Pvt. Ltd.

Affectionately dedicated to
the memory of my sister and brother-in-law

Smt. Doddabasamma
Shri. K.A. Sivarami Reddy

Advances in Integrated Pest and Disease Management in Horticultural Crops

Vol. 4: Plantation and Spice Crops

P. Parvatha Reddy

Former Director and Former Head,
Division of Entomology & Nematology,
Indian Institute of Horticultural Researc
Bangalore – 560 089

2009

Studium Press (India) Pvt. Ltd.

Advances in Integrated Pest and Disease Management in Horticultural Crops

Vol. 4: Plantation and Spice Crops

ISBN: 978-93-80012-19-3

Published by:

Studium Press (India) Pvt. Ltd.
4735/22, 2nd Floor, Prakash Deep Building
(Near Delhi Medical Association),
Ansari Road, Darya Ganj, New Delhi-110 002
Tel.: 23240257, 65150447; Fax: 91-11-23240273;
jngovil@gmail.com; jngovil@hotmail.com

Printed at:

Salasar Imaging Systems
C-7/5, Lawrence Road, Indl. Area, Delhi -110 035

Dr. M.V. Rao, MLC, Govt. of A.P.
President, Agri Biotech Foundation
Special Director-General, ICAR
(Retd) Special Secretary, DARE, GOI
(Retd) Vice-Chancellor, ANGRAU
(Retd) Ex-President, Andhra Pradesh Academy of Sciences

Plot No. 4, BHEL Enclave Akbar Road,
Bowenpally, Secunderabad 500 009,
Phone: (R) 040-27954748/27955795,
(O) 040 24002415 (Cell) 98494 77088;
Fax: 040-24002417

MVR/2008/ **10 September 2008**

FOREWORD

Plantation crops are grown over an area of 3.282 million ha with production of 11.262 million MT in India. The plantation crops contributed 6.09% of total horticultural production during 2006-07. India is the largest producer and consumer of areca nut, cashew nut, tea, third largest producer of coconut, fourth largest producer and consumer of rubber and sixth largest producer of coffee in the world. The annual income generated is more than Rs. 5,52,590 million and export earning is approximately Rs. 93,350 millions. Indian exports during 2006-2007 were US$ 866 million in tea and coffee.

India is the largest producer, consumer and exporter of spices in the world. The annual production of spices is more than 5 million MT valued at US$ 1,500 million. India's share of the world spices trade is estimated at 45 to 50% by volume and 25% by value. The present annual production of spices in the country is 5.107 million MT from over 2.857 million hectares during 2005-06. The spices contributed 2.76% of total horticulture production in 2005-2006. Spice exports from India have touched US $ one billion mark during 2007-08.

The horticultural sector in 2005-2006 has contributed around 28% of GDP in agriculture. Vegetables, fruits, plantation crops and spices contributed 59.8, 30.9, 6.5 and 2.1%, respectively of total horticulture production. The tremendous potential for export of fresh and processed plantation and spice products has made the horticulture sector as one of the thrust areas for export promotion. Processed and value added plantation and spice products are other items of potential export. For this, focus should be more on the quality and phytosanitation aspects than on quantity.

Of the several factors responsible for lower productivity of plantation and spice crops, pests (insect pests, diseases and nematode parasites) are considered as important limiting factors. Plantation and spice crops being perennial in nature, harbour and encourage the build up of pest population which forms a major constraint in the production of quality produce. Some pest problems are of national importance which causes significant losses to plantation and spice crops. The annual losses due to pests to all the crops in India was estimated at Rs. 60,000 million in 1983, which at today's prices could exceed Rs. 200,000 million. Use of pest avoidance tactics, enhancement of biological pest suppression and adoption of other non-chemical methods of pest management would help to over come much of the pest problems. Use of pesticides when the pest density reaches ETL only needs to be advocated to reduce expenses to the grower.

The information on insect pests, diseases and nematode pests on plantation and spice crops is very much scattered. There is no book at present which comprehensively and exclusively deals with the above aspects on these crops. The present book deals with integrated pest and disease management in plantation and spice crops in detail using regulatory, physical, cultural, chemical, biological, host plant resistance and integrated methods. The book is extensively illustrated with excellent quality photographs enhancing the quality of publication. The utility of the book is further enhanced by chapters like pesticide residues, useful tips in plant protection and sources of critical inputs for IPM. The book is written in lucid style, easy to understand along with adoptable recommendations involving eco-friendly components of IPM. This book is a practical guide to growers of horticultural crops. Further, it is a useful reference to students, researchers and extension workers. The book can be used for teaching undergraduate and postgraduate courses as well.

In view of greater emphasis being given by the Government of India to horticulture by establishing horticulture mission and exclusive State Horticultural Universities, there is an urgent need for good text books to teach courses on different disciplines of horticulture. In this context this book written by Dr. P Parvatha Reddy is of immense help to all those concerned with plantation and spice crops. With vast experience of more than three decades working with pests of horticultural crops, Dr. Reddy has taken up this noble task single handedly and brought out an excellent publication which I am sure will be useful to all those who are interested in these crops – the farmers, scientists, extension workers, students and policy makers. He deserves all appreciation and commendation.

(M.V. Rao)

PREFACE

Plantation crops serve a variety of human needs such as food, oil, industrial raw materials, beverages and confectionary items. They generate huge employment opportunities directly or indirectly to several million people in their production, processing, marketing and international trade sectors. Plantation crops are grown over an area of 3.282 million ha with production of 11.262 million tonnes in India. The plantation crops contributed 6.09% of total horticultural production during 2006-07. India is the largest producer and consumer of areca nut, cashew nut, tea, third largest producer of coconut, fourth largest producer and consumer of rubber and sixth largest producer of coffee in the world. The annual income generated is more than Rs. 5,52,590 million and export earning is approximately Rs. 93,350 millions. Indian exports during 2006-2007 were US$ 866 million in tea and coffee.

From time immemorial, India is regarded by the rest of the world as the "magic land of spices". Spices are natural plant products used to improve flavour, aroma, taste and colour of food products, beverages, liquors, pharmaceutical, cosmetic and perfumery products. India is the largest producer, consumer and exporter of spices in the world. The annual production of spices is more than 5 million tonnes valued at US$ 1,500 million. India's share of the world spices trade is estimated at 45 to 50% by volume and 25% by value. The present annual production of spices in the country is 5.107 million tonnes from over 2.857 million hectares during 2005-06. The spices contributed 2.76% of total horticulture production in 2005-2006. The Indian spice export for the year 1996-97 was 2,25,000 metric tonnes which was worth $350 million. In the year 2006-07, the spice exports touched the figure of 3,75,000 MT which was worth US$ 800 million. Spice exports from India have touched US $ one billion mark during 2007-08.

Of the several factors responsible for lower productivity of plantation and spice crops, pests (insect pests, diseases and nematode parasites) are considered as important limiting factors. Plantation and spice crops being perennial in nature, harbour and encourage the build up of pest population which forms a major constraint in the production of quality produce. Some pest problems are of national importance which causes significant losses to plantation and spice crops. The annual losses due to pests to all the crops in India was estimated at Rs. 60,000 million in 1983, which at today's prices could exceed Rs. 200,000 million. Use of pest avoidance tactics, enhancement of biological pest suppression and adoption of other non-chemical methods of pest management would help to over come much of the pest problems. Use of pesticides when the pest density reaches ETL only needs to be advocated to reduce expenses to the grower.

The information on insect pests, diseases and nematode pests on plantation and spice crops is very much scattered. There is no book at present which comprehensively and exclusively deals with the above aspects on these crops. The present book deals with integrated pest and disease management in plantation and spice crops in detail using regulatory, physical, cultural, chemical, biological, host plant resistance and integrated methods. The book is extensively illustrated with excellent quality photographs enhancing the quality of publication. The utility of the book is further enhanced by chapters like useful tips in plant protection and sources of critical inputs for IPM. The book is written in lucid style, easy to understand along with adoptable recommendations involving eco-friendly components of IPM. This book is a practical guide to growers of plantation and spice crops. Further, it is a useful reference to students, researchers and extension workers. The book can be used for teaching undergraduate and postgraduate courses as well. Suggestions to improve the contents of the book are most welcome (E-mail: reddy_parvatha@yahoo.com).

In view of greater emphasis being given by the Government of India to horticulture by establishing horticulture mission and exclusive State Horticultural Universities, there is an urgent need for good text books to teach courses on different disciplines in horticulture. In this context the present book is of immense help to teach courses on plant protection in plantation and spice crops.

I profusely thank Dr. K. Srinivasan, Chief Agricultural Economist, TAFE, Chennai, who has kindly made available many of the photographs included in this book. The publisher, Studium Press LLC, USA deserves commendation for their professional contribution.

Bangalore
March 15, 2009

P. Parvatha Reddy
34, UAS Layout, Ist Main, 7th Cross
Sanjay Nagar, Bangalore-560 094

CONTENTS

1 PLANTATION CROPS

1.1. INDIAN SCENARIO

Plantation crops serve a variety of human needs such as food, oil, industrial raw materials, beverages and confectionary items. They generate huge employment opportunities directly or indirectly to several million people in their production, processing, marketing and international trade sectors. Plantation crops are a large group of crops. These are grown over an area of 3.282 million ha with production of 11.262 million tonnes in India (Bijay Kumar, 2007) (Table 1.1). The plantation crops contributed 6.09% of total horticulture production, in 2006–07. India is the largest producer and consumer of areca nut, cashew nut, tea, third largest producer of coconut, fourth largest producer and consumer of rubber and sixth largest producer of coffee in the world. While the major plantation crops include coconut, areca nut, oil palm, cashew, coffee, tea and rubber, minor plantation crops are cocoa and betel vine. The major plantation crops growing states include Kerala, Tamil Nadu, Karnataka, Andhra Pradesh, Maharashtra, Orissa and West Bengal (Table 1.2). They play an important role in export as well as domestic requirement, employment generation and poverty alleviation particularly in rural sector. The annual income generated is more than Rs. 5,52,590 million and export earning is approximately Rs. 93,350 millions.

Table 1.1. Area, production and productivity of plantation crops in India (2005–06)

Plantation crops	Area ('000 ha)	Production ('000 MT)	Productivity (MT/ha)	% Share of production
Coconut	2028.9	10190.1	5.0	90.5
Areca nut	381.1	483.1	1.3	4.3
Cocoa	29.5	10.2	0.3	0.1
Cashew	843.4	579.4	0.7	5.1
Total	3282.9	11262.8	3.4	

There is a good scope for switching over the production of plantation crops *viz.*, coffee, tea, coconut and areca nut to organic farming, as it is possible to realize higher returns from the unit quantity exported when grown organically. The plantation crops are generally grown in ecologically fragile hilly tracts, adopting organic farming methods would protect environment and also prevent contamination thereby producing products of high nutritional quality. Further, the plantation crops are amenable for organic farming as they produce huge amount of waste biomass for recycling, which can meet major portion of nutrients required.

Table 1.2. State-wise area, production and productivity of plantation crops in India (2005–06)

State	Area in '000 ha	Production in '000 MT	Productivity in MT/ha
Kerala	1096.6	4545.1	4.1
Tamil Nadu	497.7	3415.0	6.8
Karnataka	735.1	1095.9	1.5
Andhra Pradesh	286.1	706.6	2.4
Maharashtra	180.0	376.1	2.0
Orissa	170.8	267.0	1.5
West Bengal	44.3	261.9	5.9
Assam	104.3	217.8	2.0
Goa	81.9	115.8	1.4
Gujarat	20.4	99.1	4.8
Andaman & Nicobar	30.0	64.7	2.1
Lakshadweep	2.7	36.5	13.5
Meghalaya	17.9	21.9	1.2
Pondicherry	5.2	20.3	3.9
Tripura	6.7	11.7	1.7
Mizoram	2.0	5.3	2.6
Nagaland	1.1	2.0	1.8
Total	**3282.9**	**11262.8**	**3.4**

1.2. EXPORT OF PLANTATION PRODUCE

India is the largest producer in the world of cashew nuts, coconuts and tea. India is largest producer, importer, processor, exporter and second largest consumer of cashew. Indian exports during 2006–07 were US$ 866 million in tea and coffee (Table 1.3).

Table 1.3. Export of plantation crops

Commodity	Apr-Mar 2006	Apr-Mar 2007	% Growth	% Share
Tea	390.92	432.00	10.51	0.34
Coffee	358.83	434.77	21.16	0.34
Cashew	585.76	553.41	–5.52	0.44

1.2.1. Tea

India is the largest producer and consumer of tea in the world. During 2006–07, India produced approximately 950,000 tonnes of tea of which around 218,000 tonnes was exported. Tea is grown mainly in three major regions in India-Assam, Darjeeling in North East and the Nilgiri region in Southern India. Other areas where tea is grown to a small extent are Karnataka, Tripura, Himachal Pradesh, Uttaranchal, Arunachal Pradesh, Manipur, Sikkim, Nagaland, Meghalaya, Mizoram, Bihar and Orissa. The northern states of India account for around 75% of the tea grown in India and the southern states accounting for around 25% of the production. Tea accounts for approx £1 billion in turn over in India.

1.2.2. Coffee

India produced 290,000 tonnes of coffee accounting for approximately 4% of the global production. Over 80% of the coffee produced in India is exported. Although over 50 varieties are produced, the popular ones are the Indian Robusta which is used for blending and Arabica. The major coffee plantations in India are in the southern states with Karnataka being the major coffee producing state. Recently some coffee plantations have come up in the north eastern states of India as well.

1.2.3. Natural Rubber

India is the fourth largest producer of natural rubber with a share of 8.8% in world production in 2006. The two major states with rubber plantations are Kerala and Tamil Nadu, with Kerala alone accounting for over 80% of the area under plantation. India produced an estimated 852,000 tonnes of natural rubber during 2006–07 and exported 56,545 tonnes while it imported 89,699 tonnes. In 2006 the yield per hectare in India at 1.879 kg per hectare, was amongst the highest in the world.

1.3. PEST AND DISEASE SCENARIO

Table 1.4. Avoidable yield losses in plantation crops due to plant parasitic nematodes

Crop	Nematode(s)	Yield loss (%)	Reference(s)
Coconut	*Radopholus similes*	30	Koshy & Geetha, 1992
Betel vine	*Meloidogyne incognita*	21.1 – 38	Saikia, 1992; Jonathan *et al.*, 1990

Table 1.5. Biological control of plantation crop pests

Plantation crop/Pest	Biocontrol agent/Dosage/ha	Frequency of application	Remarks
Coconut			
Black headed *caterpillar,*	*Goniozus nephantidis* 3000 adults.	Need based or for each generation.	Parasitoid: host ratio of 1 : 5 to be maintained. In case *G. nephantidis, Elasmus* and *Brachymeria nosatoi* are put together, the ratio should be 2 : 5

Table. 1.5. (Contd.)

Plantation crop/Pest	Biocontrol agent/Dosage/ha	Frequency of application	Remarks
Rhinoceros beetle, *Oryctes rhinoceros*	Baculovirus 10 infected beetles.	Spray as soon as infestation is noticed.	Add 0.5% jaggery and 0.1% tinopal or 0.25% boric/ tannic acid to increase the efficacy.
Arecanut			
Ischanaspis lolngirostris	*Chilocorus nigrita* at 20 50 adults/plant	Release as soon as infestation is noticed.	Identify endemic areas, mark the infested trees and release.
Coffee			
Mealybugs, *Planococcus* & *Pseudococcus* spp.	*Cryptolaemus monotrouzieri* 2–10 beetles/ infested plant.	After the blossom.	Adopt ant suppression operations.

Table 1.6. Biological control of plant parasitic nematodes in plantation crops

Crop	Nematode	Effective bioagent/s	Reference
Coconut	*R. similis*	*G. mosseae, G. fasciculatum, G. macrocorpus, G. versiforme, Acaulospora bireticulata, Sclerocystis rubiformis, Scutellospora nigra*	Koshy *et al.*, 1999
Areca nut	*R. similis*	*P. lilacinus*	Sudha *et al.*, 2000
Betel vine	*M. incognita*	*P. lilacinus*	Jonathan *et al.*, 1995 Nakat *et al.*, 1995 Hazarika *et al.*, 1998
	M. incognita, R. reniformis	*P. lilacinus*	Jonathan *et al.*, 1995
	R. similis	*P. lilacinus*	Sosamma *et al.*, 1994

2 COFFEE

Coffee (*Coffea arabica, C. canephora*) is an important commercial plantation crop mostly grown on the hilly slopes of the traditional coffee tracts in south India (Karnataka, Kerala, Tamil Nadu and Andhra Pradesh) and non-traditional Orissa, Maharashtra, Madhya Pradesh and north-eastern council states. The coffee industry is well established and flourishing enterprise. The major coffee plantations in India are in the southern states with Karnataka being the major coffee producing state. Recently some coffee plantations have come up in the north eastern states of India as well. Besides being one of the major foreign exchange earner to a tune of around Rs. 2000 million annually, coffee industry supports more than four lakh people to earn their livelihood. The two main species of coffee in cultivation are *Coffea arabica* (Arabica) and *Coffea canephora* (Robusta). *C. arabica* produces finer seeds which are more valued and grows best at higher elevations (1000 and 1500 m MSL). *C. canephora* is sturdier and better adapted to the warm climate prevailing in the lower elevations (500 and 1000 m MSL). Robusta, in general, is more tolerant to many diseases and pests including nematodes, as compared to Arabica. Coffee in India is grown under a canopy of shade trees of various species along with other plantation crops like black pepper, cardamom, banana, cocoa, etc. The present area under coffee in India is above 100,000 hectares, nearly three-fourth of which is under the fine quality Arabica species with most of the remaining area under Robusta.

India produced 290,000 tonnes of coffee accounting for approximately 4% of the global production. Over 80% of the coffee produced in India is exported. Although over 50 varieties are produced, the popular ones are the Indian Robusta which is used for blending and Arabica. There is a strong demand in the world market for production of high value, high quality, especially organic coffee. Today, the area under certified organic coffee is 2500 ha with an estimated production of 1500 MT of which 1000 MT is being exported currently.

2.1. PESTS

2.1.1. White Stem Borer, *Xylotrechus quadripes*

(i) Damage: Arabica coffee is most preferred. Infested plants show ridges round the stem.

It causes yellowing and wilting of plants. The adult beetle makes tunnel and emerges outside (Fig 2.1). Grubs feed on the wood of the stem and thick primaries and tunnel in all directions, sometimes right down to the tap root. Tunnels are packed with chalk-like excreta.

Figure 2.1. White stem borer infestation on coffee

(ii) Management

The pest can be controlled by adopting integrated measures:

- Maintenance of optimum shade.
- Regular tracing and destruction of infested plants.
- Scrubbing of main trunk and thick primaries with coir gloves to remove loose scaly bark.
- Repeated application of neem oil (2–5%) at short intervals of 1–2 weeks.
- Use of pheromone traps for monitoring incidence.
- Application of 10% lime on stem.
- Mass release of the parasitoid, *Apensia* sp. help to manage the white stem borer.
- Swab the stem with thick emulsion of lindane or carbaryl in April-May and October-December. This is also the time when adult beetles emerge, take to flight, mate and oviposit.

2.1.2. Berry Borer, *Hypothenemus hampei*

Coffee berry borer, the most serious pest of coffee, is believed to be native to northeast and central Africa. It has spread to most coffee growing countries (58) in all the continents throughout the tropics except Papua New Guinea, Hawaii and Solomon Islands. In Asia, it is found in Brunei, Cambodia, India, Indonesia, Laos, Malaysia, Philippines, Saudi Arabia, Sri Lanka, Thailand and Vietnam.

The pest entered India in 1990 when it was reported from Gudalur in the Nilgiris. The pest probably got accidentally introduced through coffee seeds brought by refugees from Sri Lanka or through illegally imported coffee seeds. It has now spread into many coffee growing areas of Tamil Nadu (Gudalur, Palani Hills, Bodinayakanur and Kilkotagiri), Kerala (Wynad and Idukki) and Karnataka (Kodagu, Hassan, Chickmagalur and Sakleshpur). It has spread to more than 1.2 million hectares accounting for about 35% of the total coffee growing area of the country.

(i) Damage: *Hypothenemus hampei* is known to feed and reproduce only in the seeds of coffee. The pest attacks both Arabica (*Coffea Arabica*) and Robusta (*Coffea canephora*) types and is also found on other species of coffee. The female enters the coffee berry by cutting an irregular hole, generally at the tip of the berry (Fig 2.2). Berries in all stages of development are prone to attack. Breeding occurs within berries with hardening or hard endosperm. The damage is mainly due to feeding inside the berries in the field. In general, only one of the two beans in a berry is damaged and hence maximum loss could be 50% of infested berries. Damage from 5–90% of the berries has been reported. Berries of the previous season left on the plant or fallen on the ground serve as pest reservoirs. This serves as a source for the next season crop.

Figure 2.2. Berry borer infestation on coffee

(ii) Biology: The adult female after entering the berry tunnels into the bean, excavating irregular galleries while feeding and laying bean shaped eggs. The brownish black female adult measures 1.5–1.9 mm. The males are smaller and generally incapable of flight. The population is usually female dominated (10 : 1). The females can lay from 30–70 eggs during its life. The eggs hatch in 4–9 days and the larvae feed and make extensive tunnels. The larval period lasts 10–26 days. The total development from egg to adult lasts for 25–35 days. On emergence, the female remains inside the berry for a few days, during which period mating takes place between siblings. The females then leave the berry to infest fresh berries. Adults are long lived, a female can live for 282 days with an average of 156 days. Females can survive up to 2 months in buried beans and mobile females can survive by feeding on immature berries. Three to four generations a year have been reported.

(iii) Ecology: The optimum temperature and RH requirements for reproduction and development was found to be around 25–26°C and 90–95%, respectively. An elevation of 500–1000 metres has been found to be optimum for infestation. Intermittent rain favours emergence and development. The incidence was found to be more under dense levels of shade as compared to open or less dense shade.

(iv) Management

(a) Cultural Methods: Cultural and phytosanitary measures that aim at disrupting the breeding are especially useful in places like India where there is a distinct gap between harvest and availability of suitable fruits in the new crop. Timely and clean harvest (removing fallen berries) helps in reducing the infestation.

Ungleaned estates recorded an average infestation of 13% as compared to 1.5% in gleaned estate. The use of picking mats while harvesting also helps in reducing the gleanings (fallen berries on the ground) to an extent of 73% and in reducing the infestation to the extent of 72% in the next crop as compared to areas harvested without picking mats.

Coffee dried to 11.5% moisture helps in killing all stages as also dipping infested fruits boiling water for 2 min or cold water for 2 days.

The beetles from migrating drying yards can be trapped by using oil coated polythene sheets and also in the berries on plants surrounding the drying yard by using them as trap plants, which can be clean harvested and destroyed.

Thinning of shade trees and training and pruning of unwieldy bushes to avoid excessive shade help in reducing the infestation.

(b) Chemical Methods: Spraying 0.07% endosulfan was found to be the most effective among all insecticides tested. It is important to spray during the period when the females are entering the berry *i.e.* during September-October. Spot application in infested patches also proved useful to prevent the spread of infestation. Spraying of chlorpyriphos, quinalphos, methomyl and malathion is also effective.

Mass trapping using chemical attractants (1: 1 mixture of ethanol and methanol is best) could be used in the management of the berry borer. A trap density of about 70 traps/ha was found optimal. Large number of beetles can be trapped during the harvest and post-harvest periods.

(c) Biological Methods: Among indigenous natural enemies, *Beauveria bassiana* was the most abundant and caused epizootics under high humidity conditions. High volume sprays gave better results than low volume sprays. The fungus is most effective at a temperature of 23–28°C and relative humidity of more than 90%. A spore load of 1×10^{10} per tree has been found effective. Trials have shown that treatment during September with 10^8 conidia/ml of spray solution resulted in 51–60% mortality of the berry borer.

Among parasitoids, only 4 are considered promising. They are *Prorops nasuta, Cephalonomia stephanoderis, Phymastichus coffea* and *Heterospilus coffeicola*, which are all native of Africa. Of these 4 promising parasitoids, three have been introduced into India. *Prorops nasuta* and *Cephalonomia stephanoderis* were imported from Mexico in 1995 and *Phymastichus coffea* and *P. nasuta* from Columbia in 1999. The parasitoid, *C. stephanoderis* has established in several areas of Kodagu district, Karnataka; Wynad district, Kerala and Lower Palanis, Tamil Nadu.

(d) Integrated Methods: The pest could be effectively tackled by adopting the following measures:

- Maintenance of optimum shade and good drainage.
- Thorough and clean harvesting of the crop in time.
- Use gunny bags/picking mats/polythene sheets while harvesting.
- Collecting the gleanings and dipping the infested berries in boiling water for 2 min or in cold water for 48 hours for killing all the stages of the borer inside the berry.
- Spreading the polythene sheet smeared with castor oil over the harvested coffee heap during the day for trapping all the beetles.
- Maintenance of trap plants around drying yard followed by clean harvesting from the trap plants and treating the berries in boiling/cold water.
- Timely spraying of entomopathogenic fungus, *Beauvaria bassiana*.
- Mass trapping of beetles using lures.
- Fumigating the storage material and transport structures effectively check the pest.

2.1.3. Shot Hole Borer, *Xylosandrus compactus*

(i) Damage: The shot hole borer is a serious pest of robusta coffee. It attacks on tertiary twigs, young primaries and suckers. The infestation starts by mid-June and increases by September. The adult female bores a hole the size of a pinhead on the lower side of a twig in the internodal region. Withered or dried branches with shot holes can be observed on infested plant. Leaves at point of attack fall prematurely. Terminal leaves wilt, droop and dry up. Extensive tunneling within the branches is noticed. The tunnel which extends 38 mm to both sides of the hole is lined with the ambrosia fungus (*Monilia* sp.) on which the larvae feed (Fig 2.3).

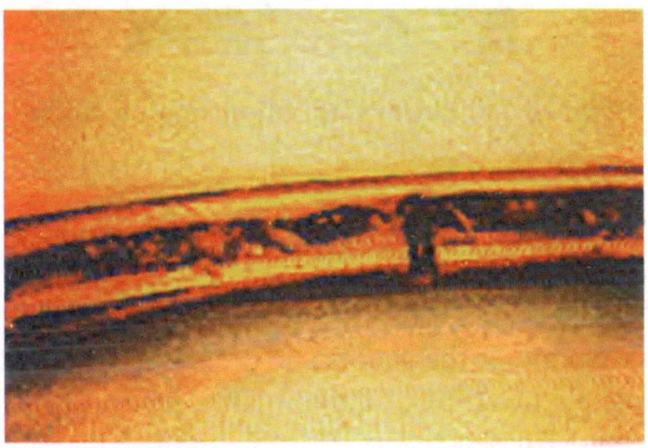

Figure 2.3. Shot hole borer infestation on coffee

(ii) Management

The incidence on Robusta coffee could be minimized by providing optimum thin shade, good drainage, regular pruning and burning of affected twigs and suckers, especially during April-May and September-December.

Swab the stem with thick emulsion of lindane or carbaryl in April-May and October-December. This is also the time when adult beetles emerge, take to flight, mate and oviposit.

2.1.4. Green and Red Scales, *Coccus viridis, Sassietia coffeae, S. hemisphaericum*

(i) Damage: Sporadic cosmopolitan; the bugs infest succulent shoot regions (Fig 2.4). All stages of the pest occur crowded on the underside of the leaves. The upper side of the leaves get blackened owing to the growth of a sooty mould on honeydew secreted by these bugs. Young plants are affected more commonly. Green scales sucks sap from tender parts, congregating down on under surface of leaves close to mid rib and veins. Infested leaves curl up and tender twigs droop. Sooty mould develops on honey dew secreted by scales. Other symptoms include curled or distorted young leaves covered with a black fungus and dropping of tender berries.

Figure 2.4. Green scale infestation on coffee

(ii) Management

Prune the affected parts and spray 0.03% dimethoate, diazinon, quinalphos, dicrotophos or monocrotophos.

Control ants by dusting ekalux 1.5% or malathion 5% around the bushes and shade trees. Destroy ant nests. Spray for the shoots and drench the roots of grown up plants with any one of the following: ekalux 25EC, folithion 1000, lebaycid 1000 in 200 litres of water or freshly prepared neutral Bordeaux mixture along with 1 kg of fish oil rosin soap with 100 ml of sandovit or vettoplant. Spray quinalphos 2.5 ml/l of water.

2.1.5. Cockchafer Beetle, *Phyllophaga (Holotrichia) conferta*

(i) Damage: Grubs (Fig 2.5) damage roots, especially in nurseries. More serious in areas full of weeds and grasses. The pest is active during the monsoon (July to September-November). They cut and eat feeder roots from May to October-November.

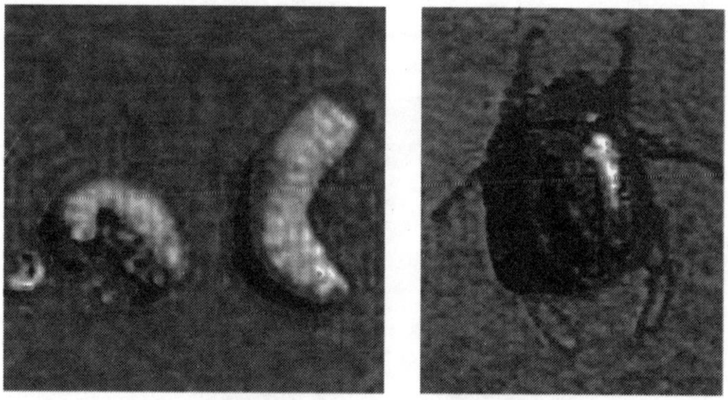

Figure 2.5. Cock chafer beetle larvae and adult

(ii) Management

- Build up good shade.
- Drench soil around the base of the plants with lindane 20 EC at 750 ml in 200 litres of water.
- Apply 5 g of phorate10G or aldicarb10G at the time of planting.
- Soil treatment with 5% aldrin or dieldrin.

2.1.6. Thrips, *Scirtothrips bispinosus, Heliothrips haemorrhoidalis*

(i) Damage: Thrips are tiny fragile insects that lacerate and feed on leaves. Affected leaves become distorted, crinkled and mottled. Severe in hot weather, though found throughout the year. Females have saw-like ovipositor. Eggs are laid inside the leaf tissue. Larvae and adults grasp and scrape the epidermis and suck the sap, resulting in local scars, and distortion of young leaves.

(ii) Management

- Maintain good shade, spray for green bug (except lannate) only in exposed fields till such time that shade trees cover up.
- Remove weeds and grasses that harbour thrips.
- Spray with 0.03% phosphamidon, dimethoate, methyl demeton or diazinon.

2.1.7. Shoot and Root Mealy Bugs, *Planococcus lilacinus, P. citri, P. perissopneum*

(i) Damage: The pest (Fig 2.6) causes yellowing, wilting and defoliation leading to stunting of plants. The other symptoms of damage include development of sooty mould, death of the feeder roots, and berry drop. Ants help to distribute the pest. The pest is active from February to June.

Leaves become chlorotic, flower buds abort and berries become small. Sooty mould develops on honey dew secreted by the bugs. Leaves of infested plants become black.

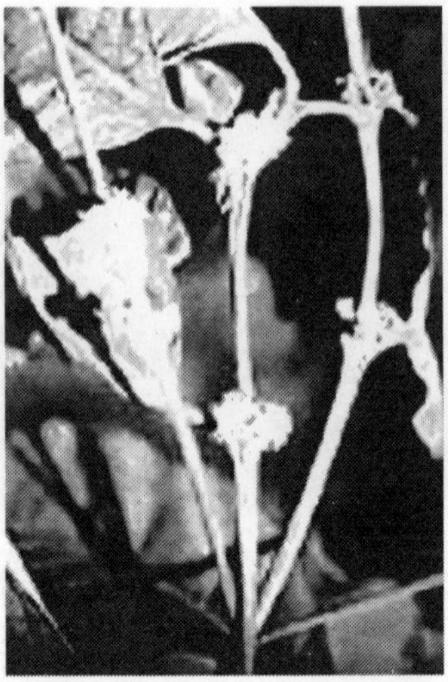

Figure 2.6. Mealy bug infestation on coffee

(ii) Management

Control ants by dusting ekalux 1.5% or malathion 5% around the bushes and shade trees. Destroy ant nests. Remove mealy bug weeds. Treat with phorate or Dasanit 5 G. Spray the shoots and drench the roots of grown up plants with any one of the following: ekalux 25EC, folithion 1000, lebaycid 1000 in 200 litres of water or freshly prepared neutral Bordeaux mixture along with 1 kg of fish oil rosin soap with 100 ml of sandovit or vettoplant.

Spray any systemic insecticide at twice the recommended concentration. After a month, release predatory Australian lady bird beetle, *Cryptolaemus montrouzieri* at 10 beetles/plant.

2.1.8. Hairy Caterpillar

(i) Damage: Appear generally on shade trees and later come down to coffee. Occur in congregations.

(ii) Management

- Collect and kill caterpillars wherever possible.
- Burn pupae from January to May.
- Install light traps for destroying moths.
- For killing caterpillars, spray ekalaux 25 EC at 350 ml or thiodan 35EC at 280 ml along with 1 kg of fish oil rosin soap.

2.1.9. Snail, *Ariophanta solata*

(i) Damage: The pest occurs from June-October on dadap and coffee. It feeds on bark and leaves causing damage to foliage.

(ii) Management

Hand-picking and clean weeding help to reduce the incidence of the pest. Broadcast snailkil (metaldehyde product) at 37.5 kg/ha.

2.1.10. Vertebrate Pests

(i) Damage: Birds *viz.* small green barbet and red whiskered Babul (*Pycnonotus jocosus*) cause economic damage to coffee. These birds feed only on ripened coffee berries. They punctured the pericarp and siphoned in the sweet contents and dropped the husk and seeds on the ground. Monkeys also cause economic damage to coffee by feeding on sweet, succulent and palatable stem tissue of young plants.

(ii) Management

Timely scaring of birds and monkeys using trained dogs effectively prevented the loss.

Table 2.1. Pest management

Pest	Nature of damage	Control measures
Shoot and root mealy bugs, *Planococcus lilacinus, P. citri, P. perissopneum*	Yellowing, wilting and defoliation leading to stunting; Sooty mould; the feeder roots die; ants help to distribute the pest; berry drop; active from Feb to June.	Control ants by dusting ekalux 1.5% or malathion 5% around the bushes and shade trees; destroy ant nests; remove mealy bug infested weeds; treat with thimet or Dasanit 5 G; spray the shoots and drench the roots of grown up plants with any one of the following: ekalux 25EC, folithion 1000, lebaycid 1000 in 200 litres of water or freshly prepared neutral Bordeaux mixture along with 1 kg of fish oil rosin soap with 100 ml of sandovit or vettoplant.
Green bug, *Coccus viridis;* Red/Brown bugs, *Sassietia coffea, S. hemispaericum*	Sporadic cosmopolitan; the bugs infest succulent shoot regions; all stages of the pest occur crowded on the underside of the leaves; the upper side of the leaves get blackened owing to the growth of a sooty mould on honeydew secreted by these bugs; young plants are affected more; curled or distorted young leaves; dropping of tender berries.	Prune the affected parts and spray 0.03% Dimethoate, Diazinon, Dicrotophos or Monocrotophos; control ants by dusting ekalux 1.5% or malathion 5% around the bushes and shade trees; destroy ant nests; spray the shoots and drench the roots of grown up plants with any one of the following: ekalux 25EC, folithion 1000, lebaycid 1000 in 200 litres of water or freshly prepared neutral Bordeaux mixture along with 1 kg of fish oil rosin soap with 100 ml of sandovit or vettoplant.
Hairy caterpillar	Appear generally on shade trees and later come down to coffee; occur in	Collect and kill caterpillars wherever possible; Burn pupae from Jan to May; light traps for destroying moths; for killing

Table. 2.1. (Contd.)

Pest	Nature of damage	Control measures
	congregations.	caterpillars, spray ekalaux 25EC at 350 ml or thiodan 35EC at 280 ml along with 1 kg of fish oil rosin soap.
Cockchafer beetle, *Phyllophaga (Holotrichia) conferta*	Cut and eat feeder roots from may to Oct-Nov; grubs damage roots, especially in nurseries; more serious in areas full of weeds and grasses; active during the monsoon (July to Sept-Nov).	Drench soil around the base of the plants with lindane 20 EC at 750 ml in 200 litres of water; apply 5 g of phorate 10G r aldicarb 10G at the time of planting; soil treatment with 5% aldrin or dieldrin; build up good shade.
Shot-hole borer, *Xylosanders compactus, X. conidens*	A serious pest of robusta; attack on tertiary twigs, young primaries and suckers; starts by mid-June and increases by September, peak infestation during Oct-Dec; the adult female bores a hole the size of a pinhead on the lower side of a twig in the internodal region; the tunnel which extends 38 mm to both sides of the hole is lined with the ambrosia fungus (*Monilia* sp.) on which the larvae feed; infested branches get discoloured around shot holes and the leaves on the distal portion turn yellow, droop and drop.	Prune and burn the affected twig, desucker during summer, maintain thin shade, good drainage, spray the crown foliage with 650 ml of lindane 20EC in 200 litres of water.
White stem borer, *Xylotrechus quadripes*	Grubs feed on the wood of the stem and thick primaries and tunnel in all directions, sometimes right down to the taproot; tunnels are packed with chalk-like excreta.	Trace, stump or uproot and burn borer-infested plants during March-April and Sept-Oct., apply lindane 20 EC at 1300 ml along with 100 ml of sandovit or vettoplant in 200 litres of water by swabbing the main stem and thick primaries during Apr-May or Sept-Dec.
Red borer, *Zeuzera coffea*	The adult moth emerges in Apr-May; larvae damage the stem of young plants and thick primaries during Sept-Nov; affected branches show wilting and reveal one or more holes	Destroy caterpillars; prune and burn infested branches, generally below pest infestation level.

Table 2.1. (Contd.)

Pest	Nature of damage	Control measures
	through which excrement of larvae hangs out in the form of pellets.	
Thrips, *Scirtothrips bispinosus, Heliothrips haemorrhoidalis, Anaphothrips* sp.	Females have saw-like ovipositor; eggs are laid inside the leaf tissue; larvae and adults grasp and scrape the epidermis and suck the sap, resulting in local scars, and distortion of young leaves; feeds on bark and leaves causing damage to foliage.	Maintain good shade, spray for green bug (except lannate) only in exposed fields till such time that shade trees cover up; remove weeds and grasses that harbour thrips; spray with 0.03% phosphamindon, dimethoate, methyl demeton or diazinon.
Snail, *Ariophanta solata*	From June-Oct on dadap and coffee.	Hand-picking and clean weeding; broadcast snailkil (metaldehyde product) at 37.5 kg/ha.
Coffee leaf-miner, *Agromyza coffeae*	Has gained importance as a major pest recently; larvae mine leaves.	Prune and destroy affected leaves; spray 0.1% BHC + 0.1% DDT.
Coffee grasshopper	A minor pest; defoliates the plants.	Dust 5–10% BHC.
Cutworm, *Agrotis segetum*	A sporadic pest; feeds on leaves.	Dust 5–10% BHC.
Coffee bean weevil, *Araecerus fasciculatus*	The weevil infests the berries in the field and dry beans in storage.	Dust 5–10% BHC.
Mites: *Oligonychus coffeae, Brevipalpus obovatus, B. phoenicis*	Mites feed in large colonies on the undersurface of the leaves; attacked leaves dry up in patches.	Dust sulphur or spray 0.05% wettable sulphur.
Nematodes, *Pratylenchus coffeae, Rotylenchulus reniformis*	General stunting and yellowing of basal leaves, decaying of roots leading to die-back, tap root may be destroyed, and tufts of adventitions roots appear from near the collar region.	Use robusta-arabica grafted seedlings, exclude nematode-infested plants, jungle soil and farmyard manure.

2.2. DISEASES

2.2.1. Leaf Rust, *Haemeleia vastastrix*

The disease was first reported from Sri Lanka in 1968 and in 1969 from Karnataka, India. Now it is widely distributed disease in all the coffee growing areas. The disease is responsible for serious crop loss and causes fluctuations in production.

(i) Symptoms: First visible symptoms are production of small brownish coloured rust pustules around 2 mm in diameter on the lower surface of infected leaves. Later, the colour turns to orange-brown and powdery spore mass appears from such surface (Fig 2.7). As the infection advances, the spore mass fades to become white, the host tissue becomes dark, turns black and dries up. Under severe conditions, the leaves drop off.

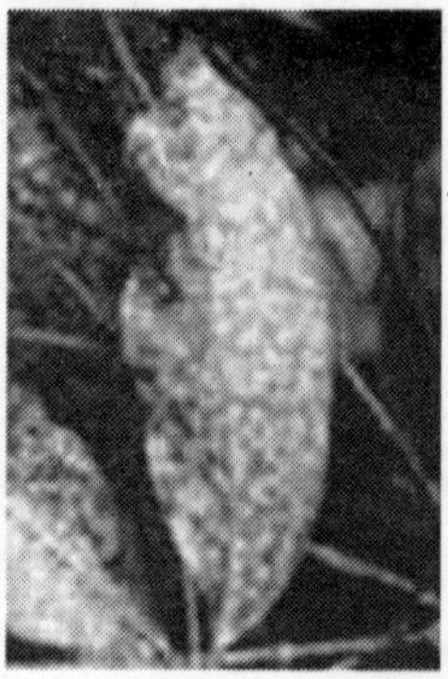

Figure 2.7. Rust on coffee leaf

If the infection is seen in early crop stage, the disease causes major damage to young flush by defoliation and thus affects the yield drastically.

(ii) Epidemiology: The fungus multiplies and spreads through uredial stage and over winter in this form only, since the pycnidial and aecial stages of the pathogen are not known. The uredospores are dispersed by wind and rain. Many insects and birds can also carry spores over long distances. Spore germinates at an optimum temperature of about 22°C and is inhibited by daylight. Hence, the disease is more severe in lower coffee growing areas where night temperature is fairly warm. Epidemics usually develop only during rainy season because of want of water. Latent infection can remain dormant over long period during dry weather. Heavily bearing coffee is epiphyotically more susceptible.

(iii) Management

- Removal and destruction of infected and fallen leaves helps to reduce spread of the disease.
- Maintaining optimum shade, judicious pruning and spraying of 1% Bordeaux mixture or 0.05% tridemefon as pre-monsoon, mid-monsoon and post-monsoon application can effectively control the disease in Arabica coffee.

- Spray tridemefon at 2 g + Teepol 0.5 ml/litre of water at fortnightly intervals at times of intermittent rainfall.
- Robusta coffee is generally more resistant and interspecific hybrids provide more durable resistance.

2.2.2. Brown Rot, *Fomes noxius*; Red Rot, *Poria hypolaterita*; Black Rot and Santavery Root Disease, *Corticium koleroga*

(i) Symptoms: Stump or brown root-rot affected bushes show a gradual yellowing of leaves and defoliation and death. Affected roots are brittle and show dark brown wavy lines of the fungus. Stem near the ground level becomes soft and spongy. Brown fungal encrustation can be seen on the affected roots (Fig 2.8).

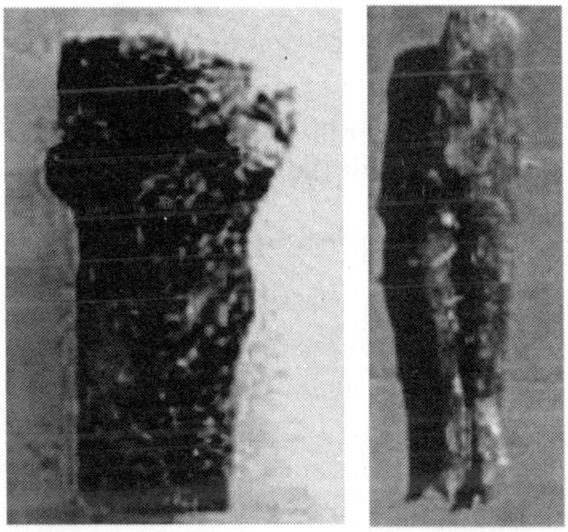

Figure 2.8. Black root and red root on coffees

The pest infestation causes blackening and rotting of affected leaves, young twigs and berries. Affected leaves get detached and hang down by means of slimy fungal strands. Defoliation and berry drop occur.

(ii) Management

Thinning out the shade, regular pruning, handling, centering (opening up the centre of the bush), cutting and burning the infected plants and plant parts followed by prophylactic sprays with 1% Bordeaux mixture in the black rot prone endemic patches would effectively check the disease.

The following measures are suggested to prevent the spread of root diseases to neighbouring areas and to recover the infected areas:

- Uprooting of shade tree stumps after timber extraction followed by uprooting and burning of affected coffee bushes.
- Making an isolation trench around the affected plants by enclosing few healthy plants and application of 1–2 kg agricultural lime to the uprooted pit and exposing the spot for at least 6 months before replanting.

- Application of 5–10 kg/plant of well decomposed FYM/compost fortified with *Trichoderma* sp. to the surrounding healthy plants.
- Soil application of 2 kg *T. harzianum* or *T. viride*/plant during June and October gave good control of brown rot and red rot diseases. Mix 2.5 kg of *Trichoderma* spp. in 100 kg of FYM. Leave it overnight. Spread these equally around plant basins twice a year.

2.2.3. Collar Rot or Damping-Off

(i) Symptoms: The pathogen is responsible for pre-emergence damping off in germinating beds and collar rot in primary or secondary beds. Seedlings show wilting, rotting extends from the collar all around and seedlings collapse.

(ii) Management

Maintaining good shade and sanitation by removing and burning affected seedlings helps in reducing the disease incidence.

Application of PCNB (brassicol) 0.4% or terrachlor or rhizoctol or Flit 406 or ferbam or dithane M 22 each at 0.4% is effective for the management of the disease.

2.2.4. Anthracnose, *Colletotrichum coffeanum*

(i) Symptoms: The pathogen is a weak parasite but becomes vigorous under favourable environmental conditions. It causes 3 diseases:

(a) Radcliffe Disease: The symptoms include yellowing of one or more functional leaves, necrosis of nodes, leaf-fall, die-back of twigs and death of many branches.

(b) Stalk Rot: It causes rotting of berries and leaves. The other symptoms include necrosis of nodes and internodes from the junction of brown wood and green wood towards apex followed by berry drop and defoliation due to decaying of stalk portion.

(c) Brown Blight: Roundish spots which are first brown, then turn grey making the leaf look blighted.

(ii) Epidemiology: The fungus occurs as saprophyte on dead tissue in outer layers of bark, which serve as major source of inoculum. It releases a large number of water-borne spores during wet season. In areas where coffee berry disease is noticed, only a small portion of virulent inoculum comes from bark, while diseased berries serve as a major source of inoculum. Young expanding berries are more susceptible to the disease. Such lesions are dormant lesions on material berries. Ripening berries are susceptible for both forms of the pathogen. In wet condition, diseased berries produce large number of spores within 2–3 weeks after infection allowing cropping season. Spores are dispersed by rain water percolating through canopy. Rain splash can disperse spores between trees. Long distance spread occurs mainly through carrying of spores on insects, birds, machinery, etc. Coffee pickers are major source of inoculum dispersal.

(iii) Management

- Providing proper drainage helps to remove excess water from the root zone.

- Spraying with 1% Bordeaux mixture or 0.3% captafol at monthly intervals prevents defoliation. Maintaining adequate shade avoids sun scald.
- Coffee *cvs*. Geisha, Rume, Sudan and Blue Mountain are reported to be resistant to anthracnose.

2.2.5. Fusarial Root Rot or Wilt

(i) Symptoms: The pathogen is responsible for sudden wilting and yellowing of leaves leading to defoliation and death of aerial parts. Roots turn brown to pinkish.

(ii) Management

Burn the plants at site, isolate affected patches by digging deep trenches, ring the bark and poison the shade trees while thinning. Remove the affected plants and treat the soil at 1 kg lime per plant to raise the soil pH. Treat the soil around the affected area with PCNB (brassicol) at 0.4%. Maintain the vigour of plants.

2.2.6. Berry Blotch, *Colletotrichum kahawae*

(i) Distribution: The disease is widely distributed in Kenya, Angola, Zaire, Cameroon, Rwanda, Uganda, Tanzania, Ethiopia, Malawi and Zimbabwe.

(ii) Symptoms: Coffee berry disease attacks young expanding berries (5–20 weeks-old) producing black sunken anthracnose lesion which cause rotting and shedding of berries. Necrotic spots on the exposed surface of green berries enlarge and cover the major portion. Fruit skin shrivels and sticks fast. Ripe berries are also susceptible and the overlapping of successive crops can produce a continuing build-up of inoculum to cause increased disease severity. The disease is favoured by wet conditions as rainfall is required for spore production, dispersal and infection. The spores can travel longer distances on passive vectors such as birds, man and machinery.

(iii) Management

Spraying 1% Bordeaux mixture during June and late august and maintaining medium shade overhead is effective.

Spraying tank mixtures of organic fungicides (captafol, chlorothalanil, dithianon, prochloraz, benomyl and carbendazim), at half the normal rates, and copper-based fungicides at a reduced rate of 5 kg/ha were effective and resulted in better yields than when using the organic fungicides by themselves (Okioga, 1978). The tank mixtures are less costly due to the reduced rates and inclusion of the much cheaper copper fungicides and hence their use improves the cost benefit ratio (Table 2.2). In addition, the tank mixtures are able to control coffee leaf rust on the same spray programme, and bacterial blight of coffee with a few additional copper sprays.

Arabica coffee *cvs*. Rume Sudan, Blue Mountain, Jackson Hybrid and Java varieties have resistance to coffee berry disease.

Table 2.2. Comparative efficacy of straight fungicides and tank mixtures on coffee yield and cost benefit ratio

Chemical	Conc. %	Yield (kg/ha)	Yield benefit (kg/ha)	Cost of chemical/ ha (Rs.)*	Cost: benefit ratio**
Captafol 80% WP	0.4	1169	735	6209	1 : 4.73
Captafol + cupric hydroxide	0.2 + 0.5	1163	729	5567	1 : 5.24
Chlorothalanil	0.4	741	307	7920	1 : 1.55
Chlorothalanil + cupric hydroxide	0.2 + 0.4	889	455	6209	1 : 2.93
Anilazine	0.4	852	418	5632	1 : 2.97
Anilazine + cupric hydroxide	0.2 + 0.4	1011	577	5236	1 : 4.41
Unsprayed	—	434	—	—	—

* Market prices for 1988

** Based on assumption that market price for coffee was £ 2000/tonne but excludes labour costs.

2.2.7. Brown Eye Spot

(i) Symptoms: Usually severe in the nursery but found on young coffee plants also. Small circular necrotic spots appear with dark brown margin and pale white centre. Leaves turn yellow and are shed.

(ii) Management

Spray nursery seedlings with 0.4% flit 406 and young coffee plants with 1% Bordeaux mixture during April-May and September-October, avoid exposing the plants to direct sun.

2.2.8. Black Rot/Koleroga, *Koleroga noxia*

The disease is more prevalent in Karnataka and Tamil Nadu in India. It is also reported from Burma, Sri Lanka and other coffee growing areas.

(i) Symptoms: The symptoms include blackening and rotting of affected leaves, twigs and developing berries. Affected leaves get detached from branches and hang down by means of slimy fungal strands. Affected leaves and berries become dry, showing a white web consisting of closely interwoven mycelium. Defoliation and berry drop takes place from infected branches in advanced stage.

(ii) Epidemiology: The fungus remains on twigs throughout the year. The dormant mycelium survives on dead and decaying matter. The secondary spread is due to basidiospores. The disease is more common during wet and humid weather. The fungus is also known to survive on several other alternate hosts and reinfects the coffee whenever environmental conditions are favourable for infection.

(iii) Management

- Shade regulation is advocated, as penetration of sunlight makes it difficult for the pathogen to infect.
- Field sanitation should be followed by cutting and burning the infected plants and plant parts.
- Periodical spraying of 1% Bordeaux mixture gives effective control of black rot of coffee.

2.2.9. Pink Disease, *Corticium salmonicolor*

(i) Symptoms: The symptoms include pink encrustation on infected branches and development of longitudinal cracks through which pink encrustation bursts. Cobweb like mycelial branches develop on affected branches. Infected branches lose leaves and die.

(ii) Management

Mix 2.5 kg of *Trichoderma viride* in 100 kg of FYM. Leave it overnight. Spread these equally around plant basins twice a year.

2.2.10. Bacterial Blight, *Pseudomonas syringae* pv. *garcae*

Bacterial blight of coffee (BBC) is confined to a few coffee growing areas in Kenya (Solai in Nakuru District and around Mt. Elgon). Recently a few cases have been reported in the East of Rift Valley. A more severe dieback occurs at high altitudes (over 1800 m) in Kenya. The pathogen has also been reported from Brazil.

(i) Symptoms: This bacterium is favoured by cool wet weather. Lesions appear on leaves with water soaked margins at start of infection. Leaves eventually dry up and roll inwards turning brown. Dead and dried leaves do not shed but remain attached to the plant (Fig 2.9). Symptoms occur on secondary or tertiary branches and in severe cases the heads may be affected. On twigs the terminal bud is attacked. Infection then extends downwards, resulting in dieback. Flowers and pin-heads shrivel and turn black.

The bacterium is an epiphyte on coffee shoots. Under wet and cool conditions, the pathogen gains entry through young tissues to produce water-soaked, dark, and necrotic lesions on the leaf, stem and berry. The infection spreads backwards from the shoot, tip of twigs causing a dieback. Affected bushes appear as if they have been scorched by a fire. The initial symptom is the blackening of a node often accompanied by blackening of the petiole and basal part of the leaf. This leads to death of the branch above the node and leaves on the branch turn brown and eventually necrotic (burnt). Severely affected trees appear scorched as if by fire. The bacterium persists on leaves, berries and on green and matured bark. During cool, wet weather, the bacterium multiplies and initiates an epidemic. The entire crop may be lost.

Figure 2.9. Bacterial blight on coffee

(ii) Management

- Desucker or cut off infected twigs (at least 5 cm from the point of infection) and collecting them in buckets for burning.
- Observe field hygiene by disinfecting tools used for pruning infected trees with a suitable disinfectant such as spirits with 70% alcohol. Implements used must be sterilized after making each cut to avoid the risk of infecting the cut surfaces of the twig with a carry-over inoculum.
- Use of copper sprays especially during wet season. Spray at 2 weeks interval during the rains and 3 weeks interval after the rains.
- Plant shade trees and wind breaks on exposed sides of BBC areas.

- Avoid planting material from known BBC areas. In emergencies spray with copper just before, during and after flowering especially when this coincides with cool wet weather.
- Spray at 2 week intervals during the rains and 3 week intervals after the rains.

The current control measures involve frequent spraying with copper fungicides and/or tank mixture of copper and organic fungicides during the rainy season. More recently, it has been observed that pruning of the infected branches before application of bactericides enhances the control (Kairu, 1991).

Table 2.3. Disease management

Disease	Symptoms	Control measures
Coffee leaf rust, *Haemeleia vastastrix*	Small pale-yellow spots on the lower surface of infected leaves, orange-yellow spore mass appears, defoliation and die-back; results in serious crop loss and causes fluctuations in production.	3 applications of 0.5% Bordeaux mixture for susceptible varieties.
Black rot, *Koleroga noxia*	Blackening and rotting of affected leaves, young twigs and berries; affected leaves get detached and hang down by means of slimy fungal strands; defoliation and berry drop occur.	Remove and burn affected parts; apply 1% Bordeaux mixture close to the south westerly monsoon if needed; centre the coffee bushes, regulate the overhead canopy.
Anthracnose, *Colletotrichum gloeosporioides*	A weak parasite but becomes vigorous under favourable environmental conditions; causes 3 diseases: 1. **Radcliffe disease:** Yellowing of one or more functional leaves, necrosis of nodes, leaf-fall, die-back of twigs and death of many branches. 2. **Stalk rot of berries and leaves:** Necrosis of nodes and internodes from the junction of brown wood and green wood towards apex followed by berry drop and defoliation duc to decaying of stalk portion.	Provide good drainage to remove excess water from the root zone; spray with 0.5% Bordeaux mixture to prevent defoliation; maintain adequate shade to avoid sun scald.

Table. 2.3. (Contd.)

Disease	Symptoms	Control measures
	3. **Brown blight:** Roundish spots are first brown, then turn grey making the leaf look blighted.	
Root rots, *Fomes* spp., *Fusarium* spp.	1. **Stump or brown root-rots:** Affected bushes show a gradual yellowing of leaves and defoliation and death; affected roots are brittle and show dark brown wavy lines of the fungus. 2. **Fusarial root rot or wilt:** A sudden wilting and yellowing of leaves leading to defoliation and death of aerial parts; roots turn brown to pinkish.	Burn the plants at site, isolate affected patches by digging deep trenches, ring the bark and poison the shade trees while thinning; remove the affected plants and treat the soil at 1kg lime per plant to raise the soil pH; treat the soil around the affected area with PCNB (brassicol) at 0.4%; maintain the vigour of plants.
Berry blotch	Necrotic spots on the exposed surface of green berries enlarge and cover the major portion; fruit skin shrivels and sticks fast.	Spray 1% Bordeaux mixture during June and late August, maintain medium shade overhead.
Nursery disease (Collar rot or damping-off)	Pre-emergence damping-off in germinating beds; collar rot in primary or secondary beds; seedlings show wilting, rotting extends from the collar all around and seedlings collapse.	Apply PCNB (brassicol) 0.4% or terrachlor or rhizoctol or Flit 406 or ferbam or dithane M 22 at 0.4%; maintain good shade and sanitation; remove and burn affected seedlings.
Brown eye spot.	Usually severe in the nursery but found on young coffee plants also, small circular necrotic spots appear with dark brown margin and pale white centre; leaves turn yellow and are shed.	Spray nursery seedlings with 0.4% flit 406 and young coffee plants with 1% Bordeaux mixture during April-May and Sept-Oct, avoid exposing the plants to direct sun.

2.3. NEMATODES

Lesion, root-knot and reniform nematodes are recognized as the major limiting factors in successful production of Arabica coffee crop.

2.3.1. Lesion Nematode, *Pratylenchus coffeae*

The lesion nematode is the most important pest of coffee in India. The annual loss of 20 million rupees was assessed in coffee due to this nematode in an area of about 1,000 ha in Karnataka alone (Van Berkum & Seshadri, 1970).

P. coffeae destroyed about 95% of *Coffea arabica* plantations in Java, which was subsequently found in other regions, including India and Central and South America.

(i) Distribution: *P. coffeae* is proved to be highly destructive to Arabica coffee in Karnataka, Kerala and Tamil Nadu. Over 3,000 hectares have been infected with this nematode in coffee growing tracts of south India (Kumar, 1984).

(ii) Symptoms: In nursery, *P. coffeae* causes death of coffee plants. The diseased trees are stunted; showing thin stems, resulting in complete destruction of secondary roots and rootlets (Fig 2.10). In some trees, severe injury caused by *P. coffeae* results in a corky region at the base of the trunk. Complete wilting of the foliage is also a typical symptom during dry seasons. Total defoliation occurs under severe attacks.

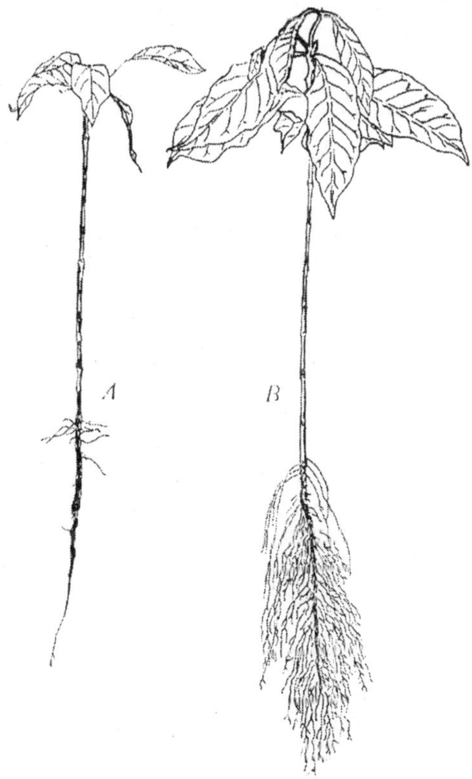

Figure 2.10. Lesion nematode on coffee. *Left*–Infected; *Right*–Healthy (Courtesy: Thorne, 1961)

The infected plants show loss of primary roots. The lesion nematode causes wounds in roots through which other pathogenic organisms, such as fungi and bacteria, enter the root tissues. The interaction of these agents results in the formation of lesions that finally destroys the root tissues. The root system of seedlings infected by *P. coffeae* is poor and the plants can be pulled out of the ground easily. Leaves and stems may show symptoms of mineral deficiencies.

(iii) Life Cycle: Lesion nematodes are typically migrant parasites of plant roots. The nematode penetrates slightly behind the elongating zone in the piliferous region. Adults and juveniles of various ages constantly migrate into and out of the roots. The first molt in the life cycle of *P. coffeae* takes place within the egg while the remaining three occur outside the egg. Eggs hatch in 6–8 days at 28–30°C. Within 6 more days, the second-stage juvenile forms. Development to the third stage occurs at 21 days and to the fourth at 28 days. Adult males emerge at 29–32 days.

(iv) Host Range: *P. coffeae* is known to attack coffee, banana, citrus, grapevine, apples, strawberry, abaca, yam, cocoa, rubber, potato, sweet potato, peas, beans, cabbage, cauliflower, tomato, chrysanthemum, caladium, dahlia, marigold, peanut, areca nut, cardamom, etc.

(v) Survival and Dispersal: *P. coffeae* is capable of surviving for about 18 months in moist soil under laboratory conditions but failed to exist beyond 12 months under field conditions.

The nematode is disseminated through working implements and running water especially across the slope. Nematodes enter the nursery through jungle soil and breed there.

(vi) Biological Races: Existence of as many as four biological races in *P. coffeae i.e.*, coffee, cardamom, pavetts and bamboo races has been recognized and coffee race is specific to coffee (Kumar, 1988).

(vii) Management

(a) Cultural Methods: Nematode control in coffee is achieved through cultural practices like maintaining heavy shade, providing mulch to individual plants, application of FYM and giving frequent irrigation or providing nutrients through foliar spray. Soil application of coffee pulp at 18 kg/sq.m. reduced the nematode population and increased the plant growth.

Lordello (1968) reported that replanting of old nematode-infested plantations preceded by fallow period of at least 2 years gave effective control of the lesion nematode.

(b) Chemical Methods: In field trials with young trees of Arabica coffee, phenamiphos at 0.5 to 1.0 g/plant gave good control of *P. coffeae* and remained effective for 90 days after application. Repeated applications were necessary to maintain continuous control (Kumar, 1982).

(c) Integrated Methods: It is advisable to change nursery site once every 7–8 years. Jungle soil and FYM should be dried and well sieved before use. The nursery site should also be dug and exposed to sun. The nematode-free plants should be used for planting. Affected plants should be pulled out and burnt. The planting site should be dug and the soil exposed to the sun for at least one summer season. Care should be taken to keep the site free from weeds. The site should be planted with either Robusta or Arabica-Robusta grafted plants. Grafting of Arabica scion on Robusta rootstock in 40 day-old 'topee stage' plants by using wedge-cleft method can be adopted (Kumar, 1988) (Fig 2.11).

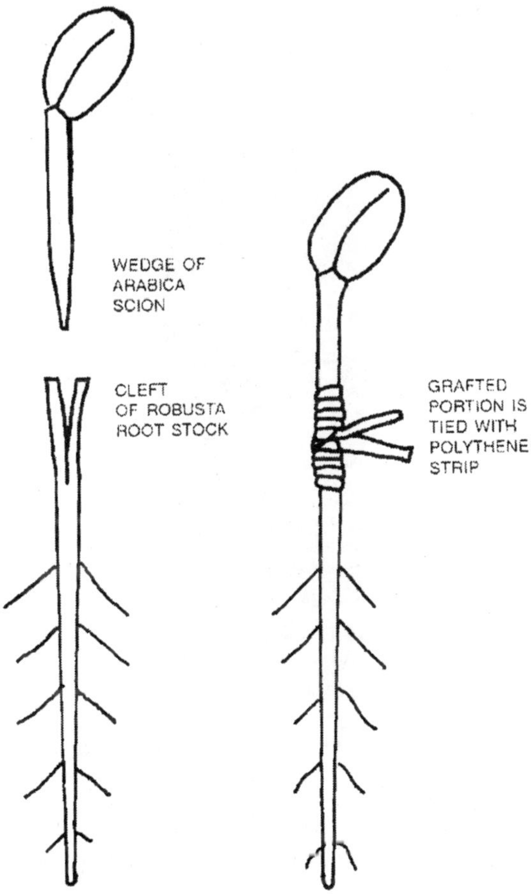

Figure 2.11. Grafting of Arabica coffee scion on Robusta rootstock (Courtesy: Indian Coffee)

2.3.2. Root-Knot Nematodes, *Meloidogyne exigua, M. coffeicola*

Srinivsan and D'Souza (1965) reported *M. exigua* on coffee from South India. *M. exigua* is a pest of *Coffea arabica* but not of *C. canephora* (robusta coffee) which is very resistant (Schieber & Sosa, 1960).

(i) Economic Importance: A 5% loss in yield due to *M. coffeicola* damage has been reported in Hawaiin coffee plantations. The yield of non-infected plants was more than that of artificially inoculated plants (104 and 50 kg, respectively). Losses due to *Meloidogyne* spp. in Brazil are estimated at 15% of that country's production.

(ii) Symptoms: Affected trees look unthrifty and have chlorotic, wilting and dying leaves. *M. exigua* produces small elongated galls that are located primarily at the root tips (Fig 2.12). In addition, roots may show cracks, necrosis and sloughing of cortical tissue. The effectiveness of root system is greatly diminished, with rootlets and root hairs practically eliminated. Trees may die, particularly during periods of drought and cold temperatures. Affected plants lack vigour and are unable to withstand adverse conditions.

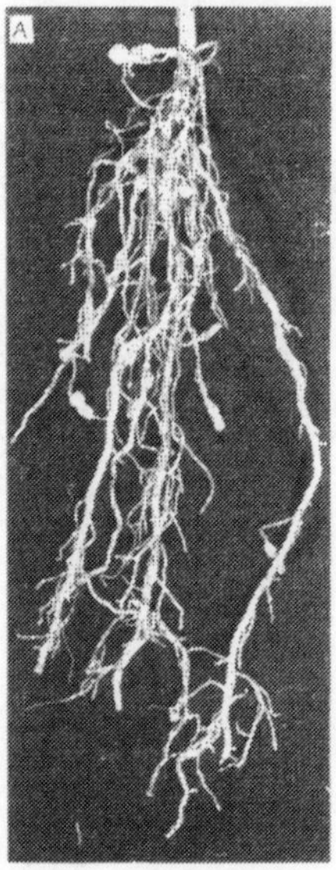

Figure 2.12. Arabica coffee roots infected with *Meloidogyne exigua* (Courtesy: Lordello, 1972)

(iii) Life Cycle: *M. exigua* takes 32–42 days at 25–30°C to complete the life cycle. The egg masses of *M. exigua* are mostly located under the epidermis of coffee roots.

(iv) Hosts: *M. exigua* is known to attack coffee, tea and sweet pepper.

(v) Survival and Spread: *M. exigua* does not survive in the soil, in the absence of the host, for more than six months after eradication of infested plants. The nematode spreads through the infested seedlings used for transplanting in the main field.

(vi) Interaction with Other Pathogens: The disease complex involving *Rhizoctonia solani* and *M. exigua* caused more root necrosis and defoliation when both pathogens were inoculated. Extensive fungal colonization within the coffee root systems was observed.

(vii) Management

(a) Cultural Methods: At the crop renovation time, the infested coffee area may be turned to pasture. Most graminicolous plants are non-hosts to *M. exigua* and *M. coffeicola*. The decision of growing forage crops and turning back again the area to coffee will certainly reduce drastically the population of *Meloidogyne* species in 6–12 months.

Rotation of coffee with cotton, soybean and maize in *M. exigua* infested areas for one year significantly reduced the nematode population, after which the farmer can return to coffee cultivation. Coffee can also be rotated with groundnut, castor (both immune), *Styzolobium deeringiatum* and *Crotalaria spectabilis* (both resistant).

(b) Chemical Methods: Application of aldicarb, carbofuran and phenamiphos each at 1.6 to 6.0 g *a.i.*/plant, in one or two applications during the year gives effective control of root-knot nematodes. The first application should be in the beginning of the rainy season followed by the second 3 months later.

(c) Host Resistance: *Coffea canephora* (line 2258 from Costa Rica) showed resistance to *M. exigua* and *M. incognita*. Planting of *Coffea arabica var.* Mundo Novo or Catuai Vermelho grafted on line 2258 in field showed good growth and production in Brazil.

2.3.3. Reniform Nematode, *Rotylenchulus reniformis*

The reniform nematode has been observed on coffee in India, Puerto Rico, Brazil and Hawaii.

(i) Symptoms: *R. reniformis* infection of coffee trees results in an almost complete absence of feeder roots, a poorly developed tap root, and the yellowing and wilting of above-ground parts. In soils heavily infested with reniform nematode, Arabica coffee cannot be grown.

(ii) Life Cycle: *R. reniformis* is a sedentary ectoparasitic nematode. Only females feed on plant tissues; males are non-parasitic. These nematodes reproduce by cross fertilization. Females lay about 120 eggs approximately 9 days after entering the root. The eggs are deposited in a gelatinous matrix from which second stage juveniles hatch within 8 days. Juveniles of both sexes develop rapidly in the soil where 3 molts occur. Apparently, individual nematodes do not feed during juvenile development.

After penetrating the root, the young female feeds, enlarges progressively, the body eventually assuming a kidney shape, hence the common name, reniform nematode. The posterior portion of the body usually protrudes from the root surface and can easily be seen in properly stained material.

(iii) Management

(a) Cultural Methods: In Puerto Rico, crop rotation with Pangola grass and sugarcane has effectively controlled *R. reniformis*. Rotation with sugarcane has also provided effective control in Brazil.

(b) Host Resistance: *Coffeae canephora var.* robusta may have some resistance to infection by *R. reniformis*.

2.3.4. Ring Nematodes, *Hemicriconemoides* spp.

(i) Symptoms: Association of large number of the ectoparasitic nematodes like *Hemicriconemoides coffeae, H. cocophilus and H. gaddi* with Arabica and Robusta coffee has been reported to cause "Crinkle Leaf" disorder of coffee in south India (Kumar, 1985).

(ii) Management

(a) Chemical Methods: Both phenamiphos (Nemacur 5G at 25 g/plant) and carbofuran (Furadan 3G at 30 g/plant) were effective in reducing the population of *H. gaddi*. The treated plants put forth more new and healthy vegetative growth.

3 TEA

India is the largest producer and consumer of tea (*Camellia sinensis*) in the world. During 2006–07, India produced approximately 950,000 tonnes of tea of which around 218,000 tonnes was exported. Tea accounts for approximately £1 billion in turnover in India. Tea is grown mainly in three major regions in India—Assam, Darjeeling in north east and the Nilgiris region in southern India. Other areas where tea is grown to a small extent are Karnataka, Tripura, Himachal Pradesh, Uttarakhand, Arunachal Pradesh, Manipur, Sikkim, Nagaland, Meghalaya, Mizoram, Bihar and Orissa. The northern states of India account for around 75% of the tea grown in India and the southern states accounting for around 25% of the production.

India is the largest producer of tea in the world. It accounts for 20% of the total area under tea in the world, 28% of world production, 22% of global tea consumption and 15% of the total global tea exports. The total area under tea is 5,07,200 ha. The tea industry provides direct employment for one million workers of which sizeable number are women. The largest producers of tea are Assam (50.7%), West Bengal (22.1%), Tamil Nadu (15.9%) and Kerala (8.3%).

More than half of the one million hectares of tea cultivation is in India and Sri Lanka, on areas ranging from 225,000 to 350,000 hectares. Between them, these two major producers account for more than three-quarters of the world's export trade of 600,000 metric tonnes of this commodity. This apparently disproportionate share in the export trade is because in these two countries, domestic consumption is far outstripped by total production. In India, tea cultivation is restricted to the north eastern and southern regions.

With the current low price trends which are highly unremunerative to the growers, that most of the growers have shifted the cultivation practices to organic farming system. Today, more than 10,000 ha have been certified under organic tea cultivation.

3.1. PESTS

3.1.1. Shot-Hole Borer, *Xyleborus fornicatus*

(i) Damage: The shot hole borer is a major pest of tea. Its infestation is characterized by small pinholes leading into the tunnel which are lined by a fungus (*Monilia* sp.). The infested branches get discoloured around shot holes and the leaves on the distal portion turn yellow, droop and drop. Peak infestation of the pest occurs during October-December.

(ii) Management

The incidence could be minimized by removing affected branches, avoiding pruning during dry weather followed by spraying of *Beauvaria bassiana* at 1.5 kg/ha during May-October when humidity levels are high and also by placing cut stems of *Montonoa bipinnatifida* in infested plantation during 3rd and 4th year.

An indigenous strain of *Beauveria bassiana* (available as a wettable powder 'Bio-Power' formulation) gives effective control of shot hole borer (Fig 3.1).

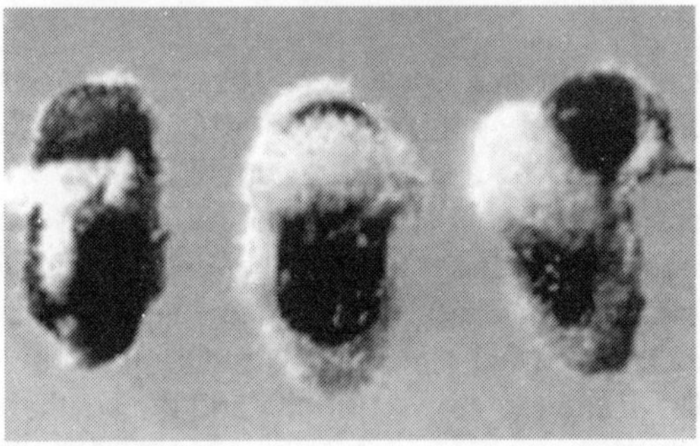

Figure 3.1. *Beauveria bassiana* infected adults of shot hole borer

While conducting studies on alternate host plants of shot hole borer, cut stems of a jungle plant, *Montanoa bipinnatifida* were found to attract shot hole borer beetles (Fig 3.2). Large scale field experiments were conducted at Valparai to find out the possibilities of using the cut stems of this plant as a trap for shot hole borer. The study showed that 22 to 28 days old cut stems measuring 90 cm long and 25 to 30 mm thickness attracted large number of adult beetles. The number of beetles attracted to the stem trap was more during June to October with a peak in July. On an average of four cut stems were required for an area of 100 sq. m. and approximately 400 cut stems are needed per hectare. The number of beetles attracted to the stems declined after 15 to 20 days indicating the need for replacement of cut stems.

Figure 3.2. Cut stems of *Montanoa bipinnatifida* to attract shot hole borer beetles

3.1.2. Mosquito Bug, *Helopeltis theivora*

(i) Damage: Tea mosquito bug (Fig 3.3) laid large number of eggs on the broken ends of plucked shoots. Leaves curl up, become gradually deformed and remain small.

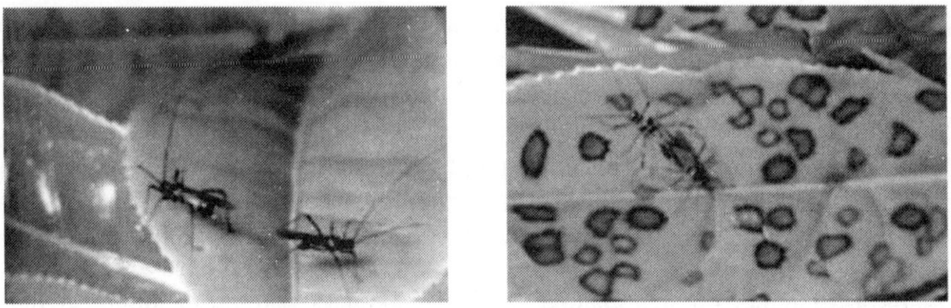

Figure 3.3. Mosquito bug infestation on tea leaf

(ii) Management

Intensive removal of stalks during plucking is recommended to reduce the incidence of this past. Weeds offer excellent hiding places and serve as alternate hosts for *Helopeltis*. Growth of weeds and wild host plants in and around tea fields may be controlled and this will help to reduce the incidence of pest.

3.1.3. Scales

(i) Damage: The most common insect pests of tea are scales. Scale insects feed on plants by piercing plant tissue and sucking sap. Scales do not look like typical insects. They are small, immobile and have no visible legs. They vary in appearance depending on species and sex. Some look like small fish scales attached to the plant. As a result of their unusual appearance, populations can reach damaging levels before they are noticed.

On tea, scales usually attach to leaves but some species also attach to stems. Their feeding weakens the plant. With a heavy infestation, symptoms include yellowing of the upper leaf surface, fewer and smaller blossoms, leaf drop, twig dieback and sometimes death.

Tea scale has an oblong shape with a ridge down the center parallel to the sides. It is a small scale with the female about 1/20 inch. The male is about two-thirds the size of the female. The females vary in colour from dark brown or gray to nearly black. Males are white. The female lays 10 to 16 eggs, which remain protected under her body until they hatch. In one to three weeks, bright yellow immature forms called crawlers hatch from the eggs.

(ii) Management

With a light infestation, scales can be scraped off the plant and discarded. If only a few leaves are infested, hand picking and destruction of infested leaves is very effective. The best time to spray with a refined horticultural oil (Bonide All Season Spray Oil, Ferti-lome

Scalecide, Green Light Horticultural Oil Spray, or Ortho Volck Oil Spray) is in spring, after the plants have finished blooming and the danger of cold weather has passed. This will kill many adults, crawlers and eggs by smothering them. Spray twice at 10 days interval when temperatures are between 40 and 85°F.

Most insecticides are effective only against the crawlers. In addition, using an insecticide against scales can result in the death of naturally occurring enemies of scales. As such, insecticides should be avoided unless required.

Monitor the crawler emergence with sticky cards, tape wrapped around a branch, or by putting an infested shoot or leaf into a bag and watching for crawler movement. Crawler activity often coincides with the flush of new plant growth in the spring. However, some scale species may have overlapping generations with an extended crawler emergence period, such as along the coast.

Insecticides labeled for use against tea scale crawlers include acephate, malathion, cyfluthrin, and carbaryl.

3.1.4. Phassus Borer, *Sahydrassus malabaricus*

(i) Damage: Phassus borer causes long cylindrical tunnels in thick branches. Entry hole is plugged with chewed wood and silk.

(ii) Management

Spraying of endosulfan or quinalphos gives effective control of the pest.

3.1.5. Thrips, *Scirtothrips bispinosus*

(i) Damage: Leaves curl and show parallel lines of feeding marks on either side of the midrib.

(ii) Management

Spraying of chlorpyrifos or dimethoate or phosalone is effective for the management of thrips.

3.1.6. Mites: Red Spider Mite, *Oligonychus coffeae;* Scarlet Mite, *Brevipalpus californicus;* Purple Mite, *Calacarus carinatus*; Pink Mite, *Acaphylla theae*; Yellow Mite, *Polyphagotarsonemus latus*

(i) Damage

(a) Red Spider Mite: Leaves turn ruddy bronze due to feeding. Scarlet mite: Causes discoluration of leaves and defoliation. Purple mite: Causes smoky grey colour on leaves.

(b) Pink Mite: It is responsible for twisting and pale colouration of young leaves. Yellow mite: Young leaves become rough, brittle and with a corky under surface.

Red spider mite (Fig 3.4) has emerged as an important pest of tea in the last few years. Studies on its population ecology showed that the incidence of RSM was high during January to May and low during June to December and reached a peak in March/April. Crop loss studies revealed that the mite can cause more than 18% loss in crop when the infestation is severe.

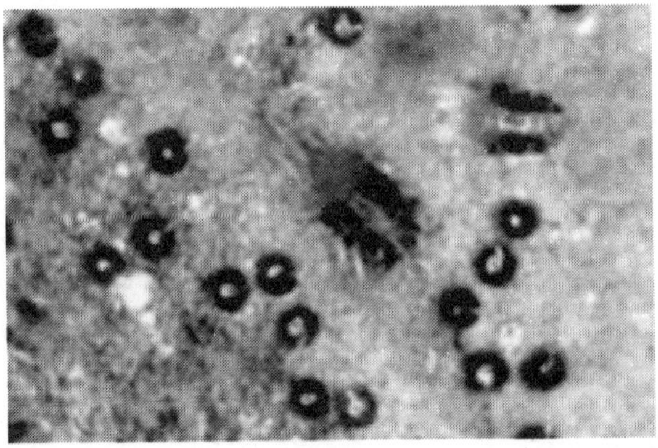

Figure 3.4. Red spider mite infestation on tea

(ii) Management

An entomopathogenic fungus, *Paecilomyces fumosoroseus* (available as a wettable powder 'Mycomite' formulation) gives effective control of red spider mites (Fig 3.5).

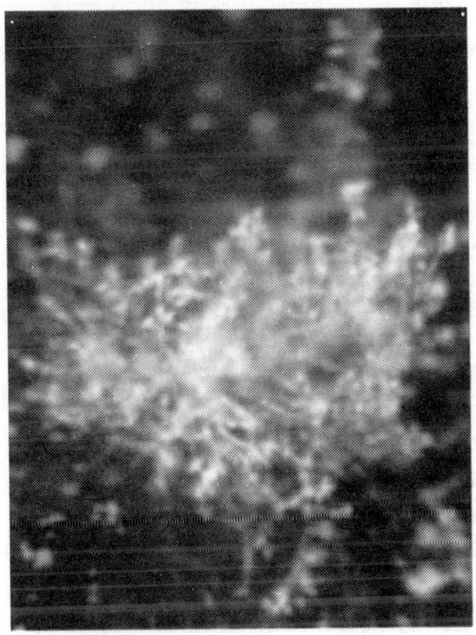

Figure 3.5. Biological control of red spider mites using *Paecilomyces fumosoroseus*

Use of hydrocarbon oil formulations of sulphur is effective against red spider mites. Recently, spray oil from paraffinic base has been found effective against red spider mites. Since this oil does not leave any residues in tea, it is incorporated into the mite control programme in tea.

IPM

- Populations of leaf folding caterpillars such as flush worms, leaf roller and tea tortrix can be reduced by their removal while plucking.
- Bushes may be pruned to reduce the damage of *Helopeltis* and shot hole borer.
- Unshaded areas are more prone to attack by thrips and mites. Proper shade management will help to prevent excessive build up of thrips and mites.
- Growth of weeds like *Mikania cordata, Bidens bifernata, Emillia* sp., *Polygonum chinense, Lantana camera* and wild host plants near tea fields may be controlled to reduce mosquito bug population.
- Application of higher levels of K (N: K_2O = 1: 2) in the pruned year helps to reduce shot-hole borer incidence.
- *Amblyseius herbicolus* and *Euseius ovalis* are 2 phytoseid mites feeding on the eryophid mites—*Acaphylla theae* and *Calacarus carinatus*. The stigmatid, *Agistemus fleschneri* is an important predator of eggs and nymphs of *Oligonychus coffeae* in north eastern India.
- Coccinellids—*Cryptogonus bimaculatus, Jauravia quadrinotata, Menochilus sexmaculatus* and *Stethorus gilvifrons* prey upon mites.
- *Chrysoperla carnea* released at 500–1000 adults/ha feeds on thrips and mosquito bugs.
- Anthocorids—*Anthocorus* and *Orius* spp. and predatory thrips—*Aeolothrips intermedius* and *Mymarothrips garuda* are important predators of thrips.
- Scale insects and mealy bugs are attacked by parasitoids—*Coccophagus cowperi* and *Encyrtis infelix* heavily parasitize *Saissetia coffeae.*
- The larval parasitoid, *Apanteles aristaeus* is efficient against flush worms.
- The leaf roller, *Caloptilia theivora* is heavily parasitized by the eulopid, *Sympiesis dolichogaster* (20 to 83%).
- The tea tortrix, *Homona coffearia* is severely affected by ichneumonid larval parasitoid, *Phytodietus spinepes.*
- The eggs of mosquito bug are parasitized by mymarid, *Erythmelus helopeltidis* (52–83%). The neuropteran, *Chrysoperla carnea* (at 500–1000 adults/ha) feeds on mosquito bugs in the field and gives effective control in north eastern India.
- *Bt* is effective for the management of looper caterpillars, cut worms, flush worms and other lepidopterous pests. *Beauvaria bassiana* is effective against shot-hole borer.
- Entomopathogenic fungi—*Verticillium lacanii, Paecilomyces fumosoroseus* and *Hirsutella thompsoni* were found effective against pink, purple and red spider mites. *Beauvaria bassiana* at 1 kg/ha is recommended in north eastern India against mosquito bugs.
- Azadirachtin is effective against pink, and purple mites and caterpillars.

Table 3.1. Pest management

Pests	Crops stages affected	Occurrence	Common chemicals recommended
Phassus borer	Thick branches	Any time	Endosulfan, quinalphos
Shot hole borer	Stem	Mainly in low and mid elevation areas	Cypermethrin, dichlorvos, quinalphos
Red spider mite, scarlet mite, purple mite	Mature foliage	Dry weather	Sulphur formulations, ethion
Pink mite, yellow mite	Pluckable shoots	Cool, humid sheltered condition	Sulphur formulations, ethion
Thrips	Young leaves and buds	Warm, humid conditions	Chlorpyrifos, dimethoate, phosalone
Tea mosquito bug	Young leaves and buds	Any time	Chlorpyrifos, dimethoate, quinalphos

3.2. DISEASES

3.2.1. Dieback and Canker, *Glomerella cingulata*

(i) Symptoms: This is one of the most serious of all tea diseases and is caused by the fungus *Glomerella cingulata*. Leaves on affected branches suddenly turn yellow and wilt. Branch tips usually die. Gray blotches appear on the bark and stem, and then sunken areas (cankers) develop, eventually girdling the stem. Parts of the plant above the stem canker lose vigour, wilt and die. Damaged plants show more symptoms during hot, dry weather.

(ii) Management

Keep tea as healthy as possible. Plant in a well-drained acidic soil, avoid wounding and fertilize properly. Remove diseased twigs by pruning several inches below the cankered areas. Disinfect pruning tools between all cuts, using a solution of one part household bleach to nine parts water. Fungicides, such as thiophanate-methyl and copper salts of fatty acids can be applied during wet periods and normal leaf drop periods to protect fresh leaf scars from infection.

3.2.2. Flower Blight, *Ciborinia camelliae*

(i) Symptoms: This serious disease of tea causes the flowers to turn brown. Flower blight appears in early spring when moisture is present and is caused by the fungus *Ciborinia camelliae*. Symptoms begin as small, brown, irregular shaped spots on the flower petals. These spots quickly enlarge to cover most of the flower. The entire flower turns brown and usually drops within 24 to 48 hours. Only the flowers of the plant are affected.

(ii) Management

Sanitation is the best control. Pull off and destroy all infected flowers. Rake up and remove all leaves, flowers and plant debris that have fallen to the ground. Replace the mulch under the plant. This fungus survives in the soil. Spores of the fungus can be wind-borne for up to 1.6 km. Therefore, control is best achieved when applied to other tea plants in the community.

Application of soil drenches, such as PCNB, around the plant in late December or early January is often helpful in reducing the intensity of disease. Fungicide sprays recommended for the flowers include mancozeb and triadimefon.

3.2.3. Root Rot, *Phytophthora cinnamomi*

(i) Symptoms: This fungal disease is caused by *Phytophthora cinnamomi*. The first symptoms are a uniform leaf yellowing, poor growth and wilting of the entire plant. Infected root systems lack small feeder roots and appear discoloured. Infected roots are reddish-brown to dark-brown in colour (healthy roots are white). Death of the plant can occur rapidly, or the plant may remain in a state of decline for several years. All varieties of common tea are susceptible and all varieties of *Sasanqua camelliae* are resistant to this root rot.

(ii) Management

This disease is difficult to control once plants are infected, so prevention is very important. In areas where this disease has been a problem, request tea cultivars grafted onto a sasanqua rootstock. Procure healthy plants that show no signs of wilting or yellowing of the leaves.

The fungus thrives in areas with poor drainage and warm soils. Always choose locations that have good drainage for planting. The drainage of existing areas can be improved by using raised beds. Fungicides can be effective on a preventative basis only, and repeat applications are required. Fungicides containing etridiazole and mefenoxam can be applied, but will not cure an infected plant.

3.2.4. Blister Blight, *Exobasidium vexans*, *E. camelliae*

The disease was reported for the first time from Assam in 1968. Now it is widely distributed in north and south India and also in Sri Lanka, Burma and Formosa. The disease occurred in epiphytotic form in 1946 and caused heavy damage to tea industry. Generally the disease appears on young leaves and buds causing significant economic damage. If the older and thick leaves are infected, they drop off. The young flush infections result in drastic reduction in the yield.

(i) Symptoms: The blister blight is the major disease affecting the tender harvestable shoots of tea resulting in enormous crop loss. Soon after the onset of monsoon, after the first showers, small, pale or pinkish translucent spots appear on tender leaves. Tender stem also gets affected. Leaf galls are most often observed during the spring flush of growth. New shoots and leaves become enlarged, thickened and fleshy, and appear abnormal. The colour of the affected areas turns from light green to nearly white or pink. The galls later rupture on the undersides of the leaves revealing a whitish mass of spores (Fig 3.6). The galls eventually harden and become brown. Plants are seldom severely damaged.

Figure 3.6. Tea leaves infected by blister blight disease

(ii) Epidemiology: The pathogen survives in tea plantations throughout the year due to favourable environmental conditions in hilly regions of tea gardens. The disease spreads rapidly if a wet spell is followed 4–5 days sunshine hours. The life cycle from spore to spore is about 3–6 weeks, during which time it builds up large masses of basidiospores, which spreads rapidly, to cause secondary infection.

(iii) Management

Remove and destroy young galls before the lower leaf surfaces turn white and spores are released, otherwise the disease will be worse during the next year. Rake up and remove fallen leaves. Avoid wetting the leaves when watering. Humid, moist, shady conditions favour gall formation. Chemical controls, such as 0.25% mancozeb, 0.05% calixin, 0.1% baycor or 0.1% bayleton are limited in effectiveness and must be applied before infection occurs. Start sprays at bud break and continue through the first of June at 7 to 14-day intervals.

The incidence of the disease can be reduced by thinning the shade, timely weed control and pruning at right time. Need based application of 1% Bordeaux mixture or 0.2% copper oxychloride also help in controlling the disease.

Hexaconazole, bitertanol, propiconazole and tebuconazole are the most effective systemic fungicides recommended against blister blight. A combination schedule of triazole fungicides (hexaconazole, propiconazole) and morpholine fungicide (tridemorph) has been recommended recently.

Spray hexaconazole (1 ml/l of water) or tridemorph (2 ml/l of water) at 15 days interval followed by copper oxychloride (3 ml/l of water) at 7 days interval to control blister blight of tea.

3.2.5. Grey Blight and Dieback, *Pestalotia theae/Pestalotiopsis theae*

(i) Symptoms: This disease is caused by *Pestalotia theae/Pestalotiopsis theae*. Mature leaves (Fig 3.7), young shoots and bare stalks are affected by this pathogen. Infection on young shoot results in dieback of shoots. Dieback of shoots became a major problem mainly due to continuous shear harvesting. Grey blight adversely affects the health of the bushes, which in turn affects yield while dieback of young shoots directly leads to substantial crop loss. Studies indicated that the disease incidence is in its peak during July to December. The crop loss due to the disease is 17%. The economic threshold level of the disease is fixed as 18% at a sale price of Rs. 50/- per kg of made tea.

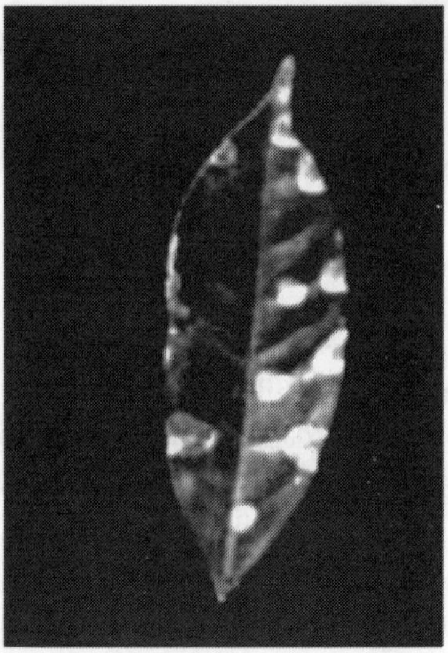

Figure 3.7. Grey blight infection on tea leaf

(ii) Management

Spraying of carbendazim, mancozeb, or thiophanate methyl is suggested to control the disease.

3.2.6. Collar Canker, *Phomopsis theae*

(i) Symptoms: A wound pathogen, *Phomopsis theae* is responsible for the disease (Fig 3.8). It is prevalent in young tea and clones are more susceptible than seedlings. *In vitro* studies indicated that the pathogen completes its life cycle in 10–13 days. Impact of predisposing factors on disease development indicated that the nature of soil is an important factor. Disease incidence was more in gravelly soil.

(ii) Management

Deep planting, application of fertilizer near to plant collar and mulching with raw coir pith close to the collar predispose the disease. Soil drenching of the systemic fungicide carbendazim and application of the same as wound dressing provided satisfactory control. However, the efficacy was superior when biological control agents like *Trichoderma harzianum* and *Gliocladium virens*, were applied both to the soil around the bush and used for wound dressing.

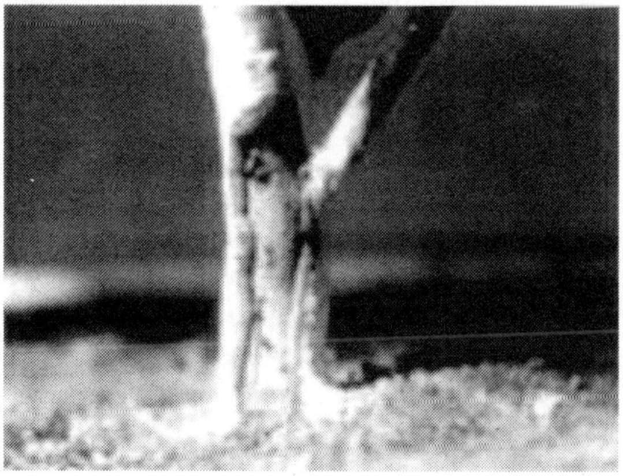

Figure 3.8. Collar canker on tea

3.2.7. Thorny Stem Blight, *Tunstallia aculeata*

(i) Symptoms: The disease is caused by the fungus *Tunstallia aculeata*. The spread of the disease was higher when the pruning timing coincided with the sporulation period. There was also a direct correlation between root carbohydrate reserves and disease development.

(ii) Management

An integrated management strategy combining replanting with resistant genotype, cultural control, chemical and biological control measures was found ideal.

3.2.8. Black Root, *Rosellina arcuata*

The black root disease generally occurs in soils that are rich in organic litter, in cool or cold habitat. It is a major problem in southern India at elevations of 1500 m above MSL and Darjeeling (West Bengal). Occasionally, it is also noticed at mid and low elevations.

(i) Symptoms: The fungal mycelium is cod-webby in appearance, whitish grey or black in colour, spreading rapidly on the surface of the soil on organic debris. When the mycelium comes in contact with tea roots, it penetrates the bark and produces typical star-shaped mycelial structures on the surface of the wood beneath the bark. Very rarely black spherical perithecia are seen at the collar region of the plant. Apart from tea, the disease also affects certain green manure and shade plants commonly grown in tea plantations.

(ii) Management

Organic debris, up to 15 m from the focus of infection should be removed by gentle scraping. Broadcasting murate of potash at 60 g/sq. m. in the disease affected area has been found to be beneficial. Soil drenching either with 0.3% mancozeb or 0.05% carbendazim has been found effective in controlling the disease (Chandra Mouli & Parthiban, 1991). Black root disease can be controlled by avoiding burial of prunings in infested field and incorporation of *Trichoderma virens/T. viride* at 200 g/plant at the time of planting.

3.2.9. Brown Root, *Fomes noxius*

(i) Symptoms: The brown root is the most common root disease of tea. This pathogen mainly spreads by root contact. Hence, the disease occurs in distinct concentric patches in tea fields (Venkata Ram, 1975). The fungal mycelium is fawny in colour. The infected root disintegrates in advanced stages of the disease and becomes spongy. When such roots are split longitudinally, typical brownish honeycomb-like reticulations are seen. The surface of the root is heavily encrusted with soil. The fructifications are typical bracket-shaped, but occur rarely.

3.2.10. Red Root, *Poria hypolateritia*

(i) Symptoms: The red root disease is extremely common in all tea areas of southern India. The fungal mycelium is initially white, then turns pink to pale red and in later stages it is dark red, almost black in colour. The mycelium is generally present as a sheet on the root surface and the typical red colour is invariably seen, when the root is rubbed vigourously under running water and held to bright light. The disease spreads both by root contact as well as by free mycelial spread to a limited extent. Roots of the affected bushes are encrusted with soil. In advanced stages, the affected root becomes spongy. It takes almost 5 years for a mature bush to succumb following the attack.

3.2.11. Root Splitting, *Armillaria mellea*

(i) Symptoms: The root splitting disease occurs in isolated areas, mainly in South India. It derives its name from longitudinal fissures observed on the bark of the invaded tea roots. It can also be observed readily by the characteristic black rhizomorphs which resemble a shoe lace and are seen in large numbers. The fungal mycelium which is white in colour grows densely under the bark and due to the pressure the bark splits. The fungus invades the host plant either by root contact or rhizomorphs which are capable of traversing long distances through the soil from the food base. The fruiting body is typical mushroom-like basidiocarp, and although it is rarely recorded. When it occurs, it is found in abundance around the collar of the diseased plant.

(ii) Management of Brown Root, Red Root and Root Splitting

These root diseases can be controlled by isolating the infected area, opening trenches of 1.3 m deep and 45 cm width and uprooting and burning of infected bushes. Rehabilitating the soil with Guatemala grass and use of biocontrol agents at 200 g/plant (*Trichoderma harzianum* for red root and root splitting; *T. harzianum, T. viride, T. hamatum, T. resei* and *T. koningii* for brown root and *T. virens* for red, brown and root splitting) effectively check the diseases.

Soil fumigation with metham sodium and carbon disulphide gives effective control of these diseases.

3.2.12. Red Rust, *Cephaleurus parasiticus, C. mycoidea*

This is an algal disease common on tea and many other plants like coffee, guava, citrus, cashew and other cultivated and jungle trees.

(i) Symptoms: Infection starts from tip of the stem, cotyledon or collar region as dark brown discolouration. The lesions are further extended to the leaves through petiole resulting in wilting and subsequent defoliation of leaves. Defoliation and dieback are the characteristic symptoms of the advanced stages of infection.

(ii) Epidemiology: The disease is more severe during south-west monsoon period (June-September). The pathogen survives in dead and decaying matter.

(iii) Management

- The entire infected and dead seedlings should be removed from nursery and destroyed.
- A combination of seed dressing and soil drenching with 0.2% copper hydroxide (Kocide) is very effective in managing the disease.
- Soil drenching with 1% Bordeaux mixture was also found effective.
- The drainage facility in the garden should be improved.

3.2.13. Camellia Yellow Mottle Virus

(i) Symptoms: Camellia yellow mottle virus is transmitted by root grafts and propagation of diseased stock. This virus causes irregular, yellow, mottled or splotchy patterns of various sizes and shapes on the leaves. Some leaves may turn entirely yellow. Irregular white blotches will appear on infected flowers.

(ii) Management

There is no chemical that will cure the virus. Plant only virus-free plants obtained from a reputable nurseries.

Table 3.2. Disease management

Diseases	Crops stages affected	Occurrence	Common chemicals recommended
Blister blight	Pluckable shoots	Cloudy weather; relative humidity 80% and temperature below 25°C	Copper oxychloride, Bordeaux mixture

Table. 3.2. (Contd.)

Diseases	Crops stages affected	Occurrence	Common chemicals recommended
Black root, red root	New area brought under cultivation after clearing jungles	Damp weather	Soil fumigation with metham sodium and carbon disulphide
Brown root	New area brought under cultivation after clearing jungles	Common in low elevation	Soil fumigation with metham sodium and carbon disulphide

Table 3.3. Maximum residue limit (MRL) of certain pesticides on tea

Name of the chemical	Pesticide residues MRL (mg/kg)	Name of the chemical	Pesticide residues MRL (mg/kg)
Acephate	0.1	Fenazaquin	—
Aldrin/Dieldrin	0.02*	Fenitrothion	0.5
Bitertanol	0.1	Glyphosate	0.1
Cartap hydrochloride	0.1	Gluphosinate ammonium	—
Carbofuran	0.2	Hexaconazole	0.05
Carbosulfan	0.1	Heptachlor	0.02*
Chlorpyrifos	0.1	Lambda cyhalothrin	1
Chlordane	0.02*	Malathion	0.5
Cypermethrin	0.5	Monocrotophos	0.1*
Diflubenzuron	0.05	Oxydemeton methyl	0.05*
DDT	0.2	Paraquat	0.1
Deltamethrin	5	Propargite	5
Diazinon	0.05	Propiconazole	0.1
Dichlorvos	0.1	Profenofos	0.1*
Dicofol	20	Phosalone	0.1
Dimethoate	0.05	Copper oxychloride	Exempted
Endosulfan	30	Sulphur formulations	Exempted
Ethion	2	Quinalphos	0.1
Fenvalerate	0.05*	Tridemorph	20
Fenpropathrin	0.02 (G)	2, 4-D	0.1

* Indicates lower limit of analytical determination

3.3. NEMATODES

Root-knot, lesion and burrowing nematodes are recognized as the major limiting factors in successful production of tea crop.

3.3.1. Root-knot Nematodes, *Meloidogyne incognita* (in nurseries), *M. brevicauda* (in field)

Barber (1901) reported *Meloidogyne* sp. in tea seedlings from south India. Seshadri and Kumarswami (1964) reported *M. brevicauda* on tea in Tamil Nadu. Root-knot nematodes are the most frequently encountered in the tea growing regions of India. Most of the species of *Meloidogyne* except *M. brevicauda* are associated only with nursery plants.

(i) Symptoms

(a) On Young Tea: The first report of root-knot nematode in young tea was from south India, where large number of seedlings were found infected (Barber, 1901). *Meloidogyne incognita, M. javanica* and *M. hapla* have been found associated with tea in India.

Young nursery plants, both seedlings and vegetatively propagated clonal plants, are severely damaged by root-knot nematodes. Seedling plants in which both the tap-root and lateral rots are severely attacked suffer greater damage than do clonal tea plants of equivalent age, probably because seedling tea plants possess less than half the root bulk of clonal plants of similar age. A marked increase in resistance is observed between 8 and 15 months.

(b) On Mature Tea: *M. brevicauda* is the only root-knot nematode that is a serious pest of mature tea. Affected bushes are unthrifty, have severely galled main roots and lack feeder roots. *M. brevicauda* occurs in southern India at altitudes of 3000 to 3500 ft. in the Wynaad area (Venkata Ram, 1963). It can attack both seedlings and mature tea.

Root-knot nematodes (*M. brevicauda*) cause severe galling on the roots of mature tea bushes with resultant decline (Fig 3.9). Tea plants of all ages are susceptible to infection by this nematode which has been recorded in only south India and Sri Lanka.

Figure 3.9. Heavy root galling of tea with *Meloidogyne brevicauda* (Sivapalan, 1972)

The infected bushes grow more slowly and the leaves tend to become smaller, dull and yellow in appearance. The roots show the characteristic swelling and pitting that is typical of *M. brevicauda.*

The effect of infection on the growth of tea bushes is most evident during the period of recovery from pruning. Healthy bushes are usually harvested about 16 weeks after pruning, but the harvesting of severely infected bushes is necessarily delayed up to a further 6 months. The production of new branches by infected bushes is very restricted, and as a consequence the cropping capacity of the bush is reduced. The most severely affected bushes fail to recover from pruning.

(ii) Life Cycle: The average size of mature female of *M. brevicauda* is about 5 or 6 times that of a mature female of *M. incognita*. Despite this massive size, the females are often observed to be empty, with only a few eggs. The mean hatch per egg mass is around 10, whilst in the other common species this is in the order of 200 to 600 juveniles/egg mass. The rest of the life cycle is very similar to the other species of *Meloidogyne.*

(iii) Management in Nurseries

(a) Physical Methods: Soil solarization in nurseries is adequate to manage root-knot and other nematodes as the plants develop complete resistance by about 9–15 months (Gnanapragasam *et al.*, 1989).

(b) Cultural Methods: Every effort should be made to raise the nurseries in nematode-free soil. Incorporation of organic matter into nematode infested soil is known to suppress the incidence of root-knot nematodes on tea.

The cultivation of marigold as a pre-plant crop considerably reduces the population of *Meloidogyne* spp. and *Pratylenchus loosi.* The reduction in the nematode population is significantly greater than when the land is left fallow. The interplanting of marigold in tea has resulted in a 7% increase in yield compared to 10% increase that followed soil fumigation with DBCP.

The rehabilitation crops such as Mana grass (*Cymbopogon confertiflorus*) or Guatemala grass (*Tripsacum laxum*) (which help to improve the soil) are resistant to *Meloidogyne* spp. The soil population of parasitic nematodes decline rapidly when these grasses are grown for one or more years before replanting tea.

(c) Chemical Methods: In field trials, aldicarb at 0.5 and 1.0 g and carbofuran at 1 g/tea seedling were found effective for the control of *M. incognita* and also caused significant increase in plant height (Basu & Gope, 1985).

(d) Integrated Methods: Combination of neem cake at 5 g/seedling + carbofuran 3G at 1.5 g/seedling improved plant growth characters and reduced number of galls, egg masses and eggs/egg mass (Kalita & Bora, 2006).

(iv) Management Under Field Conditions

(a) Cultural Methods: The first step in the management of nematodes in replanted areas is by ensuring a very thorough uprooting of all old tea roots (up to pencil thickness). In Sri Lanka, such uprooted old tea lands are routinely reconditioned by planting deep-rooted grass species including Guatemala grass and Mana grass which are non-hosts to parasitic nematodes and bring down the residual nematode population (Kerr & Vythilingam, 1966).

Shade trees and green manure crops help to provide shade to the tea plants, aid in N fixation and provide regular supplies of mulch in the form of leaf litter and loppings which help in reducing the nematode population during their decomposition. The cultivation of shade trees and green manure crops susceptible to tea nematodes should be discouraged in tea plantations.

Plants such as marigolds, *Tagetes erecta* and *T. patula*, *Vetiveria zizanoides* and *Eragrostis curvula* have been found to help reduce nematode populations.

Large inputs of organic matter including cow dung and well decomposed plant residues have been reported to suppress nematode populations. In India, neem cake has been recommended to manage root-knot nematodes in the tea field (Muraleedharan, 1991). In Sri Lanka, castor oil cake has been proved to be the most effective one for suppressing nematodes, followed by mahua, neem and karanj cakes (Gnanapragasam, 1991).

(b) Chemical Methods: The recommended nematicides for the management of root-knot nematodes in mature tea include phenamiphos (Nemacur 5%G) and carbofuran (Furadan or Curaterr 3%G) at 10 g/plant applied after pruning (Gnanapragasam, 1987a).

3.3.2. Lesion Nematode, *Pratylenchus coffeae, P. loosi*

The most destructive and most intensively investigated nematode pest of tea is the root lesion nematode *P. loosi*, which is equally destructive to both young and mature tea plants and is a problem in nurseries and in mature tea fields. Crop loss in tea infested with *P. loosi* has been estimated to average between 225–350 kg made tea/ha/year (Sivapalan, 1972). Sasser (1989) estimated the average crop loss on a worldwide scale to be 8.2%.

(i) Symptoms: The infested bushes show unthrifty tea with weak maintenance foliage. The leaves appear pale yellowish and dull mainly due to nutritional imbalances caused by altered rate of uptake of essential nutrients. The heavily infested plants show tendency for premature flowering and fruiting.

The roots of infested plants show typical marked reduction in the feeder root system. The feeder roots appear brownish in colour and dark brown necrotic patches appear on both the feeder roots, as well as the storage roots. Under greenhouse conditions, the damage threshold level of *P. loosi* to tea was estimated to be 40 nematodes/100 g soil at 24°C in a clayey loam soil (Gnanapragasam & Manuelpillai, 1984).

(ii) Host Range: The presence of crops such as sorghum, *Tephrosia vogelii*, *Sesbania* spp. and *Acacia* spp., as well as certain weeds, increase the incidence of this nematode in tea fields.

(iii) Survival and Spread: *P. loosi* is known to survive in host-free soils in the lesions of the larger old storage roots of tea that are left uncleared, following uprooting of old tea fields, for as long as three years.

One of the most important means of spread of *P. loosi* amongst tea areas, is by the dissemination of infested plants to fields from contaminated nurseries. Spread of nematodes could also occur through: (i) movement of infested soil and water; (ii) uprooting of old tea fields; and, (iii) use of contaminated irrigation water in nurseries.

(iv) Management

(a) Cultural Methods: The incorporation of large quantities of green manure such as loppings from dadaps, *Tephrosias* and marigolds, resulted in population decline of *P. loosi*. Mulching with Guatemala grass at 20 to 30 t/ha gave effective control of *P. coffeae* on tea. The use of specific potassium-enriched fertilizers to both young and mature tea plants is known to retard population build up of *P. loosi* and at the same time helps the plants to better tolerate nematode parasitism (Gnanapragasam, 1982).

The cultivation of marigold (*Tagetes erecta* and *T. patula*) as a pre-plant crop considerably reduces the population of *P. loosi* and *Meloidogyne* spp. The interplanting of marigold in tea has resulted in a 7% increase in yield. *P. loosi* is attracted to marigold roots, but upon entry the nematodes die within a week. Roots of marigold in the vegetative stage are effective in this way, but roots of plants in the full bloom or post-bloom are less effective. Consequently, it is preferable to maintain the plants in a non-flowering state by periodic lopping. Growing of grasses like Guatemala (*Tripsacum laxum*), Manna (*Cymbopogon confertiflorus*) and *Eragrostis curvula* helps to bring down the population of *P. loosi*.

The incorporation of oil cakes such as neem, coconut as well as decomposed tea waste, help to curtail pathogenicity caused by *P. loosi*. Increasing the dosage of K fertilizer brought about a decline in the population level of this nematode. Application of N in the form of urea brought about a significant suppression in nematode population.

3.3.3. Burrowing Nematode, *Radopholus similis*

R. similis has been reported attacking tea in Java. Tea is a good host of the burrowing nematode. All stages of the life cycle can be found in tea roots. *R. similis* is known to cause damage to tea at elevations ranging from 200 to 1000 meters.

(i) Symptoms: *R. similis* forms brownish lesions mostly on feeder roots and the infested roots appear dried up. Under greenhouse conditions, the damage threshold level of *R. similis* to tea was estimated to be 28 nematodes/100 g soil at 24°C in a clayey loam soil (Gnanapragasam & Hearth, 1989).

(ii) Host Range: *R. similis* also infects banana and black pepper. The latter is commonly intercropped along with tea in the mid-elevation tea areas of Sri Lanka. Guatemala grass which is often used to recondition old tea fields prior to replanting, was also found to be a host for *R. similis*.

(iii) Survival and Spread: The method of spread of *R. similis* in tea is very much similar to that of *P. loosi*. However, the survival rate in host-free soil is much shorter for *R. similis*.

(iv) Management

(a) Cultural Methods: *Eragrostis curvula*, marigold and *Vetiveria conyzoides* appear to suppress soil populations of *R. similis*.

(b) Host Resistance: In Sri Lanka, tea clone CH 13 was found resistant to *R. similis* (Gnanapragasam, 1983).

3.3.4. Reniform Nematode, *Rotylenchulus reniformis*

R. reniformis has been found to be responsible for large scale casualties in young tea fields. Damage by this nematode species is found in nursery plants and newly planted young tea fields.

(i) Symptoms: Thorne (1961) reported that in tea nurseries in Java, the young tea plants were killed by the attack of *R. reniformis* on the tap root, but not when a strong tap root had first been established.

(ii) Host Range: Perennial crops that are sometimes intercropped with tea, including black pepper, coffee and cloves, as well as grass cover crops, including Guatemala and Manna that are planted in uprooted tea fields for a process of reconditioning, are good hosts of *R. reniformis*

4 COCONUT

The coconut palm (*Cocos nucifera*), originated in South-East Asia, is now cultivated widely throughout the tropics of the world. The largest producers of coconut are the Philippines, Indonesia, India and Sri Lanka. In all the coconut producing countries, it is an important source of export earning despite being a regular constituent of the food. Coconut is an important small holder's plantation crop. The total area under coconut cultivation in India is 2.029 million ha with an annual production of 10.190 million tonnes and the average productivity of 7.0 tonnes/ha is the highest in the world (Bijay Kumar, 2007) (Table 4.1). It provides proteins, fats and some vitamins. Coconut contributes 700 billion rupees to the GDP of the country. Coir and coir products exported from India earned Rs. 2260 million during the above period. The contribution of the crop to the total edible oil pool in India is around 6%. It is supporting the livelihood of 10 million people across the country. The four southern states—Kerala, Tamil Nadu, Karnataka and Andhra Pradesh together accounts for 90% of the total area and production. Coconut is predominantly grown under rainfed conditions in Kerala and parts of coastal Karnataka, Maharashtra and Tamil Nadu. In rest of the country, it is mainly grown under irrigated conditions.

Table 4.1. State-wise area, production and productivity of coconut in India (2005–06)

State	Area in '000 ha	Production in '000 MT	Productivity in MT/ha
Kerala	897.8	4352.3	4.847
Tamil Nadu	370.5	3348.5	9.037
Karnataka	467.5	832.5	1.780
Andhra Pradesh	104.0	613.7	5.900
Maharashtra	18.0	188.1	10.450
Orissa	50.8	189.0	3.720
West Bengal	24.9	222.6	8.939
Assam	19.1	141.0	7.382
Goa	25.3	86.2	3.407
Gujarat	16.4	95.1	5.798
Andaman & Nicobar	25.6	59.9	2.339
Lakshadweep	2.7	36.5	13.518
Pondicherry	2.1	19.2	9.142
Tripura	3.3	4.8	1.454
Nagaland	0.9	0.8	0.888
Total	**2028.9**	**10190.1**	**7.0**

4.1. PESTS

4.1.1. Rhinoceros Beetle, *Oryctus rhinoceros*

(i) Damage: Rhinoceros beetle feeds on unopened fronds and spathes of coconut. Characteristic fan like geometric cuts in the newly emerged frond are observed (Fig 4.1). Infestation on spathe often destroys the inflorescence and prevents the production of nuts.

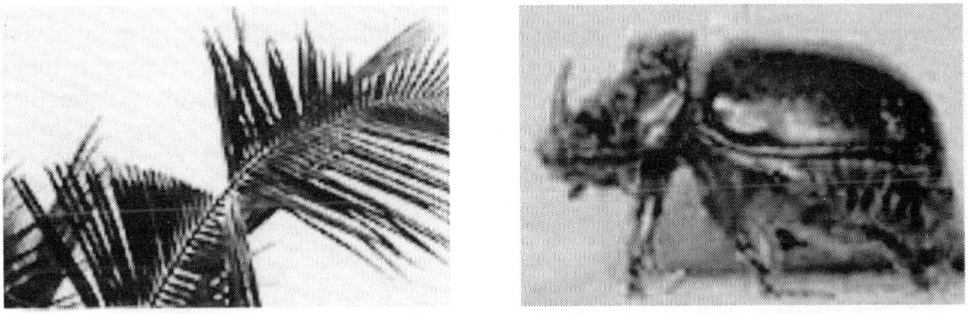

Figure 4.1. Rhinoceros beetle damage on coconut and adult beetle

(ii) Management

Releasing of 10–15 baculovirus infected beetles/ha has been found to be effective in checking the population of the rhinoceros beetle. These releases initiate a self perpetuating disease in the beetles through copulation, contamination through excreta, etc. The virus can infect all stages of the pest and the infection is passed down to the progeny also. However, re-release of the pathogen is needed after 3–5 years.

Metarrhizium anisopliae or the green muscardine fungus, is another biocontrol agent which can infect all stages of rhinoceros beetle. The fungus can be easily multiplied in coconut water, tapioca chips, etc. On inoculating the fungus in decaying organic matter, it thrives well and cause a good degree of mycosis of the grubs resulting in reduction in black beetle population.

Clerodendron infortunatum, a commonly found weed is a promising botanical. Grubs or black beetles become deformed and sterile when fed on feed containing this plant. The leaves, shoot, flowers and root can be incorporated in the composting material and the larvae that hatch out in such pits shall be affected by the contaminated feed.

Integrated pest management includes hooking out of beetle, placement of perforated polythene sachets containing phorate in leaf axil and adoption of biological control measures using baculovirus.

- Treat manure pits with carbaryl/chlorpyriphos dust to kill the grubs.
- Soak castor cake at 1 kg in 5 litres of water and keep in mud pots in the garden to attract adults. Collect the adults and allow them to swim in a tray containing *Oryctes* baculovirus and release them back in the field.

- Hook out and kill the adults. Fill the crown with the mixture of chlorpyriphos dust and fine sand (1:1 by volume).
- A commercial slow release formulation of pheromone to attract adult beetles is also available.

4.1.2. Red Palm Weevil, *Rhyncophorus ferrugenius*

(i) Damage: Red palm weevil grubs feed on the inner tissues of trees. Infected palms show holes on trunks and a viscous brown fluid oozes out. Chewed up fibre is extruded through the holes. Longitudinal splitting of leaf base and wilting of inner leaves can be observed (Fig 4.2). Gnawing sound produced by grubs feeding inside will be audible. Symptoms become clear at an advanced stage when the crown topples. Eventually the tree dies.

Figure 4.2. Red palm weevil damage on coconut and adult beetle

(ii) Management

- Injection of 1% endosulfan into the tunnels in the stem, trapping of adult weevils using baits and field sanitation are effective in controlling the pest.
- Pyrecon-E (derivative from Pyrethrum plant) has been a time tested effective chemical for the control of red palm weevil.
- Fill the crown with the mixture of chlorpyriphos dust and fine sand (1:1 by volume).
- Set up attractant traps (mud pots) containing 2.5 litres of toddy + 5 ml acetic acid + 5 g yeast.
- Scoop out the frass from the hole. Then insert one aluminium phosphide tablet inside the tunneled trunk. Plug the hole with mixture of wet clay and cow dung.
- A commercial slow release formulation of pheromone to attract adult weevils is also available known as Ferrolure (Fig 4.3).

Figure 4.3. Ferrolure pheromone for attracting adult red palm weevil of coconut

4.1.3. Black Headed Leaf Eating Caterpillar, *Opisina arenosella.*

(i) Damage: Leaf eating caterpillar eats away the green portion of the leaves giving a blighted appearance (Fig 4.4). Caterpillar feeds on green matter from the lower leaf surface, remaining in galleries made of silken web and frass. Palm present a burnt up appearance.

Figure 4.4. Black headed caterpillar damage on coconut

(ii) Management

Neem oil/soap sprays gave effective control of this pest.

Root Feeding: Select fresh and live feeder root about 60 cm away from the tree trunk. It should be pencil thick and brown in colour. Cut the root sharply at an angle and insert in a solution containing monocrotophos 36 WSC 10 ml + water 10 ml in 7 × 10 cm polybags for 3 days. Secure the bag tightly to the roots (Fig 4.5).

Effective biological control measures have been evolved using parasitoids at different stages. An exotic parasitoid, *Trichogramma embryophagum* has been found to effectively parasitize the eggs in silken galleries. *Apanteles taragamae* is an important endoparasitoid of 1st and 2nd instar caterpillars. The frass of the young larvae acts as a kairomone and attracts the females for parasitism. This parasitoid holds good potential for pest suppression. *Goniozus nephantidis* is quite hardy and gregarious. It parasitizes medium and fully developed larvae. *Bracon hebetor* parasitizes medium sized or fully developed larvae in the coastal areas. A single parasitoid parasitized 22.6 ± 7.18 late instar caterpillars during its life period. The elasmid parasitoid, *Elasmus nephantidis* is the only true prepupal parasitoid of *O. arenosella*. It is monophagous and a gregarious ectoparasitoid. It is well adopted to thrive during summer season, which is the peak period of activity of the pest. The bethylid parasitoid, *Brachymeria nosatoi* is found to possess the major attributes of an efficient biocontrol agent of the pest. It is a sturdy parasitoid prevalent in nature almost throughout the year. Its natural incidence is quite high (up to 38.8%), even during summer season, which is incidentally peak period of pest abundance. There has always been a preponderance of females, which have greater searching ability and locates and parasitizes host pupae remaining inside the cocoons in silken galleries. The 5th and 6th instar caterpillars were most readily accepted by the parasitoid. The pupal parasitoid, *Xanthopimpla punctata* is an outstanding example of a polyphagous species of parasitoid which turned out to be a better killer of the pest than even other host specific and gregarious species of parasitoids.

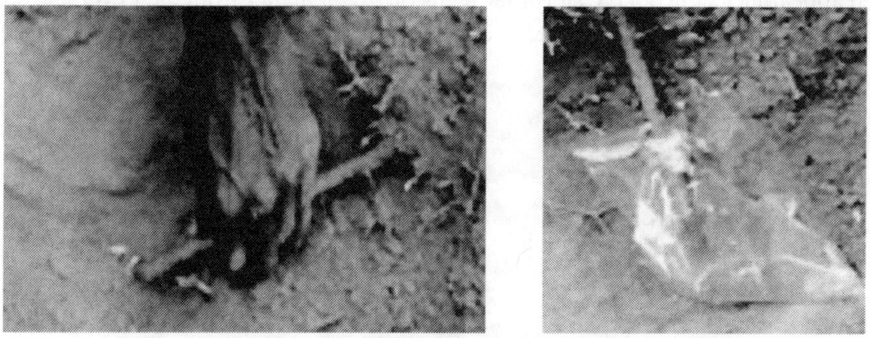

Figure 4.5. Root feeding of monocrotophos to control black headed caterpillar

The optimum doses of release of indigenous parasitoids for effective suppression of *O. arenosella* were 20.5% for *G. nephantidis,* 49.4% for *E. nephantidis* and 31.9% for *B. nosatoi*, when the parasitoids were released individually and 40.4% when a combined release was made. The release at above doses effected 81 and 83% reduction in population density of the pest in the case of individual releases of *G. nephantidis* and *E. nephantidis*, respectively. Release of the parasitoids in the early stage of pest build up was proved absolutely necessary for effective suppression of the pest.

4.1.4. Coreid Bug, *Paradasynus rostratus*

(i) Damage: Coreid bug (nut crinkler), *Paradasynus rostratus* has become a serious problem in many parts of Kerala. It causes damage to the buttons and tender nuts. Nuts are shed in severe cases of infestation and also deform the nuts with characteristic crevices on the husk just below the perianth (Fig 4.6). There will also be gummy exudation from such crevices. Mostly such nuts will be barren.

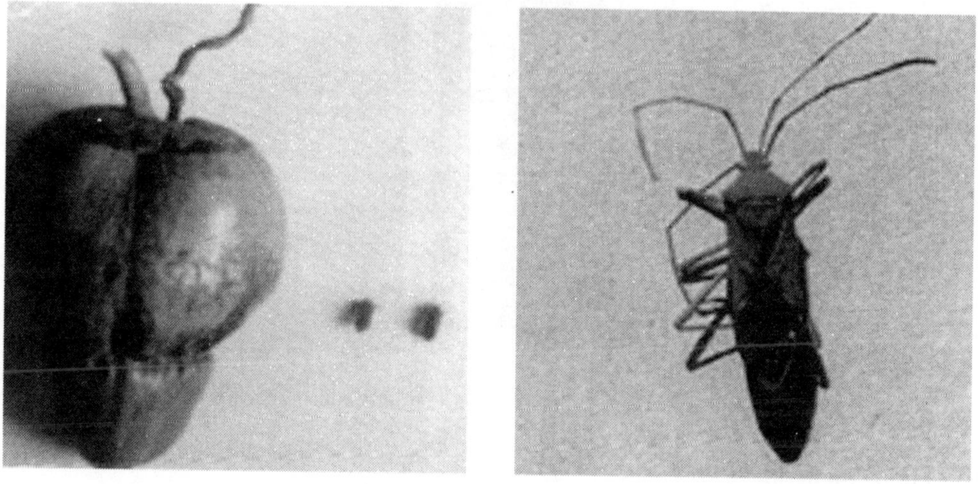

Figure 4.6. Coreid bug damage on coconut and adult bug

(ii) Management

Spraying carbaryl or endosulfan 0.1% on the unopened spathes and bunches (except the newly opened inflorescence) will control the pest.

4.1.5. Lace Bug

(i) Damage: Lace bug sucks sap from coconut foliage and acts as a vector in transmission of phytoplasma from root wilt affected palms to healthy palms.

(ii) Management

This can be controlled by spraying endosulfan or monocrotophos both at 0.01%.

4.1.6. Termites

(i) Damage: It is estimated that nearly 20% of the coconut seedlings are damaged by termites particularly in laterite soil.

(ii) Management

Drenching the nursery with 0.05% chlorpyriphos twice at 20–25 days interval and swabbing the affected trunk with the same chemical are affective.

4.1.7. Scale, *Aspidiotus destructor*

(i) Damage: The scales are severe during hot summer. In severe cases, leaflet turn yellowish, wither and dry. Scales may also congregate on green nuts and suck the sap (Fig 4.7).

Figure 4.7. Scale infestation on leaf and nut of coconut

(ii) Management

Spraying with 0.03% diazinon, dimethoate, phosphamindon or monocrotophos is effective for the management of scales.

4.1.8. Slug Caterpillars, *Latoia lepida*, *Macroplectra nararia*, *Contheyla rotunda*

(i) Damage: Caterpillars are fleshy, slug like and have a series of tufts of spines highly irritating to touch. They pupate in hard shell like grayish cocoon in trunk. Caterpillars initially feed gregariously on the leaves and then disperse (Fig 4.8).

Figure 4.8. Slug caterpillar damage on leaf of coconut

(ii) Management

Spray solution of 2 g carbaryl/litre of water when the larvae are in congregation. Root feeding also controls the pest

- Select fresh and live feeder root of about 60 cm away from the tree trunk. It should be pencil thick and brown in colour.
- Cut the root sharply at an angle and insert in a solution containing monocrotophos 36 WSC 10 ml + water 10 ml in 7 × 10 cm polybags for 3 days. Secure the bag tightly to the roots.

4.1.9. White Grub, *Leucopholis coneophora*

(i) Damage: Grubs (Fig 4.9) are found in the roots. Leaves are yellow and sick pale. Severely infested palms exhibit button shedding and tapering of crown.

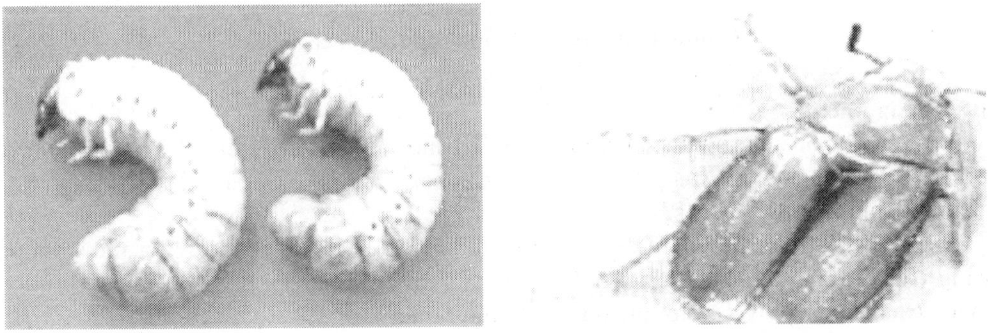

Figure 4.9. Grubs and adult of white grub on coconut

(ii) Management

Incorporate in the top soil (up to 15 cm) phorate 10 G at 100 g/palm prior to the on set of monsoon and also after the cessation of monsoon (May-June, September-October). Provide copious irrigation.

4.1.10. Eriophyid Mite, *Aceria guerreronis*

(i) Damage: Recently coconut gardens in some parts of Kerala, Tamil Nadu, Andhra Pradesh and Karnataka were seriously affected by a non insect pest called eriophyid mite. These mites infest by sucking sap from the soft tissues of buttons. In the initial stages, symptoms are seen as triangular patches close to perianth. Later, because of the continuous desapping by various stages of mites present beneath the inner bracts of perianth, brown coloured patches are formed (Fig 4.10). As the nuts grow in size, the injured patches become warts and then develop into longitudinal splits on the surface of nuts. The liquid oozing from these patches dries and as a result dried decayed matter is noticed. The damage thus caused affects the quality of husk and dehusking becomes difficult.

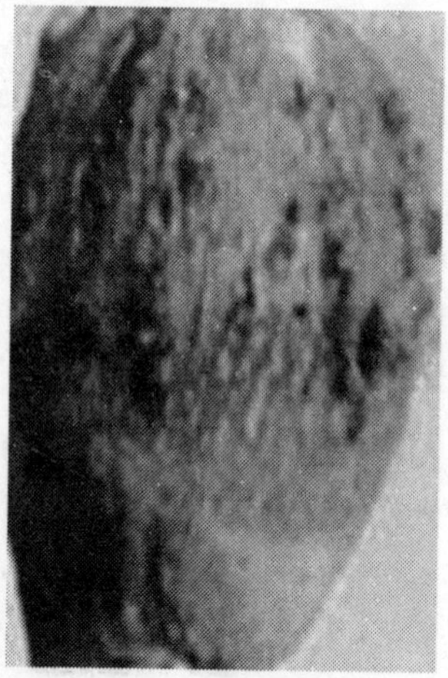

Figure 4.10. Eryophid mite damage on coconut

(ii) Management

Neem/garlic oil/soap emulsion has emerged as one of the potent measures for the control of mites. The fungus, *Hirsutella thompsonii* has been found to be an effective microbial agent for annihilating this pest.

Dissolve 1 kg biocatch in 200 litres of water and keep it overnight. Filter the solution using a muslin cloth. To the filtrate add 500 ml of Nimbecidine. Spray the resultant solution on the frond and newly developing nuts at 15 days interval.

4.1.11. Vertebrate Pests

(i) Damage: Monkey damaged the maximum number of nuts (150–200/15 days in March 1997), rodents mainly *Bandicota bengalensis* damaged 15–20 nuts/15 days and wild boar the least, 0–4 nuts/15 days in Dakshina Kannada district of Karnataka.

Rodents are a menace in coconut and cocoa gardens damaging the nuts and pods. Trapping and poison baiting are employed to control these pests.

(ii) Management

Trunk banding (metal), burrow baiting, clean basin cultivation, crown cleaning, lacing base of grown-up tree trunks with waste oils or covering base of seedlings with thorny bamboo or a mat of dry sticks, timely harvests and pick-up of fallen nuts (against porcupine, rat and squirrel) have proved effective.

(a) Integrated Method: Clean cultivation, crown cleaning, trunk banding, burrow baiting, Ibex fencing with at least 1 m of open space on either side, and covering base of seedling (up to 2–3 years) with tightly packed cover of thorny bamboo sticks or porcelain pipes wherever coconut palms were regularly prone to rodents damage saved 46% nuts compared to control against rodents, monkey and wild boar depredations. These practices to some extent afforded protection against raids by bison and elephants also.

Table 4.2. Pest management

Name of the pests	Symptoms	Control measures
Black headed caterpillar, *Opisina arenosella*	Damage the leaves by scraping green matter; caterpillar lives in galleries made of silk and excreta on the lower surface of the leaves; in case of severe attack, whole plantation presents a burnt up appearance from a distance; it results in poor coconut yield.	**Nursery:** Spray 0.15% carbaryl or 0.05% monocrotophos; if the incidence is mild, release parasites. In severely infested gardens, treat the trees by root feeding of monocrotophos at 10 ml + 10 ml of water/ palm. Before adoption, harvest all mature nuts and no harvest should be made at least for a month. **Root Feeding Technique:** Expose some roots by excavating the soil at the base of the palm (No need to trace out the tip of the root); select a matured dark brown pencil thick root near the base and cut without damaging the cut end of the root and insert it in a small (15 × 10 cm) polyethylene cover with monocrotophos at 10 ml/palm diluted with an equal quantity of water; adjust the cover so that the cut end of the root is completely immersed the solution; the solution will be absorbed by the palm within 24 hours; the pest will be controlled within 7 to 10 days after treatment.
Red palm weevil, *Rhynchophorus ferrugineus*	Grubs feed soft tissues in the trunk of 2–10 years old palms; exudation of brownish fluid and protruding of chewed fibre can be seen from the holes present on the trunk; causes very severe damage, especially when a large number of grubs bore into the soft growing parts.	Treat the affected palm by root feeding (as above) with monocrotophos 10 ml + 10 ml of water.
Rhinoceros beetle, *Oryctes rhinoceros*	Beetle bore into unopened tender fronds and spathes and throws chewed fibrous mass; this mass is commonly seen in the burrow made by the beetle at crown region and its presence indicates the pest activity; breeds in manure heaps.	Hook out beetle from the crown and kill it; fill the hole and leaf axils with 5% carbaryl + sand mixed in equal proportions.

Table 4.2. (Contd.)

Name of the pests	Symptoms	Control measuress
Slug caterpillars, *Macroplectra nararia, Latoia lepida*	Feeds on the leaves leaving only the mid rid; occurs occasionally.	Spray 0.15% carbaryl at 3 g/lit of water.
Termites, *Odontotermes obesus*	Seedlings are attacked through the nut or at the base of the young shoot; older palms through trunk to crown; affected seedlings wither and die.	5% carbaryl dust should be incorporated in to the soil; cover the germinating nuts with a layer of sand.
Cockchafer beetle, *Leucopholis coneophora*	Grubs feed on and injure coconut roots during March to May; young plants become stunted and flowering is delayed.	Deep ploughing during March-May, coupled with a soil application with 5% Aldrin or Chlordane, gives a satisfactory control.
Coconut bug, *Stapanitis typicus*	Transmits wilt (root disease).	Spray 0.1% BHC + 0.1% DDT
Leaf-rollers: *Gangara thyrsis, Suastus gremius*	Caterpillars roll leaves and feed inside; more serious in nurseries.	Collect and destroy larvae and pupae; spray 0.1% BHC + 0.1% DDT.
Coconut scales: *Aspidiotus destructor, Anonidiella orientalis, A. ficus*	The scales feed on the underside of the leaves in colonies and may cause death of leaflets; common during dry months.	Spray with 0.03% diazinon, dimethoate, phosphamindon or monocrotophos.

4.2. DISEASES

4.2.1. Basal Stem Rot or Thanjavur Wilt, *Ganoderma lucidum, G. applanatum*

Among all coconut diseases, *Ganoderma* wilt is most destructive disease reported from Karnataka, Tamil Nadu, Andhra Pradesh, Orissa, Maharashtra and Gujarat apart from Kerala (Nambiar & Rethinam, 1986, 1987). The disease was first noticed in the coastal area of Thanjavur district in Tamil Nadu, after the cyclones in 1952 and 1955 (Vijayan & Natarajan, 1972). In Thanjavur district, the disease incidence ranged between 10–15%. The disease is lethal and generally trees between the age of 10–30 years are affected (Rethinam, 1984).

(i) Symptoms: In Tamil Nadu and Andhra Pradesh, Thanjavur wilt, otherwise called Ganoderma wilt is a serious problem. The symptoms are presence of bleeding patches at the stem base, premature yellowing and drooping of outer whorl of leaves and gradual drying of spindle (Fig 4.11). The other symptoms include decay of root system, flaccid spindle leaves, browning of outer leaves, and appearance of bleeding patches on the basal region of the stem. In advanced stage, bracket of fungus will be seen at the base of the tree. Normal process of flower and bunch development is disrupted to the extent of more than 70% leading to button shedding. The leaves are initially yellow, later turn brown and drop off. The death and decay of roots is also most common symptom. In severe cases, more than 70% of root rotting was observed.

Figure 4.11. Basal stem rot in coconut

(ii) Epidemiology: Sandy loam soils are conducive for *Ganoderma* wilt. Soil moisture stress during summer, water stagnation during rainy season coupled with presence of infected stumps in field favour the development and spread of the disease. Hard subsoil impeding root penetration was found to predispose the palms to infection. Coconut plants of 10 to 30 years of age are more susceptible than other age groups. The pathogen is favoured by mean maximum temperature at soil depth of 15 to 30 cm. The pathogen is also influenced by rains suggesting its linear spread.

(iii) Management

Integrated control measures include removal and destruction of dead palms and palms in advanced stages of disease and/or stumps along with bole and root system, isolation of diseased palms from healthy palms by digging isolation trenches of 1 m deep and 30 cm wide, providing adequate soil moisture through irrigation coupled with mulches, avoiding flood irrigation and overcrowding, providing good drainage facilities, phytosanitation, root feeding of calixin, drenching the basin with 1% Bordeaux mixture and application of 50 kg FYM and 5 kg neem cake/palm/year. Raising banana as intercrop is recommended in diseased garden, since banana is resistant to the pathogen. Burying coconut husks and application of organic manure helps to check the spread of the disease. Application of manganese sulphate at 225 g/plant reduced the disease intensity.

Light irrigation coupled with Bordeaux mixture drenching (40 litres of 1% Bordeaux mixture around the trunk in a radius of 1.5 m) helps in reducing the disease incidence.

Application of tridemorph at 2 ml/100 ml water/palm at quarterly intervals through root feeding or aureofungin sol at 2 g + copper sulphate at 1 g dissolved in 100 ml water.

Application of *T. harzianum/T. viride* along with neem cake at 5 to 10 kg/plant has been found to control the multiplication of the pathogen in sick soils. When combined with phosphobacteria or plant growth promoting rhizobacteria, synergistic effects have been noticed. *T. harzianum* applied along with neem cake reduced the disease index and increased the yield of coconut.

4.2.2. Bud Rot, *Phytophthora palmivora*

(i) Symptoms: The disease is caused by a fungus *Phytophthora palmivora*. The first visible symptom of bud rot disease of coconut is withering, rotting of pale coloured spindle leaf and foul smell (Fig 4.12). Tender leaf and soft tissues of the crown rot into a slimy mass and emit foul odour. Later, the spindle turns brown and topples over. The cross section of the spindle reveals brownish to pink coloured internal tissues. Falling of younger nuts is more common in infected plants.

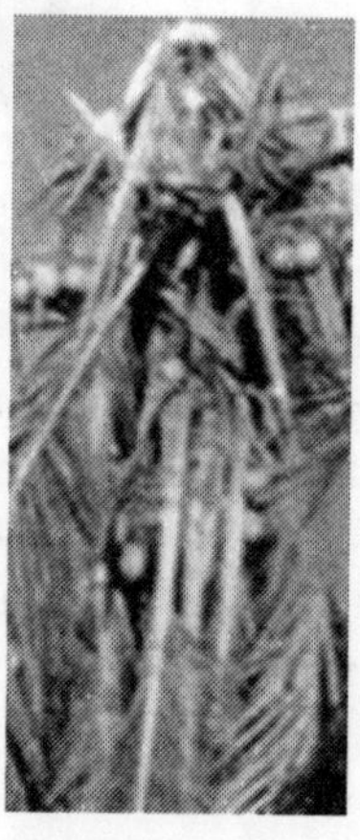

Figure 4.12. Bud rot on coconut

(ii) Host Range: Besides coconut, the pathogen also infects palms, arecanut, oil palm, rubber and hibiscus plants.

(iii) Epidemiology: The minimum temperature of 21–24°C and high relative humidity (97–100%) influences the bud rot spread and development. The young plants in low and moist situation are more vulnerable for infection. The fungus over winters in leaf bases during the dry months in the form of dormant structures such as oospores and chlamydospores.

(iv) Management

To control this disease remove decayed tissues and dress with 10% Bordeaux paste. The disease can be prevented by taking up pro-phylactic spraying of 1% Bordeaux mixture or 0.1% fosetyl AL or 0.3% metalaxyl MZ during the pre monsoon season.

Prophylactic spraying of young palms with 1% Bordeaux mixture, improved drainage, wide spacing and weed control helps to reduce the disease severity. Removal of infected spindle showing initial yellowing symptoms and application of 1% Bordeaux paste reduce the disease incidence. *Bacillus subtilis, B. macerans, B. amyloliquifaciens* and *Myrothecium roridum* have all been found to be checking the growth of the pathogen. The hybrid Malayan Dwarf x West African Tall (Hybrid PB 12) is tolerant.

Cultural practices like improved drainage, wide spacing, and weed control helps to reduce the incidence of the disease.

4.2.3. Stem Bleeding, *Theilaviopsis paradoxa*

The disease is known to occur in all the coconut growing areas of tropics like India, Sri Lanka, Philippines, Malaysia, Trinidad, Ghana and Indonesia. The disease was first reported from Sri Lanka during 1906.

(i) Symptoms: The characteristic symptom of stem bleeding is the dark gummy exudation from the trunk. The causal organism is a fungus *Theilaviopsis paradoxa*. Exudation of reddish brown liquid is observed through longitudinal cracks in the trunk, generally at the base of the trunk (Fig 4.13). Bleeding patches spread through out as the disease advances. The liquid oozing out dries up and turns black. Tissues below the lesions rot and turn yellow and then black. Premature yellowing of leaves is observed in the outer whorl. Trunk gradually tapers at apex and crown size becomes reduced.

Figure 4.13. Stem bleeding on coconut

Growth cracks on trunk, severe summer followed by sudden wetting, imbalanced nutrition, and excess salinity are the predisposing factors for the disease.

(ii) Epidemiology: This is a weak pathogen which makes entry through wounds and cracks. Poor drainage, formation of hard latent pan leading to crippled growth, imbalanced nutrition are the main predisposing factors for its infection. High humidity and moderate temperature favours the disease development. Excessive salinity with high sodium in soil during summer is associated with stem bleeding.

(iii) Management

Control measures include chipping off the infected tissues, wound dressing and root feeding with 5% calixin and application of neem cake along with antagonistic fungi like *Trichoderma*. Remove the affected tissues using a chisel and apply Bordeaux paste and coal tar after 1–2 days (Fig 4.14).

Figure 4.14. Chiseling affected potion and application of Bordeaux paste and coal tar

The integrated management practices recommended are removal of not only dead palms but also those in advanced stages of disease and/or stumps along with bole and root system and destroying; isolation of diseased palms from healthy palms by digging trenches of 1 m deep and 30 cm wide; providing adequate soil moisture through irrigation coupled with mulches; avoiding flood irrigation, overcrowding, preventing wounding of stem which predispose the palms to infection and ploughing to prevent the spread of inoculum; providing good drainage facilities; application of 50 kg FYM and 5 kg neem cake/palm/year and raising *Ganoderma* resistant crop like banana as intercrop.

Neem cake enriched with *Trichoderma virens, T. hamatum* and *T. harzianum* has been found to be very effective in reducing the population of the pathogen in the soil (Gowda & Nambiar, 1988). These antagonistic fungi thrive well in neem cake supplemented with a small quantity of rice/wheat bran thus effecting their multiplication an easy task. *Trichoderma harzianum, T. viride, T. hamatum* and *T. virens* grow very well in rice bran and neem cake (1:1 w/w) and reduced stem bleeding when applied to the soil. Soil application of neem cake and FYM mixed with. *T. virens* showed reduction of stem bleeding up to 31.3%. Soil application of *T. virens* along with neem cake and FYM showed the least disease index and the highest yield.

West Coast Tall was found more resistant to stem bleeding.

4.2.4. Fruit Rot or Mahali, *Phytophthora palmivora*

(i) Symptoms: Shedding of female flowers (buttons) and immature nuts are the symptoms of the disease. Water soaked lesions appear on the young fruits or buttons near the stalk which later develop and result in the decay of the under lying tissues (Fig 4.15). The pathogen is more active during rainy season.

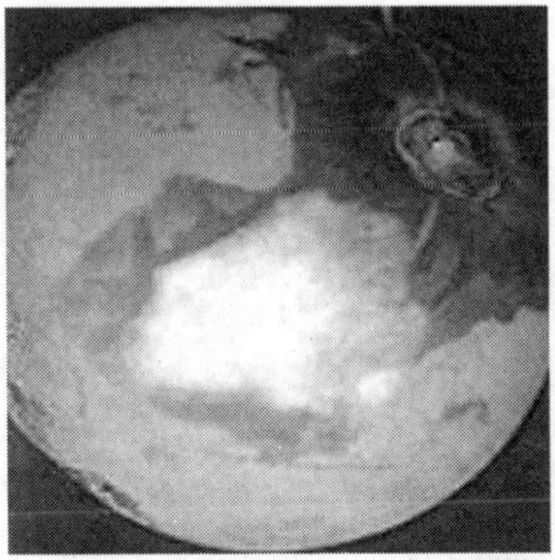

Figure 4.15. Fruit rot of coconut

(ii) Management

Pre-monsoon spraying the crown with 1% Bordeaux mixture (or any other copper fungicide such as copper oxychloride 0.5%) will control the disease. One or two sprayings at intervals of 40 days is also advised. The shed nuts should be collected and burnt.

4.2.5. Leaf Rot, *Colletotrichum gloeosporioides, Exerohilum rostratum*

(i) Symptoms: This fungal disease is seen in palms affected by root (wilt) disease. The first visible symptom is blackening and shriveling up of the distal ends of the leaflets in the spindle. Later the affected portion breaks off in bits giving the infected leaves a fan-like appearance. If not controlled, subsequent spindles also get infected.

4.2.6. Tatipaka Disease (Phytoplasma)

The disease first appeared in Tatipaka village, East Godavari district, Andhra Pradesh, following a cyclone in 1949.

(i) Symptoms: Palms in the age group of 25 to 40 years are more susceptible. Palms below 25 years of age are rarely infected (Rethinam *et al.*, 1989). Development of an abnormally large crown with dark green inner leaves and higher yield is the precursor of disease incidence. Subsequently the crown becomes smaller in size producing progressively shorter leaves (Fig 4.16). The stem begins to taper. The leaves give a fascinated appearance due to improper unfolding of leaflets. The affected tree produces smaller bunches with atrophied barren nuts. The causal agent is suspected to be phytoplasma. There will be reduction in number and size of the roots coupled with extensive root rot. In advanced stage, the affected tree become totally barren without producing even the inflorescence.

Figure 4.16. Tatipaka disease on coconut

(ii) Transmission: The pathogen is transmitted by leaf hoppers (*Sophonia greeni, T. dioscopus nigroclypealis*), lacewing bug (*Stephanitis typica*) and mealy bug (*Palmiculer* spp.).

(iii) Epidemiology: The disease is confined to black alluvial soils and incidence is low in sandy and sandy loam soils. The disease occurs in sporadic form and spreads very slowly at a rate of 3.5% in 5 years.

(iv) Management

- Destroy the affected plants to prevent the spread of the disease.
- Application of manures and micronutrients (Zinc sulphate at 2 kg/palm twice a year and subsequently at 1 kg/palm/year) to the infected plants helps in reducing the disease incidence.
- Spraying with benomyl, catechol and ascorbic acid was able to prevent further deterioration of treated plants.
- Coconut native dwarf *cv.* 'Gangabondam' is reported to be field tolerant.

4.2.7. Root Wilt (Phytoplasma)

This disease is endemic in Kerala and bordering districts of Tamil Nadu.

(i) Symptoms: The symptoms of Root (wilt) disease are flaccidity (characteristic bending or ribbing of leaves), foliar yellowing and marginal necrosis of leaflets (Fig 4.17). extensive rotting of the root system was considered as the major symptom. The capacity to regenerate new roots is drastically reduced with the progress of the disease. Radha *et al.* (1962) reported reduction in yield of nuts up to 80% in palms in advanced stage of the disease. The oil content is very much reduced and the oil loses its flavour. Research conducted at CPCRI

Regional Station, Kayangulam using electron microscope revealed the causal organism as phytoplasma. The disease is spread through insects like lace bug, *Stephanitis typica* and/or plant hopper, *Proutista moesta*.

Figure 4.17. Root wilt of coconut and its lacewing vector

(ii) Transmission: The pathogen is transmitted by the lace bug, *Stephanitis typica* and/or plant hopper, *Proutista moesta*. It is also known to be transmitted by dodder.

(iii) Management

The strategy comprises eradication of disease advanced palms, gap filling with elite disease tolerant seedlings, adoption of mixed farming/inter cropping, organic recycling, providing summer irrigation, application of Mg SO_4 along with fertilizer, control of leaf rot etc. CPCRI identified disease tolerant palms, among Chowghat Green Dwarf and local talls. Progenies of crosses involving CGD × WCT have been found to be promising.

The yield of palms can be sustained through adoption of integrated management practices like removal of disease advanced and juvenile palms, addition of organic matter, raising green manure crops in the basin and their incorporation into the soil, irrigation during summer months and adopting inter and mixed cropping (Muralidharan *et al.*, 1986; Rajagopal *et al.*, 1987). Mixed farming in diseased gardens involving the raising of fodder crops in the interspaces, maintaining milch cows and recycling of organic waste has helped in increasing the yield of palms by 26% (Sahasranaman *et al.*, 1983). Mixed cropping with cocoa also increased the yield by 30% and slowed down the decline of palms. Cultivation of tapioca, elephant foot yam and yam in the interspaces of palms in disease affected gardens for a period of 3 years increased the nut yield by 4.96, 15.57 and 8.07%, respectively (Menon & Nair, 1978).

Integrated Disease Management

- Removal and destruction of dead palms and palms in advanced stage along with roots.
- Isolation of disease palms from healthy palms.

- Avoiding flood irrigation or ploughing infected garden to prevent further spread of the disease.
- Application of 50 kg FYM/200 kg tank silt, 5 kg neem cake/palm/year and raising banana as an intercrop helps to reduce the disease incidence.
- Soil drenching with 1% Bordeaux mixture thrice a year.
- Root feeding with aureofungin 2 g/l + copper oxychloride 1 g/l alternately with calixin 2 ml/litre of water was found effective against the disease.

Table 4.3. Disease management

Name of the diseases	Symptoms	Control measures
Stem bleeding, *Ceratocystis paradoxa*	Exudation of the brownish black liquid through cracks on the trunk; this fluid turns black as it dries; tissues below the bleeding patch become decayed; in young palms tissues rot internally and on piercing the stem a yellow fluid rushes out.	Chisel out the rotten portion; treat the area with Bordeaux paste or coal tar.
Bud rot, *Phytophthora Palmivora*	The initial symptom is the withering of spindle leaf; the leaf later dries up and droops down; if the leaf is pulled, it comes out easily as the base is completely rotten; the rotting proceeds downwards killing the bud which results in the death of the palm.	Clean the rotten and drench the bud and its surrounding portions with 1% Bordeaux mixture; burn the rotten parts; spray 1% Bordeaux mixture on the surrounding palms.
Basal stem rot/ Ganoderma wilt, *Ganoderma lucidum*	Initially infects the roots; the first visual symptom is the appearance of oozing of reddish brown fluid from growth cracks at the base of the trunk; the lesions spread linearly and sideways; mean while the lower whorls of leaves start drooping and dry prematurely; as the lesions spread towards the crown, leaves dry up and ultimately the whole crown dries up and the tree dies.	Cut and remove all dead roots and a few surrounding healthy roots; drench the cut ends of roots with a mixture of 1.5 g of aureofungin. +1 g Copper sulphate in 5 lit of water and cover the root system; repeat the treatment after 15 days; also treat healthy palms in the vicinity as a preventive measure.

4.3. NEMATODES

Of the several nematode species reported on coconut, the important ones are the burrowing nematode, *Radopholus similis* and the red ring nematode, *Rhadinaphelenchus cocophilus*.

4.3.1. Burrowing Nematode, *Radopholus similis*

The burrowing nematode was reported from coconut palms in Kerala, India by Weischer (1967). *R. similis* is the most important nematode pest of coconut and is responsible for considerable amount of root rotting (Koshy *et al.*, 1978). It causes 30% yield loss in coconut (Koshy & Geetha, 1992). The threshold inoculum density required to cause significant reduction in various growth parameters of coconut is 100 nematodes/625 ml sandy loam soil over a period of 5 years under field conditions (Koshy, 1986).

(i) Distribution: *R. similis* occurs in most tropical and subtropical areas of the world and has been reported from coconut palms in Florida, Jamaica, Sri Lanka and India. Surveys conducted in coconut plantations in south India recorded 24% incidence of *R. similis* (Sosamma, 1984).

(ii) Symptoms: The burrowing nematode infested coconut palms exhibit general decline symptoms like yellowing, button shedding, reduction in number and size of leaves and leaflets, delay in flowering and reduced yield which are non-specific. Symptoms on the root are more specific. *R. similis* on infestation produces isolated elongate orange coloured lesions on tender and semi-hard roots. Consequent to nematode parasitization and multiplication, these lesions enlarge and coalesce to cause extensive rotting of roots. Tender roots on heavy infestation become spongy in texture. On semi-hard orange coloured roots, surface cracks are commonly seen. As high as 4,000 nematodes were recovered from one gram (one inch length) of main roots. The drastic reduction in the number and mass of tertiary feeder roots on parasitization by the nematode limits plant growth (Koshy & Sosamma, 1987).

(iii) Life Cycle: *R. similis* is a migratory endoparasite capable of spending its entire life cycle within the roots. Except the fourth stage male juveniles and adult males, all other juvenile stages and adult females are found to be infective. The nematode completes its life cycle (J_2 to J_2) with in 25 days at a temperature range of 25°–28°C (Koshy & Sosamma, 1977).

(iv) Host Range: Large number of crops like banana, black pepper, ginger, turmeric, sweet potato, groundnut, sugarcane, betel vine, etc. are known hosts of *R. similis*.

(v) Histopathology: Burrowing nematodes penetrate the absorbing regions behind the root cap (covered by very delicate epidermis) by the lyses of cells. Such entry points are 1–2 cells in diameter and surrounded by sclerenchymatous cells to a depth of 10–15 cells. The outer cortex harbour maximum number of cavities and nematodes. In the early stages of infestation, the roots contain independent cavities separated by several cells. Consequent to nematode multiplication and lyses of cell walls and cytoplasm, adjacent cavities merge with each other leading to almost complete destruction of the cortex. Longitudinal section of the roots showed nematodes in all stages of development (Koshy & Sosamma, 1987).

(vi) Survival and Spread: Under field conditions, *R. similis* survives for 6 months in moist soil (27°–36°C) and one month in dry soil (29°–39°C). In roots of stumps of felled coconut palms, the nematode survives for up to 6 months (Sosamma & Koshy, 1986). The infested coconut roots yielded maximum number of *R. similis* during October to November and minimum during March to July. A mean soil temperature below 25°C and light rain fall coupled with availability of tender fleshy roots are the factors favourable for *R. similis* multiplication (Koshy & Sosamma, 1978).

Infested coconut seedlings help in the dissemination of the nematode to distant places. Apart from coconut, infested planting materials of intercrops such as areca nut, banana, pepper, ginger, turmeric also serve as sources of inoculum.

(vii) Management

(a) Cultural Methods: Avoiding banana as a shade crop in coconut nurseries and use of nematode-free planting material of coconut and other intercrops are the other management approaches.

Organic Amendments: Application of oil cakes, FYM and green foliage to the basins, growing of intercrops like cocoa that enriches the soil with sizeable quantities of shed foliage which help in the build up of beneficial organisms which may inhibit nematode multiplication.

Application of 50 kg cow dung/FYM, 2 kg neem/marotti cake and 25 kg green manure to the coconut basins in November is quite effective in bringing down the nematode population. Thirty per cent increase in yield has been recorded by application of *Hydnocarpus* sp. oil cake at 8 kg/palm/year in June-July and October-November to the burrowing nematode infested coconut palms (Koshy, 1986).

Green Manure Cropping: *Crotalaria juncea* may be grown as a green manure crop in the basins and interspaces and its incorporation helps in the inhibition of nematode multiplication.

(b) Chemical Methods

Nursery Bed Treatment: In infested coconut nurseries, complete control of *R. similis* can be obtained by soil application of phenamiphos or phorate at 2.5 kg *a.i.*/ha during September, December and May (Koshy & Sosamma, 1979).

Various chemicals tried for the management of *R. similis* on coconut under nursery conditions is presented in Table. 4.4.

Bare Root-dip Treatment: Bare root dip of coconut seedlings in 1000 ppm of DBCP for 15 min eliminated *R. similis* infection.

Main Field Treatment: Application of phorate at 10 g *a.i.*/palm in June-July and in September-October is effective in reducing the population of burrowing nematodes. Maximum increase in yield was obtained with the application of phenamiphos at 10 g *a.i.*/palm.

Table 4.4. Effect of various chemicals for the management of *R. similis* in coconut under nursery conditions

Chemical	Dosage	Duration & type of treatment	Effectiveness
Fensulfothion	5.0 kg *a.i.*/ha	Thrice a year (in Sept., Dec. & March)	Reduced nematode infestation
Phenamiphos	2.5 kg *a.i.*/ha	Thrice a year (in Sept., Dec. & May)	Controlled nematodes in nursery seedlings
Phorate	2.5 kg *a.i.*/ha	Thrice a year (in Sept., Dec. & March)	Controlled nematodes in nursery seedlings

Various chemicals tried for the management of *R. similis* on coconut under field conditions is presented in Table 4.5.

Table 4.5. Effect of various chemicals for the management of *R. similis* in coconut under field conditions

Chemical	Dosage	Duration & type of treatment	Effectiveness
Aldicarb	0.5-1.0 kg *a.i.*/ha	Applied once	Reduced nematode infestation
	10-50 g *a.i.*/palm	Twice at 3 monthly intervals	Reduced nematode infestation
Phenamiphos or Phorate	10 g *a.i.*/palm	Twice a year in June-July & October-November	Increased yield by 30% & reduced nematode population

(c) Biological Methods: Introduction of biocontrol agents like *Paecilomyces lilacinus* and AMF to nursery potting mixture helps in the reduction of nematode population and at the same time impart resistance/tolerance to nematode infestation (Koshy *et al.*, 1993). Among the AMF, *Acaulospora bireticulata* was found to be more effective in reducing the nematode population. Soil application of *Cylindrocarpon effusum* reduced the rate of multiplication of the nematode and damage to coconut seedlings (Sosamma & Koshy, 1978; Koshy & Sosamma, 1987).

(d) Host Resistance: The dwarf *cvs.* Kenthali and Klappawangi and the hybrids such as Java Giant x Kulasekharam Dwarf Yellow, Java x Malayan Dwarf Yellow and San Ramon x Gangabondam recorded the least nematode multiplication and lesion indices (Sosamma *et al.*, 1980). The *cv.* Philippines Ordinary was found to be least susceptible to *R. similis* with lesion index of 1.22 (Sudha, 1998).

(e) Integrated Methods: The following measures are suggested for an integrated management schedule for R. *similis* infestation on coconut palms:

- Application of cow dung, FYM, oil cakes and green manure to the basins. *Crotalaria juncea* may be cultivated in the basins and interspaces and used as a green manure.

- Application of phorate at 10 g *a.i.*/palm twice yearly.
- Avoiding banana as a shade crop in coconut nurseries.
- Use of nematode-free planting material of coconut and other intercrops.
- Use of less susceptible/tolerant *cvs.* such as Kenthali and Klappawangi or their hybrids in infested areas.

Thirty per cent increase in yield and 5 to 10% decrease in disease indices of palms affected with root wilt disease was recorded by the application of *Hydnocarpus* oil cake and phorate granules at 10 g *a.i.*/palm in June-July and in October-November.

4.3.2. Red Ring Nematode, *Rhadinaphelenchus cocophilus*

Thee 'red ring' disease of coconut caused by *R. cocophilus* is very destructive disease on some islands of Caribbean which threatens the existence of this crop. The extensive losses caused by this nematode are a major catastrophe to natives of these areas. The nematode is transmitted by *Rhynchophorus palmarum*. The disease was first reported as occurring in Trinidad by Hart in 1905. It has not been reported from India so far.

At present, most Latin American and Caribbean countries show losses ranging from less than 1% to more than 20% of 1 to 10 year old trees. Thirty five per cent mortality of young coconut palms has been reported in Trinidad and 80% loss in single plantation of Tobago (Esser, 1969).

(i) Distribution: The disease is reported to occur in 20 countries seriously affecting their economy. At present *R. cocophilus* has been reported from the West Indian Islands of Trinidad, Tobago, Grenada and St. Vincent and from Latin America, Dominican Republic, Venezuela, Guyana, Surinam, French Guyana, Colombia, Ecuador, Peru, Mexico, Brazil, Panama, Nicaragua, Guatemala, Costa Rica, Honduras, Belize and El Salvador.

(ii) Host Range: Though *R. cocophilus* is primarily found on coconut and oil palm, it has also been found to occur naturally in the date palm and cabbage palm (Hagley, 1962).

(iii) Symptoms: The disease is present most commonly in young coconut trees (3–10 years old) with maximum incidence in trees of 4–7 years old. Chlorosis first appears at the tips of the oldest leaves and spreads towards their bases. The brown lower leaves may break across the petiole or the lower part of the rachis or they may become partly dislodged at the base and hang down (Fig 4.18). Premature shedding of the nuts may occur simultaneously with the development of the leaf symptoms or slightly before. As a result of the severe damage caused by the palm weevil internally, the crown often topples over about 4 to 6 weeks after the appearance of the first symptoms. The trunk then remains standing in the field for several months and finally it decays (Griffith & Koshy, 1989).

The most characteristic symptom of red ring disease is the internal lesions. In the beginning, about 3–5 cm beneath the stem surface, scattered reddish dots of about 1 mm diameter appear, which later coalesce to form an orange red ring of about 3 cm in width (Fig 4.18). The ring extends the whole length of the stem and roots and in petioles it assumes a crescent-like shape. Large number of juveniles are seen in the centre of the discoloured areas and adults in the periphery. Shedding of green nuts of all sizes takes place.

Figure 4.18. Red ring nematode (*Rhadinaphelenchus cocophilus*) on coconut. Infected plant and cross section of trunk showing red ring symptoms (Courtesy: Sasser, 1971)

R. cocophilus causes little leaf disease of coconut and oil palm in Surinam and Guyana. A thermostable phytotoxin was produced due to breakdown of coconut tissue. Water uptake of infested coconut palms was less due to occlusion of xylem vessels. In coconut roots, the nematodes attack cortical tissues.

(iv) The Vector: The vector of the red ring nematode is the palm weevil (*Rhynchophorus palmarum*). The adult palm weevil acts as the carrier. The female weevil makes punctures in the softer tissue of the stem of the coconut palm with the help of its proboscis to lay its eggs. Normally the punctures are made in the soft axils to a depth of 5–8 cm and the eggs are placed deep in the puncture. The nematodes are injected into the tissues of the coconut palm when the insect deposits its eggs. Larvae of the palm weevil while feeding in diseased trees get infested with the nematodes and the adult beetles which develop from them in turn become the carriers.

(v) Life Cycle: Blair (1966) carried out preliminary studies on the life-history of *R. cocophilus* in immature nuts of coconut palm. When the nuts were inoculated with eggs containing active larvae, adult nematodes and eggs were recovered 9–10 days later. The nematodes completed one life cycle in 9–10 days which is probably one of the shortest life cycles reported for plant parasitic nematodes.

(vi) Survival: *R. cocophilus* can survive for a maximum of 15 days in soil. The third stage juvenile is the most persistent form.

(vii) Management

(a) Cultural Methods: 'Cut and burn' the infested trees is one of the oldest and most effective practices used to prevent further spread of the disease.

(b) Chemical Methods: More recent methods involve the destruction of palms using arboricides, herbicides or phytotoxic nematicides applied to holes made in tree-trunks. Hoyle (1968) recommended the use of cacodylic acid ('Silvisar') to poison trees. The chemical prevented weevil breeding, killed the diseased trees within two weeks, but had no nematicidal effect. Victoria *et al.* (1970) used sodium and potassium arsenate to kill both diseased trees and nematodes.

Adult weevils are attracted to anaerobic fermentation products such as ethyl alcohol, n-butyl alcohol and to certain volatile esters extracted from diseased palm tissue (Griffith, 1987). This behaviour can be exploited to attract weevils to poisoned baits at 25/ha.

The trap or guard baskets protect plantations by attracting and killing the palm weevils which may enter the plantations from nearby diseased trees. The guard baskets are made of 2 cm mesh wire. They are cylindrical, 1 m high and 0.3 m in diameter. These baskets are filled with chunks of fresh tissue from diseased coconut trees to attract the beetle. The guard baskets are sprayed completely with about 4.5 litres of methomyl solution and distributed on the ground in the plantations at 2.5 baskets/ha of young coconut trees. This procedure is especially recommended in the dry season when the weevils are most active in the cool nights. Guard baskets remain for about two weeks, after which the tissue and insecticide in the basket should be burnt. Fresh tissue should be placed in the basket and treated as previously described.

(c) Biological Methods: Griffith (1977) reported the pathogenicity of a bacterium, *Micrococcus agilis* to the palm weevil and suggested its potential use as a biocontrol agent. The rhabditid nematodes of the genera *Steinernema* and *Heterorhabditis* can heavily parasitize the weevils by maintaining a high population of nematodes in the environment (Griffith & Koshy, 1989).

5 ARECANUT

Areca nut (*Areca catechu*), otherwise known as the betel nut, has its origin in the humid regions of Asia and Malay Islands. India continues to dominate the world in area, production and productivity of arecanut and has achieved self sufficiency in arecanut production. It is an important cash crop of India. India is the largest producer of areca nut in the world covering an area of 3.811 lakh ha with annual production of 4.831 lakh tonnes (Bijay Kumar, 2007) (Table 5.1). Most of the production is domestically consumed. The major areca nut producing states include Karnataka, Kerala, Assam, West Bengal, Meghalaya and Tamil Nadu. It is masticatory and the ripe fruits are sometimes used as an antihelminthic and astringent in Europe. It has a toning effect and stimulates the nerves. Areca nut industry forms the economic backbone of nearly 6 million people in India and for many of them it is the sole means of livelihood. The extract after boiling the nuts contains tannins which are used for tanning leather.

Table 5.1. State-wise area, production and productivity of arecanut in India (2005–06)

State	Area in '000 ha	Production in '000 MT	Productivity in MT/ha
Kerala	108.6	119.3	1.098
Tamil Nadu	4.8	10.3	2.140
Karnataka	161.6	215.7	1.334
Andhra Pradesh	0.3	0.2	0.666
Maharashtra	2.0	5.0	2.500
West Bengal	9.4	29.3	3.128
Assam	71.2	66.8	0.938
Goa	1.6	2.6	1.625
Andaman & Nicobar	4.4	4.8	1.090
Meghalaya	11.5	15.5	1.356
Pondicherry	0.1	0.1	1.000
Tripura	3.4	6.9	2.029
Mizoram	2.0	5.3	2.650
Nagaland	0.2	1.3	6.500
Total	**381.1**	**483.1**	**1.3**

5.1. PESTS

5.1.1. Spindle Bug, *Carvalhoia arecae*

(i) Damage: This is a serious pest of young areca plantations of Kerala, Karnataka and parts of Tamil Nadu. This pest multiplies rapidly with close of monsoon. Adult bugs are brightly coloured red and black. They inhabit the inner most leaf axils, usually below the spindle. The light violet brown nymphs are greenish yellow with the border of body. They suck sap from tender spindle resulting in loss of vigour and consequent death of plants. The feeding injury is caused on the lamina and petiole. The affected leaves show dry brown patches (Fig 5.1). Severe leaf damage causes stunting of palms.

Figure 5.1. Spindle bug on areca nut

(ii) Life Cycle: Eggs are laid singly between the leaflets of the spindle. The eggs hatch in 9 days. There are five nymphal stages and it is completed in 15–24 days.

(iii) Management

Filling the inner-most leaf axils around the spindle with phorate 10% granules (10 g/ palm) at 3 months intervals is effective. Conveniently this can be placed in sachets of 2 g (2 per tree).

Spray crowns with 0.1% carbaryl. The spray should reach the leaf axils. Repeat spraying after 30–35 days if pest incidence continues. Placement of 2 g phorate 10G sachets on the top most two leaf axils prevents the pest attack.

5.1.2. Root Grub, *Leucopholis burmeisteri, L. lepidophora* (Coleoptera: Melolonthidae)

(i) Damage: Root grubs or 'white' grubs (Fig 5.2) occur in low lying and clayey soils where the water table is high. They are voracious feeders on areca roots and are found in the areca tracts of Kerala and Karnataka. Adult beetles (cockchafers) emerge during May-June after few days of pre-monsoon showers *i.e.*, after 8–10 days of showers, between 6.30 to 7.30 PM. These beetles lay eggs in soil mostly up to 10 cm depth. Eggs hatch out in about three weeks. The early instar grubs feed on the roots of grasses and other humus. The grub period with three instars is completed in 7 to 8 months. The pupation is in soil in cocoons of mud. This period lasts about one month. The adult beetle is chestnut brown in colour. The second and third instar grubs of these beetles feed on tender and mature roots of the palm, in cases of severe infestation; the bole of the palm is also eaten up. They feed on roots of grasses and intercrops like banana, cocoa, tapioca, yams etc.

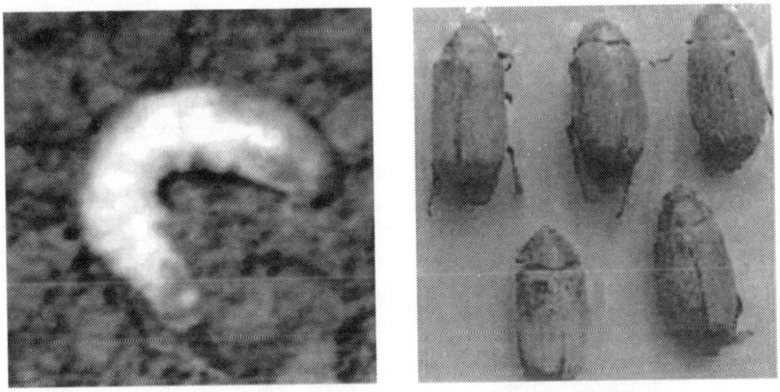

Figure 5.2. Grub and adults of root grubs on arecanut

In areca seedlings, the feeding on roots results in dropping and drying of leaves. Affected seedlings come off easily since the entire root system is usually eaten up. Palms with few years of infestation show a sickly appearance, with yellowing of leaves, tapering of stem, and reduction in yield. The palms may topple in case of severe loss of root system.

(ii) Management

Keeping the garden clean, well-drained and free of weeds reduces its attack. Loosen soil around the base of palms to a depth of 10–15 cm and drench with 0.04% chlorpyriphos suspension twice, once in May just before the onset of southwest monsoon and again in September-October towards the close of the monsoon. Repeat application for 2 or 3 years consecutively to secure a complete eradication of the pest.

Root grubs can also be controlled by soil application of phorate 10G around the palms. Apply phorate at 15 g/palm twice a year during May and September. Collect the adult beetles in the evening hours for 8–10 days after pre monsoon showers and destroy them.

5.1.3. Inflorescence Caterpillar, *Tirathaba mundella* (Lepidoptera: Pyralidae)

(i) Damage: This is reported from Karnataka and Kerala. The adult moth lays eggs in the mechanically damaged portions of the spadices and the emerging caterpillars bore into the inside of the spadices. The egg period lasts five days and the larval period for about 26 days with five instars. Pupal period lasts for 9–11 days. The caterpillars feed on the inflorescences especially the tender female flowers and rachille and web together the rachille into a wet mass with silken threads formed by the caterpillars and take shelter in it (Fig 5.3). Mature caterpillars can damage newly opened inflorescences also. In severe cases of incidence, these bore into the tender buttons and tender nuts as well.

As a result of webbing and feeding the inflorescence is unable to exert the natural pressure on the spathe necessary for its opening and thus spathe opening is delayed. Yellowing of spadices, presence of small holes with frass and drying patches on the spathe are the external symptoms of attack.

Figure 5.3. Inflorescence caterpillars on arecanut

(ii) Management

Force open the inflorescence out of the enclosing spathe and spray 0.05% monocrotophos. Control slugs, which predispose inflorescence to the attack of caterpillar, by using bait of metaldehyde.

5.1.4. Pentatomid Bug, *Halymorpha marmorea* (Hemiptera: Pentatomidae)

(i) Damage: This bug causes tender nut drop in areca. Incidence of this bug is seen from March/April to July/August. The later instar nymphs and adult bugs pierce the tender nuts of different maturity and suck the kernel sap. As a result of this feeding, the kernel dries up and the tender nuts drop. Characteristic pin prick black marks are seen at the point of feeding sites, which lead internally to the kernel.

The adult and the young bugs suck sap from the endosperm of tender nuts, causing premature shedding. The dropped nuts have one or more pin prick-like marks on the surface. The adults are bronze coloured with brown spots and in young stages they are black with white spot on the leg.

When tender nuts are not available, the insect migrates to other hosts like cowpea and bitter ground.

(ii) Management

The crops should be closely watched and the bug when noticed should be mechanically removed and destroyed. One round of spray with endosulfan (0.05%) on the bunches is effective.

Many coleopteran predators, chiefly coccinellids of arecanut mites, are used. They are *Aspects indicus, Cyleophalus semipictus, Stethrus keralicus* and *Stethorus parcepunctatus*.

5.1.5. Scale Insect, *Aonidiella orientalis* (Homoptera: Diaspididae)

(i) Damage: Scale insects colonize the leaves, spathes, leaf sheaths and bunches and suck sap from the tissues. Continuous feeding on nuts results in pre-mature yellowing of nuts and in severe infestation, the kernel may not develop and may turn black and shrivel up. The scale insects are present through out the year, but are more serious during October to February.

(ii) Management

Spraying 0.04% diazinon, dichlorvos (DDVP), phosphamidon, monocrotophos or quinalphos gives effective control of the pest.

5.1.6. Stem Borer

(i) Management

Chisel out the damaged area of the tree and swab the portion with 5% neem oil (with 0.5% teepol or soap) on the tree trunk up to a height of 1 m during April-May and October-November. Use *Metarrhizium anisopliae* or *Beauvaria bassiana* against stem borer.

5.1.7. Tea Mosquito Bug

(i) Management

Augmentative release of egg parasitoids, *Telenomus* or *Crematogaster* is effective for the management of the pest.

5.1.8. White Mite, *Oligonychus indicus* (Acari: Tetranychidae)

(i) Damage: Adults and nymphs of this spider mite colonise the lower surface of leaves (Fig 5.4). The colony is found below a white web on the leaves. Life cycle is of short duration. The total duration of the immature stages varies from 6.5 to 9.0 days. They suck sap from lower surface of leaf causing them to turn yellow or bronzed in appearance.

(ii) Management

Removal of heavily infested and dried leaves and burning help in eradicating the source of infection. Dusting sulphur or spraying 0.05% wettable sulphur is effective for the management of mites. Spray dicofol 2 ml/litre of water to the undersurface of infested leaves. Repeat after a fortnight if there is a recurrence of the pest (Fig 5.4).

5.1.9. Palm or Red Mite, *Raoiella indica* (Acari: Tenuipalpidae)

(i) Damage: Nymphs and adults of this mite are seen in large numbers on the lower surface of leaves. In severe cases of infestation, they may be seen on the leaf stalks and on the spindles. The life cycle lasts for 12.9 days. This mite also attacks palms like coconut, date and ornamentals. Population of these mites start building up after the monsoon rains and reaches its maximum in April-May. Neglected and poorly irrigated gardens and nurseries, particularly those in exposed conditions are prone to severe infestation. Colonies of these mites start declining with the onset of rains in June.

Figure 5.4. White mite damage on arecanut leaves

(ii) Management

Removal of heavily infested and dried leaves and burning helps in eradicating the source of infection. Orange coloured mites can be controlled by spraying bunches with dimethoate at 0.05%. Dusting sulphur or spraying 0.05% wettable sulphur also gives effective control of the pest. Spraying under surfaces of leaves and crown with dicofol (2 ml/litre) or dimethoate (1.5 ml/litre) is effective.

5.1.10. Vertebrate Pests

(i) Damage: Monkeys, rats, wild boars and bats damage nuts. Monkey damage nuts by irregular peeling off fruit epicarp, partially or completely. The unriped fruits are plucked-off and dropped. Bat damage was restricted to ripened or over ripened fruits leaving the fruit surface damaged. Wild boars feed only on fallen ripened nuts. They chewed the whole nut leaving the crushed nuts. Nuts damaged by monkeys are collected and sold by farmers as second quality nuts at 20% reduced rates than healthy nuts.

(ii) Management

Growers protected the crop by gun patrolling, Ibex fencing, cracker bursting, scaring and by lighting and repellent odours like waste tire burning, bonfire, etc. Damage to areca seedlings from wild boar and rat could be avoided by using locally available porcelain pipes. Monkey damage can be reduced by disfavouring the habitat like debranching shade trees, keeping dog for chasing monkeys, etc.

(a) Integrated Method: Clean cultivation, crown cleaning, trunk banding, burrow baiting, Ibex fencing with at least 1 m of open space on either side, and covering base of seedling (up to 2–3 years) with tightly packed cover of thorny bamboo sticks or porcelain pipes wherever arecanut palms were regularly prone to rodents damage saved 46% nuts compared to control against rodents, monkey and wild boar depredations. These practices to some extent afforded protection against raids by bison and elephants also.

Table 5.2. Pest management

Pest	Damage	Control
Scale insect, *Aonidiella orientalis*	The scale insects are seen feeding on nuts, rechille, and leaves- the damage is done by sucking the sap from the fronds and sheaths affecting the vitality of plants adversely; as a result of continuous sucking, the tissues become yellow in colour and severe feeding leads to withering and shedding of buttons/fruits-the damage is heavy during drought situation.	Spray 0.04% diazinon, dichlorvos (DDVP), phosphamidon, monocrotophos or quinalphos; conserve the beneficial biocontrol agents which are killed by the insecticide spray; ladybird beetles, *Chilocours nigrita* and *Chilocorus circumdatus* are found to be effective biocontrol agents against the scale insects attacking arecanut; these can be released in the affected areca garden to control the scale insects.
Aphid, *Cerataphis tetaniae*	The aphids suck the sap from the tender parts of the plants and affect their vitality.	Same as above.
Arecanut beetle, *Araecerus fasciculatus*	Cosmopolitan; small pale-grey beetles feed on the leaves and ripening fruits.	Spray with 0.2% BHC or 0.04% endosulfan.
Mites (Red and white), *Raoiella indica* and *Oligonychus indicus*	Large colonies, individuals minute in size, suck the sap and devitalize the tender plants parts; feed on lower surface of arecanut leaves; the colony is found under white webs; the attack is severe in summer months.	Dust sulphur or spray 0.05% wettable sulphur. Spray dicofol (Kelthane) 2 ml/litre of water or formothion (Anthio 25 EC) at 1.5 ml/litre of water; repeat spraying at an interval of 15–20 days if there is recurrence of pest.
Spindle bug, *Carvalhoia arecae*	Inhabit the inner leaf axils–suck sap from the tender leaflets and spindle-linear lesions turn necrotic at later stages.	Placement of 2 g phorate granules (thimet 10G) in perforated polybags in the innermost leaf axils of areca palms during April is an effective practice for maintenance of areca gardens free of spindle bug infestation; efficacy of the insecticide lasts for 8 months; as new leaves emerge,

Table 5.2. Pest management

Pest	Damage	Control
		polybags are to be shifted to the innermost leaf axils.
Root grub, *Leucopholis burmeisteri*	Grubs feed on growing roots–infested palms show a sickly appearance with yellowing of leaves, tapering of stem and reduction in yield.	Phorate (thimet 10G) applied at 15 g/palm gives effective control of the pest; apply phorate to the soil around the plant twice a year, once in May before the onset of south-west monsoon and again in September-October after the monsoon; repeat the treatment for 2–3 years continuously; collect the adult beetles from 18.30 to 19.30 hours, after 8–10 days of pre-monsoon showers and kill them.
Tender nut drop, *Allomorph marmorea*	Many reasons are attributed for this-A bug, *Allomorph marmorea* is associated with this problem.	To control the bug, spray endosulfan 0.05% or fenvalerate 0.02% to the bunches of the affected palm and few palms around it.
Inflorescence caterpillar, *Tirathaba mundella*	Causes damage to areca inflorescence.	Affected spadices may be force-open and if all the female flowers have been damaged, the inflorescence should be removed and burnt; if damage is partial, remove affected portion and spray malathion (0.125%).

5.2. DISEASES

5.2.1. Koleroga/Mahali/Fruit Rot, *Phytophthora arecae*

The koleroga disease is one of the destructive diseases reported from Karnataka and Kerala. The disease is responsible for 10–75% loss in Karnataka, Kerala and Maharashtra.

(i) Symptoms: Koleroga caused by *Phytophthora arecae* is one of the major diseases of arecanut causing serious losses. This occurs as an epidemic in the heavy rainfall areas of Karnataka and Kerala. The disease first makes its appearance after monsoon period. The first symptom is the appearance of water-soaked lesions on the nut surface near the calyx. The patches enlarge and nuts darken and they shed in large number. The fallen nuts soon develop whitish mycelial mass all over. Nuts of all ages are attacked and if unchecked, the pathogen invades crown causing the leaves and bunches to whither. Sometimes, the infected nuts may not be shed and remain mummified in the bunches. Such type of infection is known as 'dry mahali'.

(ii) Epidemiology: The disease is prevalent during south-west monsoon period and rain plays a key role in disease development. Usually the disease makes its appearance

15–20 days after the onset of monsoon (May-June) and persists till the end of monsoon (September-October). High rainfall, low temperature and lesser duration of sunshine hours are congenial for the growth of the fungus. Zoospores released from the sporangia germinate in the film of water and penetrate nut surface through stomatal openings. The pathogen survives in the form of dormant mycelium or resting spores on tree top, on infected debris and on upper surface of the soil. The disease affected dead palms remaining in the garden may also act as potent source of primary inoculum.

(iii) Management

Removal and destruction of infected debris, shed nuts, dried bunches and dead crowns help to minimize the disease spread and development. Covering the bunches with polythene (Fig 5.5) or leaf sheath or grass before the onset of monsoon helps to reduce the disease. Pre and post monsoon spraying with 1% Bordeaux mixture gives effective control of koleroga. Spray Bordeaux mixture 1% on all bunches three times in a year, once just before the onset of southwest monsoon and the rest at 40 days intervals. If monsoon season is prolonged give a third spray. Use rosin soda adhesive to ensure tenacity of the spray deposit on treated substrate.

Figure 5.5. Covering the bunches with polythene film to control koleroga of arecanut

5.2.2. Foot Rot or Anabe Disease, *Ganoderma lucidum*

The disease was reported for the first time from areca palm of erstwhile Mysore state by Coleman (1911). It is widely distributed in Tamil Nadu, Kerala, Assam and West Bengal. The disease is responsible for 5.7% loss in yield. In endemic areas, 20–25% incidence of the disease has been reported.

(i) Symptoms: This disease is caused by the fungus, *Ganoderma lucidum*. It is more prevalent in neglected gardens, causing 7% losses. Symptoms of the disease are akin to that of the draught. The initial visible symptom is yellowing of outer whorl of leaves which gradually extends to inner whorls followed by wilting and drooping. The development of inflorescence and nuts are arrested. Nuts already formed are shed. At later stages, the weakened crown topples off leaving a bare trunk. All around the base of the palm, brownish patches appear which exude a brown liquid (Fig 5.6). The interior of the stem at the basal region is discoloured and rotten, emitting a foul smell. The infection extends to roots and gets discoloured, brittle and dry. The fungal invasion interrupts uptake of water and nutrients by the palm, leading to yellowing and wilting.

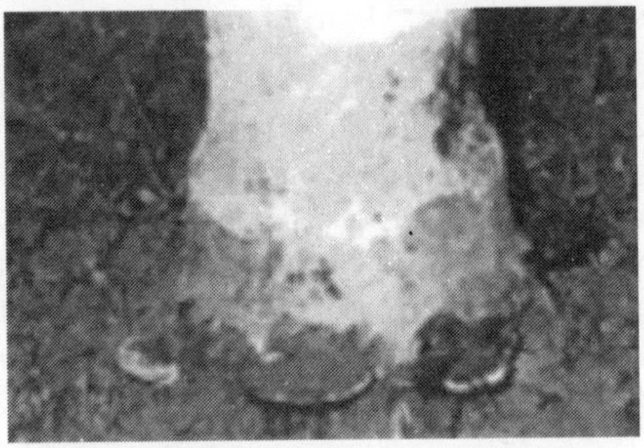

Figure 5.6. Foot rot of arecanut

(ii) Host Range: The pathogen has wide host range infecting plants from 19 families, 16 genera and 48 species of crop plants. Some of the important hosts include coconut, mango, citrus, grapevine, *Cassia siania, Pongamia glabra*, and *Casurina equisetifolia*.

(iii) Epidemiology: The pathogen is soil-borne and remains in dead and decaying plant material. The fungus spreads from plant to plant through root contact or through plant debris left over in the garden. The pathogen also spreads through water, excessive irrigation, repeated ploughing and digging. Presence of alternate hosts near arecanut gardens serve as source of inoculum.

(iv) Management

Since the infected stumps and roots act as the main foci of infection, strict phyto-sanitary measures are to be adopted by removing and destroying the stumps along with roots. Isolate affected palms by digging trenches 60 cm deep and 30 cm wide around, away from the base and drench with 0.3% captan. Drench the soil with 1% Bordeaux mixture before planting. Discourage growing of collateral hosts of the fungus like *Delonix regia* and *Pomgamia glabra*.

- Isolate affected palms by digging trenches 60 cm deep and 30 cm wide around, one metre away from the base and drench with captan (0.3%), calixin (0.1%) or copper oxychloride (0.3%).
- Remove and destroy all severely affected palms and stumps of dead palms.
- Drench the soil with 1% Bordeaux mixture before planting healthy seedlings.
- Follow recommended spacing of 2.7 × 2.7 m.

- Discourage growing of collateral hosts of fungus such as *Delonix regia* and *Pongamia glabra* in the vicinity of gardens.
- Apply organic manure at 15–20 kg/palm and 15–20 kg green leaf along with 2 kg neem cake per palm to reduce spread of the disease.
- Avoid flood irrigation and water flowing from infected palms to healthy palms.
- Drench the root zone of affected palm with calixin (3 ml/l) at 15–20 l/palm + root feeding with calixin (15 ml/l) at 125 ml/palm at quarterly intervals.
- Avoid repeated ploughing and deep digging which helps to reduce the disease incidence.

5.2.3. Bud Rot, *Phytophthora palmivora*

The disease was reported for the first time by Coleman (1910). The disease is more severe in Karnataka.

(i) Symptoms: Bud rot is a fatal disease of areca palms caused by *Phytophthora palmivora* and characterized by rotting of terminal bud and surrounding tissues and ultimately killing the palm. Affected spindle appear yellow, later turning brown (Fig 5.7) and finally the whole spindle rots emitting a disagreeable odour. Severe infection leads to death of palm.

Figure 5.7. Bud rot of arecanut

(ii) Management

Early detection of the disease and prompt removal of infected tissues will help in recovery of palms and also prevents the spread of the disease. In early stages of infection, scoop out affected rotten tissues by making longitudinal side splits and apply Bordeaux paste on the exposed healthy tissues or drench with 1% Bordeaux mixture. Apply 1% Bordeaux mixture to the crown of healthy palms in the vicinity of the affected ones. Avoid heavy intercropping, cultivation in low lying areas and closer planting to reduce spread and development of the disease.

5.2.4. Inflorescence Die-back and Button Shedding, *Colletotrichum gloeosporoides*

(i) Symptoms: Die-back of inflorescence due to micro-organisms is reported to be associated with the low fruit set in arecanut. About 60% of the palms in the states of Karnataka and Kerala are infected by this disease causing severe shedding of buttons. The symptoms include yellowing and drying of the rachis from the tip towards the base followed by shedding of female flowers (buttons) (Fig 5.8).

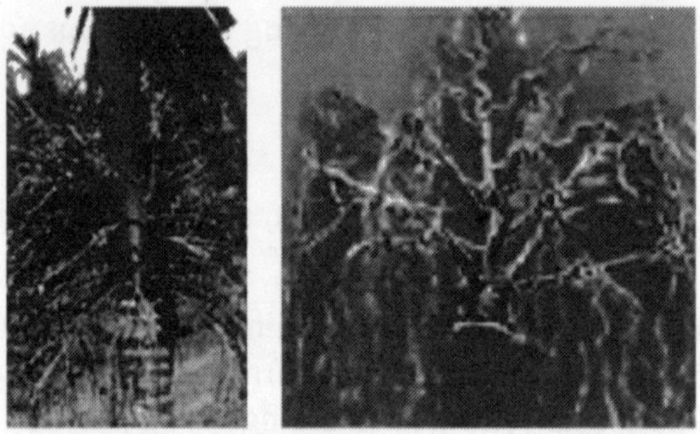

Figure 5.8. Button shedding in arecanut

(ii) Management

Removal and burning of inflorescence helps to reduce the load of inoculum in the field. Spray zineb (4 g/1) or mancozeb (3 g/l) twice, once just after female flowers are set and again 15–28 days later. Aureofungin solution at 50 ppm concentration is also effective in controlling the disease.

5.2.5. Stem Bleeding, *Thielaviopsis paradoxa*

(i) Symptoms: Stem bleeding caused by *Thielaviopsis paradoxa*, is prevalent in isolated pockets in all arecanut-growing States of south India. Younger and middle-aged palms are more susceptible. Symptoms appear mostly on lower portions of the stem as small discoloured depressions which later coalesce and cracks develop with the progress of the disease. The fibrous layer disintegrates which hollows up to varying depths and brown gummy exudates oozes out. Crowns of affected palms get reduced in size followed by reduction in yield. The disease is serious in gardens with poor drainage.

(ii) Management

Improving the drainage may help in minimizing the disease incidence. Scooping out the affected portions and application of coal tar or Bordeaux paste is effective to reduce the disease incidence.

5.2.6. Bacterial Leaf Stripe, *Xanthomonas campestris* pv. *arecae*

(i) Symptoms: The disease is caused by *Xanthomonas campestris* pv. *arecae*. The symptoms are purely parenchymatous in nature causing water-soaked linear lesions parallel to the midrib of the leaflets. The lesions are covered with abundant creamy-white bacterial exudates on the under surface which is a striking feature of the disease. The entire leaflet in a frond may be affected resulting in complete or partial blighting. In severe cases, entire crown may be affected. When growing buds are affected, death of palm takes place. The disease is aggressive during monsoon. Younger palms (3–5 years old) are highly susceptible.

(ii) Management

Spraying or stem injection with tetracycline group of antibiotics at 500 ppm concentration is effective.

5.2.7. Yellow Leaf Disease (Phytoplasma)

(i) Symptoms: The disease is rampant in Kerala and Karnataka. Yellowing of leaves begin in the inner whorl, gradually spreading to the outer parts of the crown (Fig 5.9). Chlorosis is finally observed on almost all leaves in the whorl from the edges of the individual leaflets to the midrib region. Withering of the tips starts and gradually spreads to the older portions of the leaf. The freshly formed leaves grow shorter and their laminae show unequal growth and flaccidity. In a few cases, wilting and shedding of leaves are also observed. The nuts are reduced in size, shriveled with their kernels often turning black and there is severe reduction in yield. Stem of the affected palms becomes spongy and friable, the conducting strands getting destroyed. In advanced stages, the stem breaks off at the top. Rotting of the roots is also observed. There is a gradual decline in yield. Association of mycoplasma-like organisms (phytoplasma) with the disease has been confirmed. It is spread through plant hoppers. Water logging is a predisposing factor in the incidence of the disease. Lack of balance nutrition and unscientific cultivation practices make the palm susceptible to the disease.

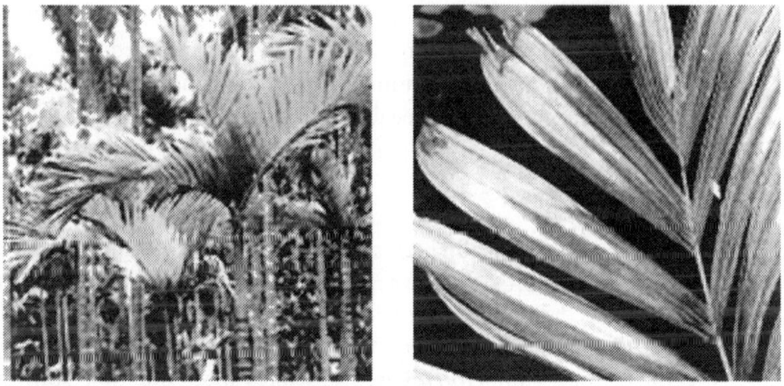

Figure 5.9. Yellow leaf disease of arecanut

(ii) Management

Destroy affected plants to prevent the spread of disease. Apply FYM at 12 kg/palm/year with provision of summer irrigation and drainage to improve the condition of the palm.

Maintain the garden properly to keep affected palms in a healthy condition by adopting recommended manurial, cultural, plant protection and other management practices. Improve drainage conditions in the garden.

Since the disease is not amenable to management by conventional plant protection measures, other means of containing the disease should be looked into. They are:

- Physical condition of the soil should be improved by deep digging and adding sand/ organic matter. In heavy sticky soils, gypsum at 500 kg/ha may be incorporated, once in two years prior to normal fertilizer application.
- Adequate drainage should be provided, especially during monsoon season.
- Phyto-sanitary and plant protection measures should be adopted to control Anabe, bud rot, spindle bud and mite infestation.
- Sun-scorching of the stem should be avoided by covering stem with arecanut leaves or painting with lime slurry.
- Application of NPK fertilizers as per schedule along with lime and zinc at 8.5 g each per palm.

Table 5.3. Disease management

Disease	Symptoms	Control
Koleroga or Mahali, *Phytophthora arecae*	Heavy shedding of nuts during rainy season–water soaked lesions are formed near the perianth end and gradually covering the entire surface of the nut–the nut becomes dark green–finally leads to nut shedding.	Spray 1% Bordeaux mixture to the bunches at least two times at an interval of 45 days; the first spray should be given immediately after the first few monsoon showers; if the monsoon prolongs, a third spray is essential; collect all the infected nuts and other plant parts and destroy it; covering the bunches with polybags gives a complete control.
Bud rot, *Phytophthora arecae*	Yellowing of spindle leaf–rotting of growing bud and surrounding tissues–palm emits a disagreeable odour.	Remove the infected tissue completely and treat the wound with Bordeaux paste; spray 1% Bordeaux mixture to the crown of healthy palms which are in the vicinity of the affected palm; remove koleroga affected bunches and destroy them.
Inflorescence dieback and button shedding, *Collectotrichum gloeosporioides*	Yellowing and drying of rachis from the tip towards the base followed by shedding of female flowers or buttons–caused by several factors; *Collectotrichum gloeosporioides*, a fungus is also associated with this disease.	To control the fungus, spray mancozeb at 3 g/litre of water at the time of opening of female flowers in most of the inflorescences; if necessary, second spray should be given 25 days after the first spray; remove the fully affected inflorescences and destroy them by burning to prevent the spread and severity of the disease.
Anabe roga or foot rot, *Ganoderma lucidum*	Yellowing of outer whorl of leaves which gradually extends to the inner whorls.	Proper management of the garden is the best way to check the occurrence of the disease; improve the drainage; drench the

Table. 5.3. (Contd.)

Disease	Symptoms	Control
		root zone of the affected palm with 0.3% calixin at quarterly intervals at 15–20 litres/palm + root feeding of calixin (125 ml/palm) during January, March, July and October; apply 2 kg neem cake/palm/ year–phytosanitory measures like cutting and burning of the dead palms along with bole and roots should be followed strictly.
Yellow leaf	Yellowing of leaves is the main symptom-yellowing starts from the tip of leaves in the mid whorl which spreads gradually extending from the margin to the middle of lamina-portions near the midrib remain green in advanced stages, yellowing spreads to all leaves completely and they dry and fall off- kernel of the nuts of affected palms become soft, show blackish discolouration and assumes a spongy texture.	Yield of palms in the disease affected garden can be sustained by adopting the recommended management practices such as balanced fertilizer application, application of additional dose of superphosphate with lime (1 kg/palm), application of organic manure, provision of summer irrigation and provision of drainage; remove the diseased palms in the mildly affected areas to prevent the spread of the disease and adopt need based plant protection measures against other pests and diseases.

5.3. NEMATODES

Although a number of nematodes have been reported from the rhizosphere of arecanut, the endoparasite reported as an important pest is only the burrowing nematode, *Radopholus similis*.

5.3.1. Burrowing Nematode, *Radopholus similis*

Radopholus similis was first reported from soil around areca nut roots in Mysore, Karnataka, India by Kumar *et al.* (1971). The threshold inoculum level causing significant damage to growth of areca nut was found to be 100 nematodes per seedling or one nematode in 800 g laterite soil (Koshy, 1986). Ten-fold increase in yield was recorded by treatment with nematicides (Sundraraju & Koshy, 1986).

(i) Symptoms: The burrowing nematode infested areca palms exhibit non-specific above-ground symptoms like general yellowing and visible reduction in growth, vigour and yield. The characteristic symptom of *R. similis* infestation is the appearance of lesions and rotting of roots. The nematode produces small, elongate, orange-coloured lesions on the young, succulent creamy-white to light orange coloured portion of the main and lateral roots. Later these lesions coalesce and cause extensive rotting. The thick primary roots arising from the bole of the palm exhibit large, oval, sunken dark lesions. In areca nut, the tips of lateral and tertiary roots on infestation becomes black.

(ii) Life Cycle: At a temperature range of 21–31°C, *R. similis* takes 25–30 days to complete one life cycle (J_2 to J_2) on areca nut seedlings.

(iii) Ecology: Maximum population density of *R. similis* in roots was recorded during October-November and minimum during March-June.

(iv) Histopathology: The burrowing nematodes are found in inter and intracellular positions in the cortex. They do not enter the stellar region (Sundraraju & Koshy, 1988a).

(v) Management

(a) Cultural Methods: Use of nematode-free planting materials of areca nut and other inter/mixed crops and avoiding *R. similis* susceptible inter/mixed crops like black pepper and banana in infested areas helps in reducing the nematode population.

Complete control of *R. similis* was obtained on areca nut palms affected by yellow leaf disease (YLD) after four years of continuous application of neem cake at 1.5 kg/palm thrice a year (June, September, January). Significant increase in yield and decrease in disease indices of YLD-affected palms were recorded only in the fifth year.

(b) Chemical Methods: Application of aldicarb or fensulfothion at 1 g *a.i.*/seedling thrice a year for three years controlled *R. similis* infestation in the seedling stage. In adult palms, reduction in nematode population and 10-fold increase in yield was obtained by application of fensulfothion at 50 g *a.i.*/palm in May-June, September-October and December-January for five years (Sundararaju & Koshy, 1986).

Complete control of *R. similis* was obtained on areca nut palms affected by yellow leaf disease (YLD) after four years of continuous application of fensulphothion at 50 g *a.i.*/palm or aldicarb 10 g *a.i.*/palm or DBCP at 10 ml *a.i.*/palm thrice a year (June, September, January). Significant increase in yield and decrease in disease indices of YLD-affected palms were recorded only in the fifth year.

Various chemicals tried for the management of *R. similis* on areca nut under nursery and field conditions is presented in Table 5.4.

Table 5.4. Effect of various chemicals for the management of *R. similis* in areca nut under nursery and field conditions

Chemical	Dosage	Duration & type of treatment	Effectiveness
(i) Nursery level			
Aldicarb or Fensulfothion	1 g *a.i.*/seedling	Thrice a year (in Sept., Dec. & March) for 3 years	Controlled soil and root population

Table. 5.4. (Contd.)

Chemical	Dosage	Duration & type of treatment	Effectiveness
(ii) Field level			
Aldicarb	1 g *a.i.*/palm	Thrice a year (in June, (September & January)	Reduced nematodes increased the yield
DBCP	10 ml *a.i.*/palm	Thrice a year (in June, September & January)	Reduced nematodes increased the yield
Fensulfothion	50 g *a.i.*/palm	Thrice a year (in June, September & January)	Reduced nematodes populations & increased the yield

(c) Host Resistance: The areca nut hybrid VTL-11 × VTL-17 has been found to be highly resistant to *R. similis*. The *cvs*. Indonesia-6, Mahuva B and Andaman-5 are tolerant to *R. similis* and hence can be recommended for infested areas (Koshy *et al.*, 1979; Sundararaju & Koshy, 1982). The *cvs*. Indonesia-6 and Singapore (VTL-17) are known to yield 50% more nuts over local South Kanara variety.

(d) Integrated Methods

- Use of nematode-free planting material of areca nut and other inter/mixed crops.
- Avoiding *R. similis* susceptible inter/mixed crops like black pepper and banana in infested areas.
- Use of tolerant/resistant varieties.
- Application of 5–10 kg of green manure/palm preferably *Glyricidia* or *Crotolaria.*
- Application of 1 kg neem oil cake/palm/year.
- Application of phorate at 3 g *a.i.*/plant in the root zone of areca nut, banana and black pepper in June-July and in September-October in areca nut based farming systems.

Integration of neem oil cake application at 1 kg/plant with phorate at 15 g *a.i.*/plant gave effective control of *R. similis* in areca nut. In areca nut + banana + black pepper cropping system, integration of phorate at 15 g *a.i.*/plant with neem oil cake at 500 g/plant was found most effective in reducing the nematode population (Sudha & Sundararaju, 1998).

6 BETEL VINE

Betel vine (*Piper betel*) is a perennial, dioecious, evergreen creeper grown in India. It is a chewing stimulant of brain, lungs and heart. The leaves contain B and C vitamins. It is an important cash crop in Andhra Pradesh, Assam, Bihar, Karnataka, Kerala, Madhya Pradesh, Maharashtra, Orissa, Tamil Nadu, Tripura, Uttar Pradesh and West Bengal with an annual turnover of about Rs. 700 crores. Its leaves are exported to Pakistan, Bangladesh, Indonesia, Malaysia, Burma and Thailand.

6.1. PESTS

6.1.1. Mirid Bug, *Disphincuts politus, D. maesayum*

(i) Damage: The mirid bug (Fig 6.1) de-saps the tender leaves and shoots. The nymphs and adults damage the leaves by puncturing and sucking the juice causing the leaves to shrivel, fade and dry up. Adult bugs are slender, active and reddish brown; both nymphs and adults suck the sap from tender leaves which develop black spots. The market value of the leaves is reduced. The pest is active from June to October. Warm and humid climate is favourable for development of the pest.

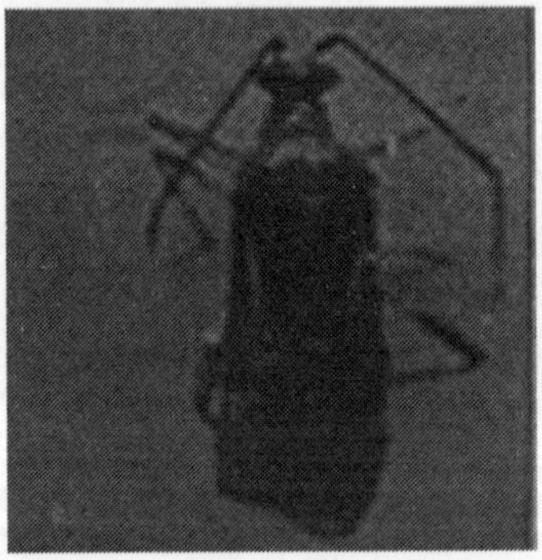

Figure 6.1. Mirid bug on betelvine

(ii) Management

Spraying 0.1% malathion or 0.002% pyrethrins is effective. In serious infestations, apply fish oil soap at 1.5% against the mirid bugs. The leaves should be harvested only after 15 days of insecticide application. The treated leaves should be consume/marketed after thorough washing in water. Spraying 4% neem seed kernel extract also gives effective control of the pest.

6.1.2. Scale Insects, *Lepidosaphes* sp., *Pseudococcus* sp.; Mealy Bugs, *Ferrisiana virgata*

(i) Damage: Scale insects and mealy bugs occur on the stems. Adults and nymphs infest tender shoots and suck the sap and reduce the vitality of the vines.

The leaves are covered with white masses of mealy bugs which sucks the sap from leaves and shoots.

(ii) Management

Pruning and destroying the affected parts helps in reducing the incidence of the pests. Spraying 0.025% quinalphos or 0.5% malathion is effective against the pests. Against scale insects, restrict insecticide application on the stem only. The leaves should be harvested only after 15 days of insecticide application. The treated leaves should be consumed/marketed after thorough washing in water.

Table 6.1. Pest management

Pest	Damage	Control
Bugs, *Disphinctus politus*, *D. maesayum*	Adult bugs slender, active and reddish brown; both nymphs and adults suck the sap from tender leaves which develop black spots; market value reduced; active from June to October; warm and humid climate favourable	Spray 0.1% malathion or 0.002% pyrethrins
Scales and Mealy bugs, *Ferrisiana virgata*, *Lepidosaphes cornutus*, *Pseudococcus* spp.	Adults and nymphs infest tender shoots and suck the sap and reduce the vitality of the vines	Prune and destroy the affected parts.

6.2. DISEASES

6.2.1. Foot and Leaf Rot, *Phytophthora parasitica*

(i) Symptoms: The plant loses luster, leaves droop while still green, turn yellow and drop. Leaves turn brown and later turn black and rot (Fig 6.2). Blackish brown marks will be visible on the stem at ground level or slightly above. The stem is discoloured and later rots. Stem becomes disintegrated.

(ii) Management

Drenching the soil with 4 g copper oxychloride/litre of water gives effective control of the disease. Soil application of *T. harzianum* reduced the vine mortality (11 to 21% compared to 15 to 55% in control) and increased the disease control (46 to 75% compared to 0% in control).

Figure 6.2. Foot rot of betelvine

6.2.2. Collar/Basal Rot, *Sclerotium rolfsii*

(i) Symptoms: The plants are usually attacked at ground level. Dense white cotton like mass of threads (mycelium) are seen on stems. This causes rotting of affected portion causing wilting and ultimate death of plants.

(ii) Management

Spraying 1% Bordeaux mixture or 0.2% zineb, or drenching with 0.1% wettable ceresan is effective.

Soil application of *T. harzianum* (strains MTCC Nos. 3841 to 3843) and *T. virens* (strain MTCC No. 794) increased vine height (199 to 254 cm compared to 67 to 69 cm in control), decreased plant mortality (7 to 33% compared to 73 to 87% in control), increased the disease control (64 to 83% compared to 0% in control) and increased the yield significantly. Consortium of 2 strains of *Pseudomonas fluorescens* (NBRI-N6 and NBRI-N) showed significant increase in yield of betelvine leaves.

6.2.3. Anthracnose/Leaf Spot, *Colletotrichum piperis*

(i) Symptoms: Small black circular irregular lesions appear on leaves expanding rapidly in humid conditions. Leaf spot appears as brownish black centre with yellowish halo around (Fig 6.3) and in serve causes the leaves drop owing to shrinkage of tissues.

(ii) Management

Destroying affected leaves helps to control spread of the disease. Spraying of 1% Bordeaux mixture or 0.2% zineb gives effective control of anthracnose.

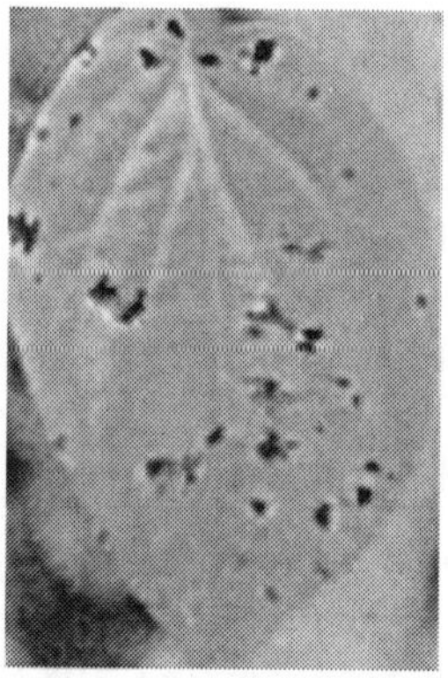

Figure 6.3. Leaf spots on betelvine

6.2.4. Powdery Mildew, *Oidium piperis*

(ii) Management

Destroy affected leaves to control the spread of disease.

6.2.5. Bacterial Leaf Spot, *Xanthomonas betlicola*

(i) Symptoms: The bacterial leaf spot caused by *Xanthomonas betlicola* is most serious in southern Kerala. The initial symptoms appear as small water-soaked regions on the under surface of leaves, which enlarge and turn dark brown in the centre with yellow halo. Later these appear on upper surface also. The centre of spots are mottled brown and later turn black and in severe conditions leaves turn yellow and fall. Profuse bacterial ooze may be seen on the lower side of the lesion. In severe cases, defoliation and stem injury occur, leading to wilting of the plant.

(ii) Management

Drenching with wettable ceresan 0.1% or Agallol 0.1% or spraying 1% Bordeaux mixture at every month, staring from August to January gives effective control of the disease.

N.B: Save the standards (Avisi) from the insect pests by applying 37 kg heptachlor/ha basally to the soil before sowing Avisi. The leaf eating caterpillars can be controlled by spraying malathion 0.5%.

6.3. NEMATODES

Root-knot nematodes are recognized as the major limiting factors in successful production of betel vine crop.

6.3.1. Root-knot Nematode, *Meloidogyne incognita*

M. incognita was responsible for 21–38% loss in leaf yield of betel vine (Saikia, 1992; Jonathan *et al.*, 1990).

(i) Symptoms: Blackening and drooping of the growing tip and yellowing of leaves occurs due to root-knot nematode infestation. Galls of varying sizes and shapes develop in roots leading to quick root decay.

(ii) Interaction with Other Pathogens: Association of *M. incognita* with severe wilt symptoms of betel vine was reported from India (Mammen, 1974) and *M. incognita* is known to predispose betel vine to root-rot caused by *Phytophthora palmivora* (Sivakumar *et al.*, 1987).

(iii) Management

(a) Physical Methods: White polyethylene mulching for 15 days on beds prepared for planting betel vine revealed high reduction in nematode populations due to increased soil temperature of 44.1°C (Sivakumar & Marimuthu, 1987).

(b) Cultural Methods: Nematode-free planting material is a cardinal principle in the management of betel vine nematodes. Summer ploughing, summer ploughing and solarization or rice cropping have been beneficial. Application of FYM at 25 t/ha or neem cake at 1 t/ha before planting or crop rotation with rice helps to reduce the nematode population in soil. Considerable reduction in nematode populations in the soil and number of galls on roots has been reported after application of 50–75 kg K_2O/ha (Jagdale *et al.*, 1985a).

Crop Rotation: A pre-rice crop reduces populations of *M. incognita, Rotylenchulus reniformis* and *Helicotylenchus incisus* in the subsequent betel vine crop. Crop rotation with non-host crops like marigold, sesame and groundnut also suppress root-knot nematode populations. Support plants such as *Erythrina indica, Moringa oleifera* and *Bombax malbaricum* which are not susceptible to root-knot nematodes can be used in betel vine gardens.

Intercropping: In pots having 5 seedlings of *Tagetes erecta* around a betel vine plant, the population of *M. incognita* was reduced by 41.4% and root galling by 54% (Medhana *et al.*, 1985).

Botanicals: Green leaf manuring with African marigold, neem or *Xanthium* have also been quite effective against root-knot, lesion and other nematodes.

Soil application of neem oil cake at 1 tonne/ha and sawdust at 2 tonnes/ha can reduce nematode populations, number of galls and increase the number of leaves harvested significantly (Jagdale *et al.*, 1985b). Significant reduction (60%) in the nematode population has been observed in beds amended with chopped and shade dried leaves of *Calotropis gigantea* at 2.5 t/ha followed by neem oil cake and poultry manure (44.4 and 40.9%, respectively). Beds amended with *C. gigantea* leaves yielded 14.2 kg/4840 leaves and with neem oil cake yielded 12.1 kg/4220 leaves. Soil amendment with sawdust at 2 t/ha + NPK (3638 leaves) and neem cake at 2 t/ha is effective in reducing nematode numbers and increasing yields (Sivakumar & Marimuthu, 1986b).

(c) Chemical Methods: Application of aldicarb and carbofuran at 0.75 kg *a.i.*/ha reduces nematode populations by 71 and 55%, respectively, resulting in increased yields. The granules degraded to non-detectable levels 41 days after application (Sivakumar *et al.*, 1987). Aldicarb, carbofuran and benfurocarb applied at 1.5, 3.0 and 5.0 kg *a.i.*/ha, respectively, in furrows on either side of the rows can reduce *M. incognita* populations in soil and galling on the roots significantly (Dethe & Pawar, 1987).

(d) Biological Methods: Three applications of *Paecilomyces lilacinus* grown on neem cake at 10 g/vine at 60 days interval gave a significant reduction in nematode infestation and increased vine growth and leaf yield under field conditions (Jonathan *et al.*, 2000). Nakat *et al.* (1995) reported that the application of *P. lilacinus* at 8 g/kg soil as the best treatment for the management of root-knot nematodes and increasing the leaf yield by 38.3 to 56.4% in betel vine.

(e) Host Resistance: The *cvs.* Kakair, Bangla, Karapaku, Gachipan, Aswani pan and Berhampuri are reported to be tolerant to root-knot (Anon, 1987). The *cv.* Kuljedu had the lowest root-knot index and number of egg masses per plant (Jagdale *et al.*, 1985a; Sivakumar *et al.*, 1987).

(f) Integrated Methods: *Trichoderma viride* mixed with mustard oil cake applied in three split doses reduced the nematode infestation and enhanced the leaf yield.

Because of the residue problems on leaves, it is preferable to manage root-knot nematode infestation on betel vine by adopting the following non-chemical measures:

- Crop rotation wherever possible.
- Use of resistant/tolerant cultivars.
- Use of non-living standards or nematode-resistant live standards for supports.
- Soil solarization by mulching the land with clear polythene (100 gauge) before planting.
- Application of organic amendments such as leaves of neem and *Calotropis*, and sawdust at 2 t/ha.
- Supply of nitrogen through neem oil cake at 2 t/ha.

6.3.2. Burrowing Nematode, *Radopholus similis*

The burrowing nematode has been reported to cause yellow/slow wilt disease of betel vine in India.

(i) Symptoms: The symptoms produced on betel vine are akin to the symptoms caused by *R. similis* on black pepper vines (Eapen *et al.*, 1987).

(ii) Management:

The integrated management schedules suggested for the control of nematodes on black pepper, other than application of nematicides, can be largely adopted with modification to suit the local conditions for controlling *R. similis* on betel vine.

6.3.3. Reniform Nematode, *Rotylenchulus reniformis*

Acharya and Padhi (1987) found *R. reniformis* to be pathogenic to betel vine. At inoculum levels of 1,000 and 20,000 nematodes/cutting, the reduction in number of leaves was 20 and 60%, respectively.

6.3.4. Root-knot Nematode, *Meloidogyne incognita* and Foot Rot, *Phytophthora capsici* Disease Complex

(i) Management

(a) Integrated Methods

- Summer ploughing and exposing the field to sunlight during May prior to sowing of *Sesbania* sp. (live standard) minimizes the initial load of inoculum of both the nematode and the fungus.
- Selection of healthy seed vines from nematode-free and disease-free mother plants for planting.
- Dipping seed vines in 0.25% Bordeaux mixture solution for 5 min before planting.
- Application of FYM at 30 t/ha to promote multiplication of antagonistic microbes which in turn kills nematodes.
- Spot application of neem cake enriched with *P. lilacinus* at 3 t/ha in 3 split doses- 1st dose at 45 DAP and remaining 2 splits at 45 days interval during North-East monsoon season (Oct-Dec).
- Rotation of betel vine crop with rice.

Combined application of *Pseudomonas* sp. (Pfbv 22) and *Bacillus* sp. (Bbv 57) gave significant reduction in nematode infestation (gall index, no. of egg laying females and soil population), wilt disease incidence and increase in leaf yield. The treatment also enhanced the biochemical markers responsible for induced systemic resistance such as peroxidase, polyphenol oxidase and phenylalanine ammonia lyase (Jonathan *et al.*, 2004).

7 OIL PALM

Oil palm (*Elaeis guineensis*) is the highest oil-yielding plant (4–6 tonnes oil/ha) among perennial oil-yielding crops, producing palm oil and palm kernel oil. These are used for culinary as well as industrial purposes. It occupies an area of 44,413 ha in Andhra Pradesh (80% area), Karnataka, Assam, Gujarat, Goa, Kerala, Maharashtra, Orissa, Tamil Nadu,Tripura and West Bengal. Average productivity in oil palm plantations is now 4–5 tonnes/ha which compares favourably with other countries.

7.1. PESTS

A number of insects are potentially damaging to oil palm in various parts of the world including: palm weevils (*Rhynchophorus* spp.), rhinoceros beetles (*Oryctes* spp.), weevils (*Strategus aloeus*, *Temnoschoita quadripustulata*), leaf-miners (*Coelaenomenodera elaeidis*, *Hispolepis elaeidis*, *Alurunus humeralis*), slug caterpillar (*Parasa viridissima*), nettle caterpillar (*Setora nitens*) and bagworms (*Cremastophysche pendula*, *Mahasena corbetti*, *Metisa plana*).

7.1.1. Rhinoceros Beetle, *Oryctes rhinoceros*

(i) Damage: *Oryctes rhinoceros* is a severe pest of oil palm. The pest causes severe damage to emerging fronds and spindle (Fig 7.1). The adult beetle feeds on the softer tissue of the rachis, resulting in snapping off of the fronds and spears at the feeding sites.

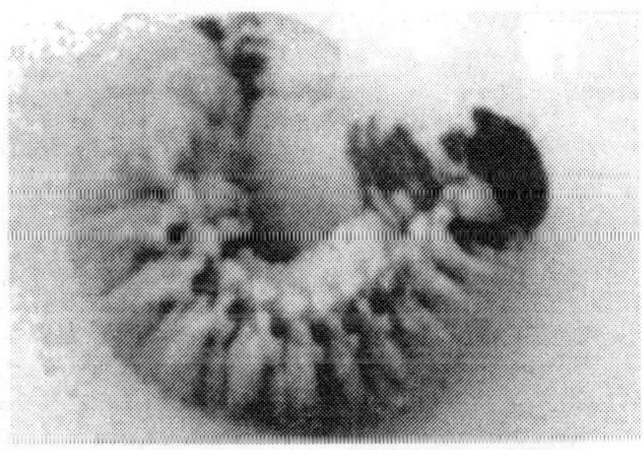

Figure 7.1. Pheromone trap for rhinoceros beetle and virus infected larva

(ii) Management

A combination of cultural, chemical and biological control is now being used in an integrated manner to successfully manage outbreaks of the pest. Field sanitation and elimination of breeding sites are essential components of the pest management operation. This pest can be suppressed by using the virus *Baculovirus oryctes*.

7.1.2. Bag Worms, *Cremastophysche pendula*, *Mahasena corbetti*, *Metisa plana* and Nettle Caterpillar, *Setora nitens*

(i) Damage: Bag worms and nettle caterpillars are opportunistic pests that can devastate large areas of oil palm if not detected and treated early. Early detection is achieved by carrying out regular inspection and counting of pest numbers in the field. Action is taken to control the pests by use of selective insecticide application, *e.g.*, trunk injection or spot spraying with pyrethroid or *Bacillus thuringiensis* formulations once when an economic threshold level of pest population is exceeded.

(ii) Management

More proactively, methods of direct or indirect biological control are utilized. Direct biocontrol involves the use of non-occluded spherical and granulosis viruses and *Cordyceps* entomopathogenic fungi for control of nettle caterpillars. Predatory pentatomid bugs such as *Cantheconidea furcellata* and *Platynopus melacanthus* are also mass bred for release into fields.

Indirect biocontrol is achieved by the cultivation of nectariferous plants like *Euphorbia heterophylla*, *Cassia cobanensis*, *Antigonon leptopus* and *Turnera subulata* in and around infested fields. These plants encourage populations of parasitoids and predators of bagworms and nettle caterpillars by providing them with nectar and shelter. The increased populations of natural enemies would in turn exert biocontrol on the pests.

7.1.3. Red Palm Weevil, *Rhynchophorus* spp.

(i) Damage: Red palm weevil (Fig 7.2) is a major pest of oil palm in India. These weevils lay their eggs at the cut end of petioles or other wounds. The emerging larvae tunnel into the crown and feed on the growing tissues. Palms infested by red palm weevil start wilting and leaves show gradual increase in chlorosis and fracture when strong wind blows.

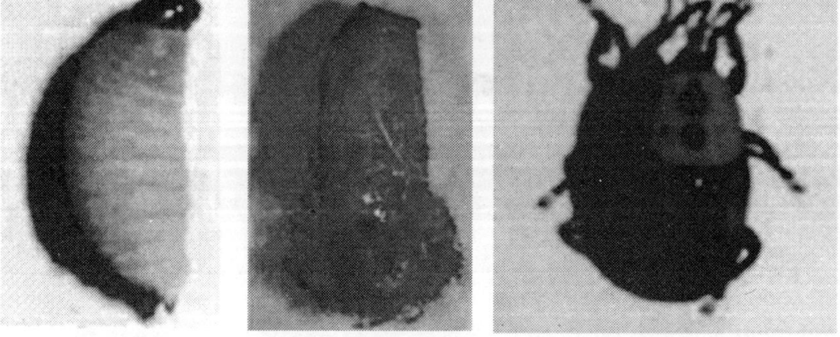

Figure 7.2. Larva, pupa and adult of red palm weevil

(ii) Management

If detected early, treatment of affected palm with 0.2% solution of endosulfan or 0.1% carbaryl would save the palms.

7.1.4. Vertebrate Pests

7.1.4.1. Rats

(i) Damage: Rats (Fig 7.3) are also implicated in heavy losses of seedlings. Rats, if uncontrolled, can bring about at least 5% loss of crop in oil palm. These pests were traditionally controlled with poison baits with potential undesirable secondary effects.

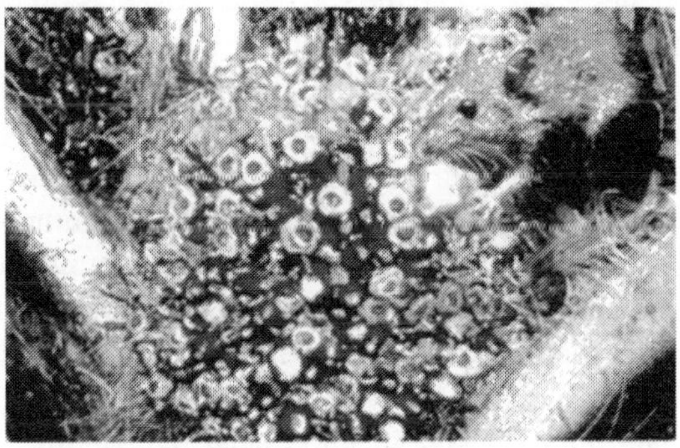

Figure 7.3. Rat damage on oil palm fruit bunches

(ii) Management

For biocontrol of rats the use of the barn owl *Tyto alba* has been implemented. By setting up owl boxes at an intensity of 1 box per 10 hectares of oil palm in a regular grid, success in biocontrol was achieved (Fig 7.4). Owls have been recorded to suppress fresh fruit damage by rats to below the economic threshold of 5% for the past 12 years without having to use rodenticide baits.

Figure 7.4. Biological control of rats by barn owl

7.1.4.2. Birds

(i) Damage: Many birds such as the forest crow, the house crow and the common Indian myna cause severe damage to oil palm fruit bunches. Parakeets, crows and common myna are the major pests in Karnataka. The roseringed parakeet cause severe damage to oil palm when fruit bunches are borne above 3 m, *i.e.*, when the palm is over 5 year old. These birds feed on the mesocarp of the oil palm fruits.

(ii) Management

The damage can be minimized by scaring the birds and covering the ripe bunch with wire net, 150 days after fruit set. Wire mesh covered bunches recorded 10.9% damage, plastic bag covered bunches 9.2%, while the k.k. bunch cover recorded the least fruit damage of 1.6% as compared to 48.95% damage in control bunches (Chakravarthy *et al.*, 1998). Covering the bunches with used bags proved effective at Hyderabad, Andhra Pradesh.

(a) Integrated Management: Covering ripening bunches with gunny bags or thatch of farm wastes or paddy straw, timely harvests and scaring away vertebrates by odours, watch and ward saved 62% fruits from birds, rodents and wild boar compared to control.

A sustainable approach to crop production relies on protecting the crop by an integrated system of pest management (IPM). There are many examples of how managing species other than the crop, rather than attempting to eliminate them, has benefits in oil palm. For example:

- *Elaeidobius kamerunicus*, a weevil introduced into South East Asia in the 1970's eliminated the need for costly and inefficient hand pollinating, resulting in a sharp increase in oil yields.
- Owls are encouraged for rodent control.
- Leaving less competitive species of weeds provides a habitat for predators of insect pests and helps to prevent soil erosion.

7.2. DISEASES

Oil palm diseases include: blast (*Pythium splendens*, *Rhizoctonia lamellifera*), freckle (*Cercospora elaeidis*), anthracnose (*Botryodiplodia palmarum*, *Melanconium elaeidis*, *Glomerella cingulata*), seedling blight (*Curvularia eragrostidis*), yellow patch and vascular wilt (*Fusarium oxysporum*), basal trunk rot (*Ceratocystis paradoxa*, *Ganoderma* spp., *Armillaria mellea*); crown disease and fruit rot (*Marasmius palmivorus*). Spear (bud) rot is caused by the bacterium *Erwinia* spp., which has been devastating in Central Africa.

7.2.1. Basal Stem Rot, *Ganoderma boninense*

(i) Symptoms: The basal stem rot (BSR) of oil palm caused by *Ganoderma boninense* is the most serious disease of oil palm in Malaysia. Under severe infestation situations, more than 50% of oil palm stand can be lost to the malady (Fig 7.5).

(ii) Management

Arbuscular mycorrhizal fungi (AMF) have recently been experienced to be able to confer protection to nursery inoculated palms that were subsequently planted in *Ganoderma* endemic fields. Wide-scale use of AMF is expected to further delay if not prevent infection of palms replanted on fields previously devastated by *Ganoderma*.

Figure 7.5. Basal stem rot on oil palm

7.2.2. Anthracnose, *Botryodiplodia palmarum*, *Melanconium elaeidis*, *Glomerella cingulata*

(i) Symptoms: This disease occurs in the nursery. It is recognized by regular or irregular brown to black leaf blotches surrounded by yellow haloes, which develop along the margin, centre or tip of the leaves. It causes heavy seedling loss.

(ii) Management

The disease can be controlled by spraying mancozeb or captan at the rate of 200 g/ 100 litres of water. Copper fungicides should not be used because of the extreme susceptibility of oil palm seedlings to copper burn (scorching).

7.2.3. Spear Rot, *Erwinia* spp.

(i) Symptoms: This bacterial disease is noticed to affect oil palms of all ages. The incidence is less than one per cent. Yellowing starts from tip of the innermost whorl of leaves. Small lesions occur at the distal portions of spear and rotting extends downwards. As the disease advances, new leaves become rudimentary and show rotting. General decline in vigour and production is then noticed. Occurrence of spear rot without yellowing has also been noticed. Distinguishable marginal yellowing of leaflets and sudden drying of leaves showing yellowing are other symptoms.

(ii) Management

Rouging of all the affected palms may be adopted to prevent further spread of the disease. In early stages of the disease, the affected portions of leaves may be removed and burnt.

7.3. NEMATODES

7.3.1. Red Ring Nematode, *Rhadinaphelenchus cocophilus*

The red ring nematode is recognized as the major limiting factor in successful production of oil palm crop. The red ring disease caused by *R. cocophilus* is the most important disease of oil palm. *R. cocophilus* causes economic loss in oil palm in South Africa. Malaguti (1953) observed that in a 7-year old oil palm grove suffering from little leaf, many trees had died.

(i) Symptoms: Diseased palms in young groves are initially found clustered which gradually expand. In older groves, such palms have a random distribution. Little leaf is the prominent symptom in diseased palms which have erect, short and deformed leaves. The top part of the main vein bore suberized patches. The pinnae are shorter, wavy and necrotic at the tips and yellow patches are present on the petioles and leaf bases. The diseased palm has a brownish discolouration. *R. cocophilus* is found in discoloured tissue and thrived more in the petioles of young folded leaves than in stems and roots. A notable feature is that the band of necrotic tissue is always very small. In diseased oil palms, an average of less than 500 nematodes is found in one gram of infected tissue.

(ii) Management

Same as in coconut.

8 CACAO

Cacao (*Theobroma cacao*) is an important commercial crop grown the world over. It is native of Amazon Valley of South America. The total area under cacao cultivation in India is 0.295 lakh ha with an annual production of 0.102 lakh tonnes and the average productivity of 3.0 tonnes/ha (Bijay Kumar, 2007) (Table 8.1). In India, it is being cultivated in Kerala, Karnataka, Andhra Pradesh and Tamil Nadu. Kerala account for about 76% of the area and 78% of the total production in the country.

Table 8.1. State-wise area, production and productivity of cacao in India (2005–06)

State	Area in '000 ha	Production in '000 MT	Productivity in MT/ha
Kerala	10.2	6.5	1.5
Tamil Nadu	1.4	0.2	7.0
Karnataka	6.0	2.8	2.1
Andhra Pradesh	11.8	0.7	16.8
Total	29.5	10.2	3.0

8.1. PESTS

8.1.1. Mealy Bug, *Planococcus lilacinus*

(i) Damage: This cosmopolitan group of insects with exceptional powers of dispersal is often seen colonizing cacao in most of the cacao growing tracts of South East Asia like India, Sri Lanka and Papua New Guinea. Though few species of mealy bugs colonize cacao trees, two species of *Planococcus* and one species of *Planococcoides* are the most important and most abundant of cacao insects. They are the cacao mealy bug, *P. lilacinus*, the citrus mealy bug, *P. citri* and the West African mealy bug, *Planococcoides njalensis*. In India, the cacao mealy bug, *Planococcus lilacinus* is reported as a serious pest causing damage to cacao and is present in all cacao tracts of the country. It occurs consistently on the plant, and is present throughout the year. This colonizes the tender parts of the plant such as the growing tips of the shoots, the terminal buds, the flower cushions, the young cherelles and mature pods (Fig 8.1).

Mealy bug poses major problem for cacao. During summer months mealy bugs cause damage to spindle leaves, spathes and bunches. The infested portions turn yellow and finally dry up.

Mealy bugs suck the sap from developing pods and tender shoots. Retarded growth and excessive branching at undesired height are the initial symptoms of damage. Later, it causes abortion and wilting of cherelles.

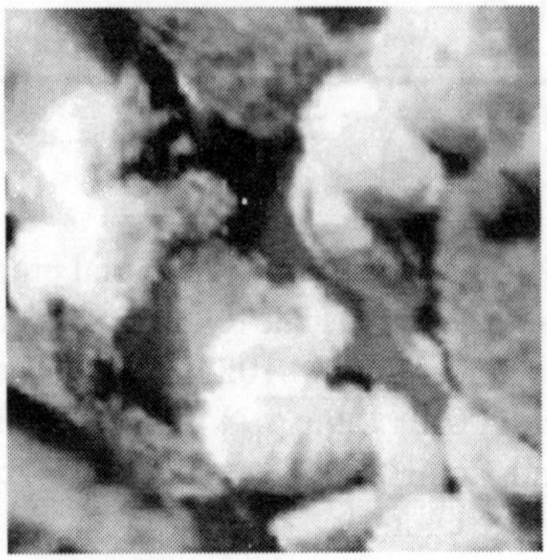

Figure 8.1. Mealy bug infestation on cacao

(ii) Management

The mealy bugs can be controlled with two rounds of spray of 0.1% fenthion or monocrotophos. Spot application of monocrotophos at 125 ml in 100 litres of water is also effective against mealy bugs.

8.1.2. Aphids, *Toxoptera aurantii*

(i) Damage: Two species of aphids occur on cacao. The abundant and cosmopolitan one is the citrus black aphid, *Toxoptera aurantii*. The Adults and nymphs of aphids colonize the flower buds, flowers, flush leaves, tender stems and very small cherelles. Leaf crinkling and shedding of flowers are the outcome of infestation.

(ii) Management

A number of natural enemies feed on the aphids and reduce the population. These include syrphids, cecidomyiids, coccinellids, hemerobiids etc. In severe cases of infestation, spraying with dimethoate at 1.5 ml/litre of water is effective for the management of aphids.

8.1.3. Ash Weevil, *Myllocerus viridanus, M. maculosus*

(i) Damage: Ash weevils, *Myllocerus viridanus* and *M. maculosus* (Curculionidae) are recorded in the Nilgiris and Yercadu regions of Tamil Nadu as early as in 1967. *M. viridanus* occur as a major insect in all cacao growing tracts of Kerala and Tamil Nadu. Adults are seen in large numbers on the underside of leaves and feed on the interveinal tissues leaving the veins intact. The flaccid tender flushes are not preferred for feeding. Population peak is observed in July-September. Infestation is relatively severe on young plants and quite often, the entire foliage of such plants is skeletonised leading to growth retardation. The incidence of ash weevils is more serious in coconut-cacao plantations.

(ii) Management

Spraying of quinalphos or chlorpyriphos at 2 ml/litre of water is effective against weevils. Drenching the soil with a suspension of carbaryl (Sevin 50%WP) at the rate of 20 g in 10 litres of water is effective in controlling the pest.

8.1.4. Stem Borer, *Zeuzera coffeae*

(i) Damage: The red borer of coffee, *Zeuzera coffeae* (Cossidae) is a pest of cacao in many countries. Caterpillars of this leopard moth bore into young branches and make unramified hollow tunnels inside the stem. The symptoms of attack are a round hole on the stem (Fig 8.2), drying up of the upper portions of above the hole, and excreta and chewed up fibres strewn out on the ground. If the main stems of young plants are attacked, the plants die. When the branches are attacked, only the branches dry up and simple pruning will save the trees.

Figure 8.2. Stem borer infestation on cacao

(ii) Management

Prune and destroy the affected branches. Smear the affected portion with an emulsion containing carbaryl wettable powder.

8.1.5. Leaf Feeding Caterpillars, *Lymantria* spp., *Euproctis* spp., *Dasychira* spp.

(I) Damage: Cacao lymantriids, are the most commonly encountered leaf feeding insects of cacao in India. The caterpillars of this moth feed on the tender leaves and on the pericarp of cherelles and unripe pods. Three genera of lymantriids, *viz.*, *Lymantria, Euproctis* and *Dasychira* are so far reported on cacao. The caterpillars of these moths feed on the tender leaves, and on the surface tissues of cherelles and green pods.

L. ampla is commonly and abundantly observed leaf-feeding insect of cacao in India. These are seen more in the field after the monsoon rains. The caterpillars cause severe damage on leaves in young plants. The early instars of this moth feed on leaves or the surface tissues of growing pods during day and night, but later instar caterpillars are nocturnal in habit. During day time they hide on fallen leaves at the base of the tree or the basal surface of the main stem.

Caterpillars of *Euproctis* spp. are also defoliators of cacao. These caterpillars feed on tender leaves, surface of cherelles and green pods. The sporadic high population of *Euproctis* spp. can cause severe damage of tender leaves. The period of attack is seen from June-July.

(ii) Management

If the damage is very severe, spraying could be given by mixing 16 ml of dimethoate in 10 litres of water.

The indigenous natural enemies of this lymantriid include the insect pathogen of the mature larvae, *Paecilomyces fumosoroseus*; a pupal parasite, and a braconid larval parasite, *Apanteles* sp.

8.1.6. Vertebrate Pests

8.1.6.1. Rats, *Rattus* spp., *Bandicota* spp.

(i) Damage: Rodents are the major problem in cacao interplanted with coconut in all the countries since they have the ability to exploit new food sources. Rats and squirrels are the main groups of rodents that damage cacao in all the cacao cultivating tracts of the world. These damage the pods and the attack follows a fairly constant pattern. Ripe pods are usually chosen and a large hole is bitten through the pod husk (Fig 8.3). The beans are then extracted and after the sweet mucilage is eaten, these are discarded. Many species of rats damage both young and mature pods. Rat damage is reported from most of the tropical countries including the islands. The most important rat species observed is the black rat *Rattus rattus* which is very serious pest of coconuts in all coconut growing areas. The black rat is recorded in India as a major pest occurring in areca-cacao and coconut-cacao mixed gardens. Black rats damage cacao grown as monoculture also since in any South Indian condition the vicinity of preferred nesting hosts like coconut is assured.

Figure 8.3. Rat damage to cacao pod

Rats inhabit the crown of nearby coconut palms. They descend during night and damage the cacao pods.

(ii) Management

Only the cooperative efforts of plantation owners/small farm holdings can achieve management of rodent population. Otherwise the local trials done by individual farm owners have no impact on the population levels of rodents and the damage they inflict on cacao and other crops. This is more pronounced in the case of rats. Baiting and trapping methods are available. The rats can be controlled by keeping 10 g bromadiolone (0.005%) wax cakes on the branches of cacao trees twice at an interval of 10–12 days. Squirrels are best controlled by trapping with wooden or wire mesh single catch live trap with ripe coconut kernel as the bait. The success is more if trapping is carried out during the lean periods of the crop (September-November) and when the alternate foods such as paddy, cashew apples and jack fruit are not available. The traps are to be inspected daily and the trapped animals are to be killed to minimize the chances of communication with other animals. Timely harvest of the pods will help in increasing the efficiency of poison baiting as well as trapping.

8.1.6.2. Striped Squirrel, *Funambulus palmarum*

(i) Damage: Two species of squirrels are reported as pests of cacao in India. They are the western Ghats squirrel, *Funambulus tristriatus*; the south Indian palm squirrel, *F. palmarum*. Of these the western Ghats squirrel is the most serious pest of cacao in India. Squirrels usually make oval holes on the central or terminal portion of the pods for feeding (Fig 8.4). Squirrels are diurnal in habit and damage ripe pods. Striped squirrel gnaws the bronzing pods and extracts beans along with mucilaginous pulp.

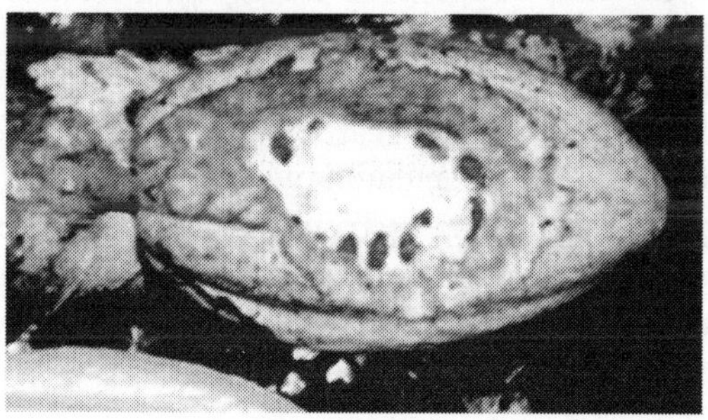

Figure 8.4. Squirrel damage on cacao pod

(ii) Management

Cover pods with punched polybags of 150 gauge smeared with bitumen kerosene mixture.

8.1.6.3. Palm Civet, *Paradoxurus hermaphroditus*

(i) Damage: The palm civet, *Paradoxurus hermaphroditus* (Viverridae) also known as toddy cat, is a nocturnal tree climber. It is present in many cacao-growing tracts of India and Malaysia. This carnivore damages the pods by biting and breaking the husk of cacao pods. The damage caused by this to cacao is reported from south India. The palm civet bite and break the husk of cacao pods. The pieces of broken chunks are 2.0 to 3.0 cm in diameter. There is no distinct pattern for the damage. The terminal half or one side of the pod is broken. While feeding the beans are swallowed and as such no trace of beans is visible directly under the tree. Piles of defecated beans are seen scattered around the cacao plantations.

(ii) Management

Palm civets could be easily controlled by poison baiting with 0.5 g of carbofuran granules using ripe bananas as baits. Two such banana fruits have to be tied to the trunk /jorquettes of five to six cacao trees per hectare.

8.1.6.4. Other Vertebrate Pests

(i) Damage: Jungle cat (*Felis chaus*), squirrel (*F. palmarum*), rat (*Rattus rattus*), monkey (*Macaca radiata*), jackal (*Canis aureus*), bison (*Bibos gaurus*), stag (*Cervus canis*) and birds constitute vertebrate pest complex. Squirrel damage commenced at the beginning of the fruiting season (July-August), jungle cat during pod ripening stage (October-November) and rat damage throughout the year. Monkey, jackal and birds were found damaging pods during ripening stage. Jungle cat is the most important pest incurring over 20% loss in pod yields in hill region of Karnataka and loss due to all vertebrates exceed 35% (Chakravarthy, 1993).

(ii) Management

Thyagaraj *et al.* (1996) conducted a baiting trial with carbofuran 3G and warfarin 0.05% simultaneously with wooden snap trapping (15–18 traps/ha) at 3 locations and banana fruits of *cv.* Puttabale served as a poison carrier. The number of healthy pods/plant increased significantly in the treated plots compared to untreated plots. In northern Karnataka, wet jaggery and dry sea fish are used as poison carriers.

Integrated Management: Cacao plantations were effectively protected from jungle cat, rodents, jackal, monkey and stag damage by baiting carbofuran 3G in local banana fruits or wet jaggery, intercropping with areca or coffee, timely harvest, clean cultivation and snap trapping at peak production period (August/September and December/January) (Thyagaraj *et al.*, 1996).

Table 8.2. Pest management

Pest	Damage	Control
Caster capsule borer, *Dichocrocis punctiferalis*	See under castor	Same as in the case of castor
Mealy bug, *Planococcus citri*	Colonies of nymphs and adults infest pods and de-sap the plants	Spray 0.03% diazinon, dimethoate, phosphamindon or monocrotophos
Cacao thrips, *Selenothrips rubrocinctus*	See under cashewnut	Same as in the case of cashewnut

8.2. DISEASES

8.2.1. Seedling Dieback, *Phytophthora palmivora*

(i) Symptoms: The infection starts from tip of the stem, cotyledon or collar region as dark brown discolouration. The lesions are further extended to the leaves through petiole resulting in wilting and subsequent defoliation of leaves. Leaves turn yellow and fall leaving a bare length of shoot. Defoliation and dieback are the main symptoms during advanced stages of infection. Xylem is discoloured by brown streaks.

(ii) Epidemiology: The disease is more severe during south-west monsoon (June-September) period. The pathogen survives in dead and decaying material.

(iii) Management

The entire infected and dead seedlings should be removed from the nursery and destroyed. Seed treatment in combination with soil drenching with copper hydroxide has been very effective in managing the disease. Improvement of drainage facilities in the garden followed by drenching with 1% Bordeaux mixture was also found effective.

8.2.2. Black Pod Rot, *Phytophthora* spp.

Black pod disease of cacao is an economically serious problem in all areas of the world where cacao is grown (Brasier & Griffin, 1979; Evans, 2001; Evans & Prior, 1987; Fulton, 1989; Gregory, 1974). Annual losses due to black pod may range from 30–90% of the crop. This disease is especially severe in West and Central Africa which contributes 60–70% of the world production of cacao beans. The disease is caused by a complex of species of *Phytophthora*.

P. palmivora has been recognized as one of the most important pathogens in the tropics attacking many plantation crops, including cacao, rubber, black pepper, coconut, pineapple, and papaya, as well as citrus, avocado, and many ornamental plants. *Phytophthora palmivora* is present in most countries and is an important part of the black pod complex, while *P. megakarya* is present only in several countries in West Africa (Gregory & Maddison, 1981). However, *P. megakarya* appears to be more virulent than *P. palmivora* and is becoming the dominant species in West Africa, moving from Nigeria and Cameroon into countries where it has not been reported previously. In the major cacao-growing region of Bahia, Brazil, three species have been implicated: *P. palmivora*, *P. capsici*, and *P. citrophthora* (Kellam & Zentmyer, 1986; Lawrence *et al.*, 1982; Luz *et al.*, 1984). *Phytophthora capsici* appears to be the dominant and most important species attacking cacao in Brazil, and also has been reported to occur in other countries in Central and South America, as well as in the West Indies, Indonesia, and India (Erwin & Ribeiro, 1996). *Phytophthora citrophthora* is the least common species found in Brazil, but is the most virulent (Kellam & Zentmyer, 1986; Luz *et al.*, 1984). *Phytophthora heveae* also has been found to cause black pod in some countries (Erwin & Ribeiro, 1996).

(i) Symptoms: Although *Phytophthora* species attacks all parts of the cacao plant, the major economic loss is from infection of the pod (Fig 8.5) (Evans & Prior, 1987; Gregory, 1974).

Pods or cherelles (immature pods) may be infected at any place on the surface, but infection is most often initiated at the tip or stem end. The disease causes a firm, spreading, chocolate-brown lesion that eventually covers the whole pod. The beans inside the pod may remain undamaged for several days after initial infection of the husk, thus frequent harvests may prevent much yield loss. In advanced infections, *Phytophthora* invades the internal pod tissues and causes discoloration and shrivelling of the cacao beans. Diseased pods eventually become black and mummify. The pathogen also will cause seedling blight in cacao nurseries as a result of infection of the stem and young leaves (Fig 8.6).

Figure 8.5. Black pod disease of cacao caused by *P. capsici*

Figure 8.6. Seedling blight of cacao caused by *P. megakarya*

The inoculum that initiates black pod can come from the soil and/or infected roots, stems, and leaves (Erwin *et al.*, 1983; Evans, & Prior, 1987). Root infection from residual soil inoculum usually is not an economic concern; however, the pathogen is capable of producing spores on all infected plant parts. Thus, infected roots may serve as a source of inoculum for infection of the pods. Bark and stem cankers may function in the same manner. Once a pod is infected and sporulating, it may then provide a massive source of inoculum to infect other pods.

(ii) Epidemiology: Under humid conditions, sporangia (asexual reproductive structures and a source of secondary inoculum) may form on the surface of infected roots, cankers, or diseased pods (Fig 8.7) (Gregory *et al.*, 1981). This type of propagule can be dispersed by rainfall, splashing water, and water moving over the surface of the soil (Erwin *et al.*, 1983). When sporangia are in free water, zoospores are formed inside the sporangia and released into the water (Fig 8.8). Zoospores are motile spores that actively swim towards an infection site and are the primary infective propagules. Thus, windborne rain is a primary factor in the spread of the disease. Ants also have been reported to transport inoculum.

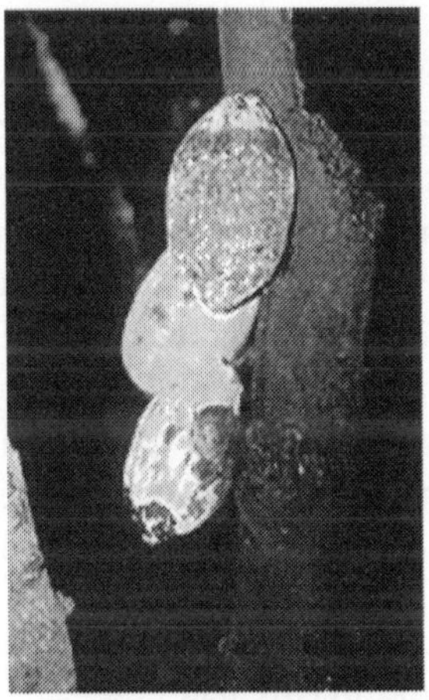

Figure 8.7. Sporangia of *P. megakarya* forming on the surface of an infected pod

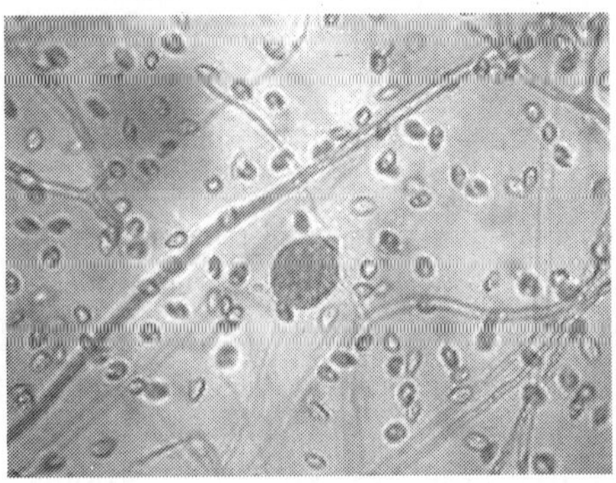

Figure 8.8. Sporangium and zoospores of *P. megakarya*

(iii) Management

Control of black pod is difficult because *Phytophthora* can persist in soil and debris for several years (Erwin *et al.*, 1983). Also, since susceptible pods may be present on the trees most of the year, the pathogen may always be present in the canopy, ready to cause major epidemics when environmental conditions become favourable for sporulation and dispersal (Evans & Prior, 1987). Frequent harvesting will lessen the danger of spread of the disease from infected pods (Soberanis *et al.*, 1999). Regular pruning to remove infected chupons (small suckers at the base of the tree) and increase air circulation (to reduce the humidity under the canopy) is an important disease management tool. Other measures, such as the removal of infected pods and husk piles, may have some effect on inoculum levels. However, under high rainfall conditions, it is difficult to effectively suppress the inoculum to attain good control (Evans, & Prior, 1987). Chemical control relies on the use of copper and metalaxyl-based fungicides. However, these are not entirely effective, are expensive for the small farmer, and not economically feasible (Fulton, 1989; Pereira, 1985). Additionally, there are environmental concerns regarding the heavy use of chemicals in the tropical rain forests, and there may be problems with non-target effects and resistance of the pathogen. At present, there is no acceptable genetic resistance in cacao to control black pod. Active research is underway throughout the world on this important disease of cacao. Control measures being investigated are aimed at integrated pest management (IPM) strategies, and include biological control micro-organisms, genetic and induced resistance, cultural practices, natural products, and limited use of chemicals (Krauss & Hebbar, 1999).

All the diseased pods should be removed and destroyed at weekly interval especially during rainy season. Regulation of shade by proper pruning and allowing the sunlight to penetrate helps in minimizing the disease incidence. The disease can be prevented by phytosanitation and prophylactic spraying with 1% Bordeaux mixture before the onset of monsoon.

8.2.3. Witches' Broom, *Crinipellis perniciosa*

Witches' broom disease, caused by the fungus *Crinipellis perniciosa* (formerly *Marasmius perniciosus*), was first reported in 1895 from Surinam and is currently found in several countries of South America and the Caribbean islands (Purdy & Schmidt, 1996; Wheeler & Suárez, 1993).

(i) Symptoms: The disease is initiated by basidiospores produced and released from pinkish mushrooms called basidiocarps (Fig 8.9) that are dispersed by wind and rain onto leaves, flowers, and fruit (pods) of cacao plants. It is estimated that a single basidiocarp can release 80 to 90 million basidiospores. In the presence of free moisture (rain and dew) and high relative humidity, basidiospores germinate and penetrate young meristematic tissues in vegetative and floral buds through stomata, epidermis, or trichomes. The colonized tissues undergo several physiological and hormonal changes leading to swelling and formation of numerous succulent vegetative branches, known as brooms, within flower cushions (Fig 8.10) and on vegetative apical (Fig 8.11) or auxiliary buds (Orchard *et al.*, 1994). The brooms are usually formed within 5 to 6 weeks following infection. The fungus also infects pods causing necrotic lesions, uneven ripening, and various deformations. Brooms and infected pods become progressively brown and dry, and within 3 to 8 months following broom and pod drying, basidiocarps (Fig 8.12) are formed following several alternating wet and dry periods.

Figure 8.9. Basidiocarp of *C. perniciosa*

Figure 8.10. Vegetative branches (brooms) arising from a flower cushion infected by *C. perniciosa*

Figure 8.11. A dry broom formed from infection of an apical vegetative bud by *C. perniciosa*

Figure 8.12. Basidiocarps (pinkish structures) formed on a dry broom and pod infected with *C. perniciosa*

Witches' broom debilitates cacao trees by diverting energy to the numerous brooms that do not form flowers, thus reducing yield potential. Further yield loss is incurred by the prevention of seed formation in pods infected early in their development. If seeds are formed prior to infection they may be unusable depending on the extent of pod colonization by *C. perniciosa* (Fig 8.13).

In its natural habitat of the Amazonian forest, cacao is found as scattered under story trees which are genetically variable, a characteristic that hampers the development of epidemics of witches' broom. The cultivation of cacao has intensified over the years as cacao has progressively become a global trade commodity. Production environments that encourage the planting of cacao varieties with little genetic diversity ensure a constant supply of susceptible tissue for infection by *C. perniciosa*.

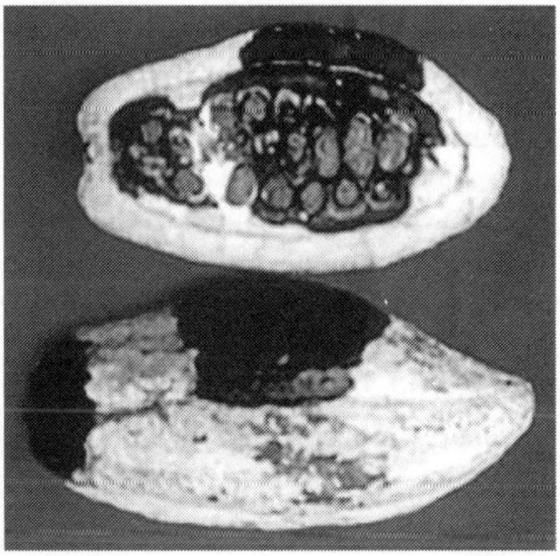

Figure 8.13. Cacao pod infected with *C. perniciosa* showing lesion on the surface and extensive rotting of the seeds

(ii) Management

Management of witches' broom has received considerable attention from the beginning of the 20th century (Evans & Prior, 1987; Fulton, 1989; Pereira, 1996; Purdy & Schmidt, 1996; Wheeler & Suárez, 1993). Broadly, there are four major strategies that may be adopted: phytosanitation, chemical control, genetic resistance, and biological control (Krauss & Hebbar, 1999). Phytosanitation, by removal and destruction of diseased plant parts, has been shown to reduce pod loss and delay disease epidemics (Rudgard & Butler, 1987; Soberanis *et al.*, 1999). However, this strategy is tedious and in one study it has been shown that 95% removal is required to achieve 50% reduction in pod loss (Rudgard & Butler, 1987). Chemical control of witches' broom with protectant and systemic fungicides is not a routine practice in cacao production because of high costs and risks associated with cacao bean contamination and environmental health (Evans & Prior, 1987). The development of genetically resistant cacao cultivars is an on-going endeavour in many countries, and it is expected that the use of these cultivars would reduce the incidence of the disease.

Prospects of managing witches' broom disease of cacao through biological control have been investigated for over 20 years leading to the isolation of a new species, *Trichoderma stromaticum*, a parasite on the mycelium and basidiocarps of *C. perniciosa* (Samuels *et al.*, 2000). In Brazil, commercial formulations of *T. stromaticum* are currently used in managing witches' broom. However, the inconsistent performance of *T. stromaticum* indicates the need to further understand the relationship between the cacao plantation environment and the survival and establishment of this mycoparasitic fungus. Collaborative research efforts by scientists at research institutions in the United States and Central and South America are currently underway to optimize the use of *T. stromaticum*.

8.2.4. Frosty Pod Rot, *Moniliophthora roreri*

Frosty pod rot (or *Moniliophthora* pod rot), caused by *Moniliophthora roreri* originally described in 1933 as *Monilia roreri*, is a devastating disease of cacao pods (Evans, 1981; Evans *et al.*, 1978). Frosty pod rot occurs on all *Theobroma* and related *Herrania* species. The first report of the disease was by J.B. Rorer (1917–1925) while on a scientific trip to Ecuador to investigate serious losses in yields due to diseases (Evans, 1981). Frosty pod rot currently is restricted to northwest South America (Ecuador, Peru, and Colombia) and south Central America (Nicaragua, Costa Rica and Panama). However, it is an immediate threat to Brazil, which is one of the largest cacao growing areas in South America. Yield losses have been estimated between 25% to total loss of the crop if frequent (7–10 days schedules) phytosanitary measures are not implemented (Soberanis *et al.*, 1999).

(i) Symptoms: The conidia (the only known infective propagules) infect by penetrating the surface of the pods. The pods are highly susceptible during the first 90 days of their growth. Early symptoms include discoloured areas of swelling on the pods followed by a dense formation of cream-colored spores, the so called frosty pod, developing on the pod surface within 2 weeks after infection (Fig 8.14). The area of the pod with sporulation spreads rapidly, and spore densities have been estimated to be as high as 44 million conidia per square centimetre (Evans, 1981; Fulton, 1989). The spores are later released by wind or by water droplets during rainy periods.

Figure 8.14. Frosty pod rot of cacao caused by *M. roreri* with whitish to creamy-colored spores on the pod surface

(ii) Management

The most economical way of controlling the spread of this disease is to eliminate the inoculum sources, which are dead sporulating pods (Fig 8.15), with frequent harvests on regular cycles (Rudgard & Butler, 1987). Other practices should include reduction of tree height (maximum 3.5 m) to facilitate removal of diseased pods, biological control, and planting of resistant or tolerant cultivars as they become available.

Figure 8.15. Trees in abandoned plantations where the presence of dead pods covered with spores of *M. roreri* is a constant source of inoculum

In view of the difficulty in finding a practical and cost-effective fungicidal treatment, alternative control strategies, such as biological control are being investigated in Peru, Panama, and Costa Rica to manage frosty pod rot (Krauss & Hebbar, 1999; Krauss & Soberanis, 2000; Krauss *et al.*, 1999; Soberanis *et al.*, 1999). This involves the use of naturally occurring fungal agents (mycoparasites) capable of parasitizing the pathogen and preventing the dispersal of fungal spores. Saprophytic fungi belonging to the genus *Trichoderma* and *Clonostachys*, selected for their parasitizing capacity, are being tested in field trials (Piper *et al.*, 2000). Bi-weekly to monthly sprays of biocontrol agents have been shown to reduce the production of conidia on diseased pods (Fig 8.16) (Krauss & Soberanis, 2000). Biocontrol treatment along with phytosanitation has reduced pod loss and increased yields in previously abandoned cacao plantations (Krauss & Soberanis, 2000).

Figure 8.16. Infected pod sprayed with biocontrol agents (left) reduce spore dispersal by parasitizing the pathogen mycelium and spores (note the cream-colored, parasitized spores). The pod on the right was not sprayed

8.2.5. Charcoal Pod Rot, *Botryodiplodia theobromae*

(i) Symptoms: Infection appears as dark brown to black coloured spots. Affected pods turn black and remain as mummified fruit. Internal tissues rot and affected beans turn black. Spores appear in masses forming a soot.

(ii) Epidemiology: The disease is more severe during summer months.

(iii) Management

- Since the pathogen enters through the wounds caused by insects and rodents, management of these pests is important.
- Spraying with 1% Bordeaux mixture was found effective for management of the disease.

8.2.6. Brown Pod Rot, *Diplodia natalensis*

(i) Symptoms: The pathogen produces water soaked chocolate brown or dark lesions on pods at tip or stalk end. Development of black sooty mould takes place on affected pods.

(ii) Management

Spraying of copper oxychloride or captafol at 2 g/litre of water is effective for the management of disease.

8.2.7. Pink Disease, *Corticum salmonicolor*

(i) Symptoms: The pathogen causes pinkish powdery coating on stem of the affected plants. It also causes wilting of shoots, shedding of leaves and ultimately drying of branches.

(ii) Management

Spraying of copper oxychloride or captafol at 2 g/litre of water is effective for the management of disease.

8.2.8. Vascular Streak Die-back, *Oncobasidum theobromae*

(i) Symptoms: The pathogen is responsible for characteristic yellowing of 1 or 2 leaves on the 2nd or 3rd flush behind growing tip. Other leaves on the shoot soon show similar symptoms. Brown streaking of woody tissue extending well beyond the region of yellow leaves is also observed.

(ii) Management

Soil drenching with copper oxychloride at 2 g/litre of water is recommended against wilt and dieback.

8.2.9. Cherelle Rot, *Colletotrichum gloeosporoides*

(i) Symptoms: The infection starts at the stalk end as it becomes highly sunken and can be easily detached. As the infection spreads, the internal tissue of the pod also becomes discoloured. The dark brown sunken lesions coalesce and form bigger lesions. At later stage, the pod turns dark brown to black and remains on the tree as mummified fruit.

(ii) Epidemiology: Usually the cherelle affected pods are seen during January-May. The critical period for infection is February-May (Chandra Mohan *et al.*, 1989).

(iii) Management

Spraying of 0.05% carbendazim and 0.2% mancozeb was found effective against cherelle wilt (Chandra Mohan & Kaveriappa, 1984).

8.2.10. Stem Canker, *Phytophthora palmivora*

(i) Symptoms: Generally round to oval, grayish brown or dark brown, water-soaked lesions appear on main stem and branches. The oozing of reddish brown liquid is seen from such lesions. Rotting of the tissue beneath the outer lesion always appears as reddish brown discolouration referred as 'stem canker'.

(ii) Epidemiology: The disease is usually observed during December-February after south-west monsoon period.

(iii) Management

Wound sealing with Bordeaux paste or copper oxychloride effectively controls the disease.

8.3. NEMATODES

Root-knot nematodes are recognized as the major limiting factors in successful production of cacao crop.

8.3.1. Root-knot Nematodes, *Meloidogyne* spp.

Root-knot nematodes are the most important nematodes of cacao due to their pathogenicity and wide distribution in cacao producing regions. *Meloidogyne* spp. have been found in cacao since 1990 (Sosamma *et al.*, 1980a), and they have been reported from Zaire, Sao, Java, Cotterel, Ghana, Malawi, Ivory Coast, Nigeria, Venezuela, Brazil and India (Sosamma *et al.*, 1980b). *M. incognita* seems to be the most frequently found in cacao. *M. incognita* and *M. javanica* have caused losses in cacao areas around the world including yield decrease, sudden death of trees and growth retardation of seedlings in nursery.

(i) Symptoms: *M. incognita* causes dieback, stunting, wilting, and yellowing of leaves and small leaves. On roots tiny galls and females with egg masses can be observed. The leaf tips and margins first turn brown and become dried. This spreads to the entire leaves which are eventually shed. The infested plant exhibit the syndrome of sudden death.

(ii) Survival and Spread: The commonly used shade plants such as banana may become a source of inoculum in the cacao plantation (Sosamma *et al.*, 1980). Nursery soil infested with the nematodes will allow the production of infested seedlings which will spread nematodes into plantations. Run off water may also spread nematodes.

(iii) Management

(a) Cultural Methods: The most efficient control strategies include production of seedlings free from major pathogenic nematodes and to cultivate in soils or areas from which the nematodes are absent. Soil to be used in the nursery should be collected from areas that are not infested by root-knot and *Pratylenchus* spp. Care must be taken with the selection of shade plants, avoiding trees susceptible to root-knot or lesion nematodes such as *Leucaena glauca* and banana (Sosamma *et al.*, 1980a).

(b) Chemical Methods: Application of granular nematicides such as phenamiphos (Nemacur 10G), aldicarb (Temik 10G) at 50 mg of commercial product/plant (Sharma & Ferraz, 1977) helps in reducing the nematode population. Tarjan *et al.* (1973) reported increased yield of 11 to 96% after field application of ethoprophos (Mocap) or fensulfothion (Terracur) at 34 kg of the commercial product/ha and phenamiphos (Nemacur) at 22 kg of the commercial product/ha.

9 RUBBER

India is the fourth largest producer of natural rubber (*Hevea brasiliensis*) with a share of 8.8% in world production in 2006. The two major states with rubber plantations are Kerala and Tamil Nadu, with Kerala alone accounting for over 80% of the area under plantation. India produced an estimated 852,000 tonnes of natural rubber during 2006–2007 and exported 56,545 tonnes while it imported 89,699 tonnes. In 2006 the yield per hectare in India at 1879 kg was amongst the highest in the world.

9.1. PESTS

9.1.1. Bark Feeding Caterpillar, *Aetherastis circulata*, *Ptochoryctis rosaria*

The pest is distributed in Nagercoil, Nedumangad, Punalur, and Thrissur localities.

(i) Damage: The caterpillars build galleries with fecal matter and silk all over the trunk region and branches of trees. Generally larvae feed on dead bark (Fig 9.1) and occasionally on live bark causing exudation of latex. Deep scars are found at the regions of feeding.

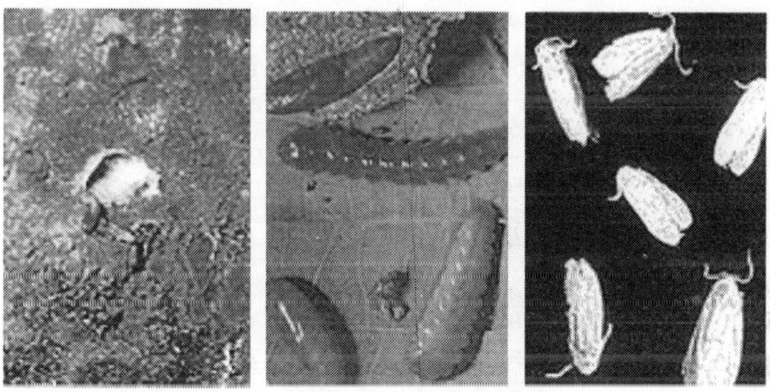

Figure 9.1. Bark eating caterpillar on rubber

(ii) Management

Remove the mud galleries along with the larvae. Mix 100 g carbaryl in a litre of water. Dip a paint brush in this solution and swab around the affected portion.

When the infestation is severe, apply carbaryl 5% dust at the rate of 10 kg per hectare or fenval 0.4% dust at the rate of 7 kg per hectare (provisional recommendation) with a power duster. Spraying the trunk with fenvalerate 0.02% (Tatafen 20 EC, Arfen 20 EC at 1 ml/L) is also effective.

9.1.2. Cockchafer Grub, *Holotrichia serrata*, *H. rufoflava*, *Anomala varians*

The pest is more prevalent near forest areas and in loose soils.

(i) Damage: Cockchafer grub (Fig 9.2) feeds on the roots of seedlings in the nursery and young plants. The affected plants droop and fall over.

Figure 9.2. Cockchafer grubs and adults

(ii) Management

Incorporate 25 kg per hectare of phorate 10 G in the soil at the time of bed preparation.

9.1.3. Mealy Bug, *Ferrisiana virgata*

The pest is seen mostly in nurseries in all rubber growing areas.

(i) Damage: Mealy bugs are soft bodied small insects with white mealy outer covering. Occurrence and damage are similar to scale insects. Occur on tender shoot (Fig 9.3), suck the sap and devitalize the plant. Association with sooty mould is also observed.

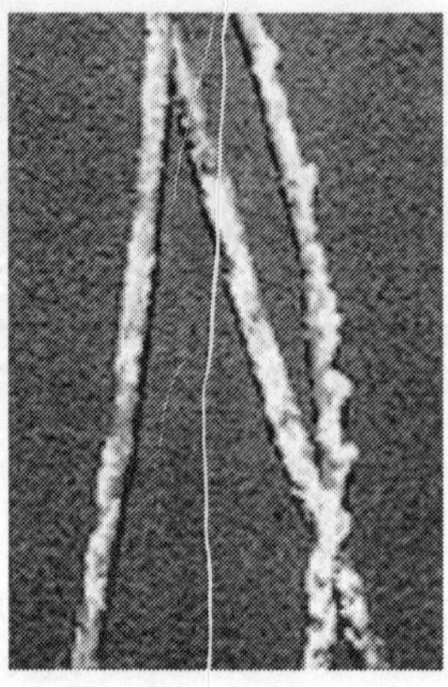

Figure 9.3. Mealybug infestation on rubber

(ii) Management

In the normal course, its own parasite will keep this pest under check. But, when the population increases beyond a level, then spraying with monocrotophos at 2 ml/litre of water is recommended. Spraying with organophosphorus insecticides like malathion 0.1% (2 ml/l) or quinalphos 0.075% (Ekalux 25EC at 3 ml/L) also gives effective control of the pest.

9.1.4. Scale Insect, *Saissetia nigra*

The pest is seen generally in young plantations and nurseries in almost all rubber areas.

(i) Damage: Scale insects are small insects with an outer black, dome-shaped covering. Occur on leaflets, petioles and tender shoot portions and suck the sap (Fig 9.4). Severely affected portions dry up and die. Ants and sooty mould are associated with this insect.

Figure 9.4. Scale infestation on rubber stem

(ii) Management

Natural enemies like insect parasites and entomopathogenic fungi keep this pest in check. When severe infestation is noted, spray organophosphorus insecticides like malathion at 0.05% concentration.

9.1.5. Slugs and Snails, *Mariaellae dussumieri*, *Crytozona (Xestina) bistrialis*

The pest is distributed randomly in rubber areas in Kanjirappally, Thrissur and Calicut.

(i) Damage: Feed on latex by lacerating the tender leaves and buds. Growth of affected buds is arrested and side shoots develop giving a bunchy appearance. Slugs drink latex from the tapping cut and collecting cup also (Fig 9.5). Nursery seedlings are more susceptible.

Figure 9.5. Snail infestation on rubber

(ii) Management

Broadcast 2.5% Metaldehyde bait pellets (snailkill 2.5) at the base of the infested plants or seedling beds or paint the base of the stem with metaldehyde. To repel slugs and snails, brush 10% Bordeaux Paste around the stem above the bud union to a length of 30 cm. Repeat the application after 30–45 days if the attack continues. Spraying metaldehyde 0.05% or 0.1% of one day old suspension in the infected area is more effective than broadcast.

9.1.6. Termite (White Ant), *Odontotermes obesus*

The pest is distributed in dry regions of Central Kerala (Thrissur and Palakkad) and non-traditional areas like Dapchari in Orissa.

(i) Damage: Feeds on the dead bark of trees and young plants (Fig 9.6). Builds covered passageways of soil on the tapping panel and collection cup. Sometimes young plants dry up due to attack.

Figure 9.6. Termite infestation on rubber

(ii) Management

Drench the soil at the base of affected plants with chlorpyriphos 0.1% solution. When mulch is present, spray the mulch also.

9.1.7. Mite, *Hemitarsonemus dorsalis*

(i) Damage: Mites are responsible for crinkling and shedding of leaves.

(ii) Management

Spraying sulphur 50WP at 2 g/litre of water gives effective control of mites.

9.2. DISEASES

9.2.1. Abnormal Leaf Fall, *Phytophthora palmivora, P. meadii, P. nicotianae* var. *parasitica, P. botryosa*

This is the most severe disease of rubber plants occurring in India, Malaysia, Burma, Java and other far Eastern countries. The disease recurs annually during southwest monsoon period. Prolonged wet weather coupled with humid atmospheric conditions favour the disease. Disease incidence is less in Kanyakumari District, but increasingly heavy northwards along the southwestern coast.

(i) Symptoms: First the fruits rot, later infected leaves fall in large numbers prematurely, either green or after turning coppery red. A black lesion may develop on the petiole with a drop of latex, often coagulated, in the centre. Lesion may develop on the midrib and leaf blades also (Fig 9.7). Heavy defoliation may lead to considerable loss of crop and die-back of terminal twigs.

Figure 9.7. Leaf fall in rubber

(ii) Epidemiology: The disease occurs during rainy season, since the pathogen requires high humidity. The fungus over winters inside the bark or soil and infects during the onset of monsoon with production of abundant sporangia and zoospores for secondary spread.

(iii) Management

Prophylactic spraying of the foliage prior to the onset of southwest monsoon with either 1% Bordeaux mixture using high volume sprayers or oil-based copper oxychloride dispersed in diluents spray oil employing either low volume air blast sprayers from the ground or through aerial application. For micron spraying, based on the tree spread, foliage intensity, planting material used and age of plants, two rounds of spray using about 17 to 22 litres of fungicide oil mixture (1: 6 proportion) per hectare per round with a gap of 10 to 15 days or a single round of spray with about 30–40 litres of fungicide oil mixture (1: 5 proportion) per hectare may be necessary. For aerial spraying, 6.2 litres of 40% oil based copper oxychloride paste in 37 litres of diluents oil or 8 kg oil dispersible copper oxychloride powder in 40 litres of oil is used per hectare. Spraying should be done as close to the monsoon as possible. Rubber clones RRII 105, PB 217, GT 1 and GL 1 are showing some tolerant reaction to the disease.

9.2.2. Birds' Eye Spot, *Drechslera heveae*

This is a hot weather disease causing serious damage in the nursery during April-May. Weaker plants and plants growing under exposed situations are more susceptible.

(i) Symptoms: Symptoms appear as small necrotic spots with dark/brown margins and pale centre (Fig 9.8). Severe infection leads to premature defoliation and dieback. Nursery seedlings are susceptible.

Figure 9.8. Bird's eye spot on rubber leaf

(ii) Management

Repeated spraying with 1% Bordeaux mixture or 0.2% mancozeb (Mancozeb/Indocile M 45 at 2.5 g/l) or carbendazim 0.02% (Baiting at 0.4 ml/l). Shading the nursery plants reduces the disease incidence. Seedlings should be maintained in vigorous condition through adequate balanced nutrition.

9.2.3. Black Stripe, Black Thread or Black Rot, *Phytophthora palmivora, P. meadii*

The disease is prevalent during the rainy season.

(i) Symptoms: In the renewed bark region, small depressions are formed due to localized rotting and drying of bark, which gets pressed to the wood (Fig 9.9). When scraped, deep vertical black lines are noticed running downward into the tapping bark and upwards into the renewed bark. Bark rot often reduces yield. The renewed bark becomes highly uneven.

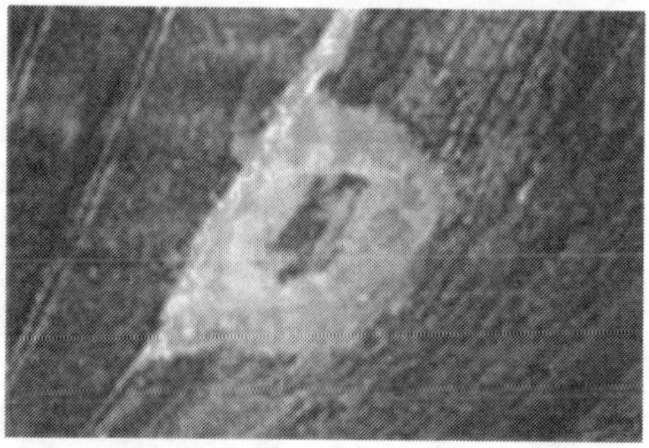

Figure 9.9. Black stripe on rubber

(ii) Management

If tapping is regularly done during rainy season, the tapping panel should be disinfected at weekly intervals by brushing with mancozeb 0.375% *a.i.* (dithane/indofil M-45 at 5 g/l). Phosphoric acid formulations at 0.08% (akomin and phosjet at 2 ml/l) are also effective. In the infected cases, scrape off the affected tissues and apply the fungicide. When the fungicide dries up, a wound dressing compound may be applied (Fig 9.10).

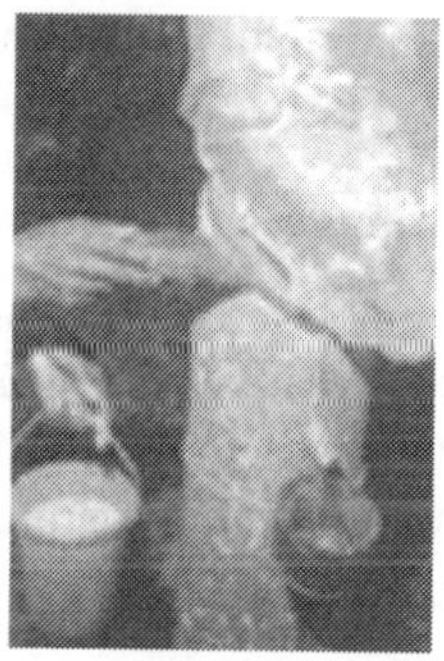

Figure 9.10. Fungicidal application for the control of black stripe disease

9.2.4. Dry Rot, Stump Rot, Collar Rot or Charcoal Rot, *Ustulina deusta*

Infection of the pathogen takes place during the wet weather.

(i) Symptoms: Affects roots, collar, trunk and branches with copious exudation of latex from the lesions (Fig 9.11). Affected bark and wood become soft and powdery with double black lines in the wood. Grey, crust-like flat fructifications are seen on the bark, which later coalesce and turn black. Affected trees or branches are killed or blown over. *Ustulina* root infections are also noticed.

Figure 9.11. Dry rot disease on rubber

(ii) Management

Scrape off the fructifications, affected bark and wood showing black lines. Apply a wound dressing compound in which thiram 0.75% *a.i.* (Thiride at 10 g/l) incorporated. Hexaconazole 0.02% *a.i.* (Contaf at 4 ml/l) is also effective. Avoid accumulation of rubber at the base of the tree. For root infection, clean up affected areas by washing with Emisan solution. Scrape out the fructifications. Wash the wound again with fungicides. When it dries up, apply a wound dressing compound.

9.2.5. Leaf Spot, *Corynespora cassiicola*

Earlier the disease was confined to nurseries only. From 1996 onwards severe incidence of the disease was observed in mature plantations in the Dakshina Kannnada District of Karnataka State and in the adjoining districts of Kerala. In nurseries, disease incidence is noticed during November to May period. On mature trees, disease is observed during the refoliation period from February to May.

(i) Symptoms: Appearance of large spots, with brown margins and pale centre is the common symptom (Fig 9.12). The centre may fall off forming shot holes. On mature trees, light green leaves during refoliation are more susceptible. Several lesions coalesce to form large blighted area. Disease spreads along the veins leading to a brownish "railway track" or "fish bone" like appearance. Even a localized infection on midrib or base of a leaf causes leaf abscission. Defoliation leads to die-back of branches.

(ii) Management

Repeated spraying with 1% Bordeaux mixture or 0.2% mancozeb or 0.02% carbendazim is recommended for nursery. Shading the nursery reduces the disease incidence. Seedlings should be maintained in vigorous condition through adequate balanced nutrition. High volume spraying with mancozeb 0.2% (Dithane/Indofil M-45 at 2.66 g/l) carbendazim 0.05% (Bavistin at 1 g/l) at 2–3 weeks interval during refoliation is effective in mature plantation. Micron spraying with oil dispersible copper oxychloride 56% (8kg) or oil dispersible mancozeb 70% (7 kg) dispersed in 40 litres spray oil per ha is also effective.

Figure 9.12. Leaf spot on rubber

9.2.6. Patch Canker or Bark Canker, *Phytophthora palmivora*, *Phytophthora meadii*, and *Pythium vexans*

Infection mostly occurs during the wet weather.

(i) Symptoms: Infection occurs on the tapping panel region, or anywhere on the stem including the collar region and occasionally on the roots causing swelling and bark burst with amber-coloured liquid oozing out (Fig 9.13). This may not be evident in all cases. In most cases oozing of latex is observed. The bark rots, and a coagulated rubber pad, emanating a foul smell is seen in between the wood and the rotting bark. When this is removed, a slight discolouration of the wood in this region is also noticed.

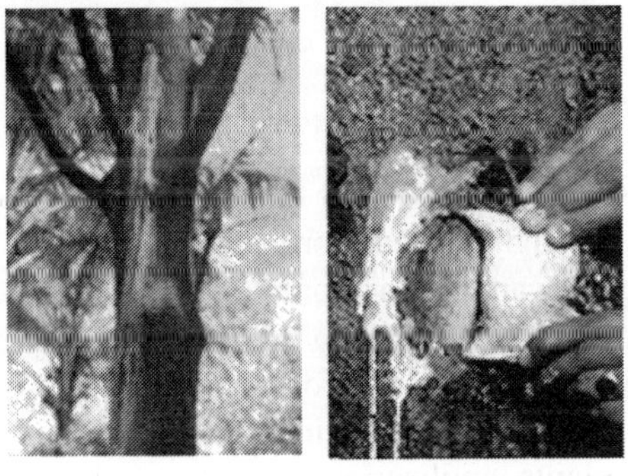

Figure 9.13. Bark canker on rubber

(ii) Management

The affected region may be scraped to remove all the rotting bark and the coagulated rubber and the wound washed well with mancozeb 0.75% (dithane/indofil M-45 at 10 g/l). When the fungicide dries up, apply wound dressing compound.

9.2.7. Pink Disease, *Corticium salmonicolor*

The disease is noticed during southwest monsoon period in the traditional rubber growing areas in south India. The disease incidence is very mild in Kanyakumari District due to the weak southwest monsoon and absence of prolonged wet conditions. Disease has been noticed in Assam, Tripura and Meghalaya in north east.

(i) Symptoms: The disease is more damaging on plants in the age group of 2 to 12 years.

Main seat of infection is usually the fork region. White or pink coloured cobweb mycelial growth on the bark surface with streaks of latex oozing out from the lesions can be observed (Fig 9.14). Rotting, drying up and cracking of the affected bark follows. Sprouts develop from below the affected portion. The distal portion of branches dry and dried leaves stick to the dead branches.

Figure 9.14. Pink disease on rubber

(ii) Management

(a) Prophylactic: Prophylactic treatment for pink disease can be taken up for highly susceptible clones planted in highly disease prone areas. Two rounds of spraying with 1% Bordeaux mixture (during May and August) onto the fork and branches reduce the disease incidence. For 2-year-old plants, the top most brown portion has to be applied with Bordeaux paste in a 30 cm wide band all around. The forking region also has to be applied with the paste. In leaning plants, in addition a 30 cm wide band on the convex side of the bend has to be applied. In 3 years old plants, application has to be made on all major forking regions except the lower most one and also on all the leaders of the top most brown regions. Only one application is required. No scraping is necessary. The application can be made from ground using a long handled-brush before the onset of monsoon. During October, a single tracing and treatment need be made for infection, if any. No sticker should be added to the paste. Rubber clones like PB 86, RRIM 513, GL 1, PR 107, GT 1 and PB 260 are less susceptible.

(b) Curative: Frequent tree inspection during June to October for detecting infection and application of Bordeaux paste in early stage on infected part and up to 30 cm above and below. In advanced cases, Bordeaux paste may be initially applied and then the affected portion scraped to remove mycelium and decayed bark, followed by a second Bordeaux paste application. Dried branches should be pruned after disinfection. Thiram 0.75% [thiride (TMTD) at 10 g/l] is incorporated in a wound dressing compound or tridemorph 2% (Calixin at 25 ml/l) in 1% ammoniated field latex or thiride 0.75% or tridemorph 1% or propiconazole 0.1% (Tilt at 4 ml/l) in pidyvyl, china clay and water (1: 2: 4 by volume) is also effective as rain wash is prevented.

9.2.8. Powdery Mildew, *Oidium heveae*

The disease is predominantly noticed on newly formed tender flush during the refoliation period from January to March. The disease is severe in Kanyakumari, Idukki and Wynad district of south India and north eastern states. Cloudy days with light rains and/or misty nights with dew formation during refoliation favour serious disease outbreaks.

(i) Symptoms: Tender leaves with ashy coating curl, crinkle, and edges roll inwards and fall, leaving the petioles attached to the twigs like a broom-stick. After a few days, the petioles also fall. Die-back of twigs follows. On older leaves, white patches later causing necrotic spots reduce photosynthetic efficiency. Infected flowers and tender fruits are shed, affecting seed production.

(ii) Epidemiology: The fungus over winters as resident candidate in the host bark. It infects during warm weather with low humidity. It is active during the entire season under shade in higher elevation. The fungus infection takes place during conducive weather conditions, and the secondary spread is through new flush.

(iii) Management

Dusting of 11 to 14 kg 325-mesh fine sulphur dust/round/hectare should be taken up during the refoliation period commencing from bud break in about 10% of the trees, giving 3 to 5 rounds at weekly to fortnightly intervals. Sulphur mixed with an inert material like talc (70: 30) is commonly used. Wettable sulphur or carbendazim at 2.5 g/l is also effective in nurseries and for young plants as a spray. Spraying 0.05% *a.i.* carbendazim (bavistin at 1 g/l) is more effective than sulphur for nurseries and young rubber plantations. Alternate use of carbendazim and sulphur is recommended to avoid resistance problem. For efficiency, dusting may be carried out in the early morning hours when the leaves are moist and the atmosphere calm. Rubber clones PB 86, GT 1, GL1, PR 107, PB 5/139, RRIM 703, RRII 208 and PB 310 show some tolerance.

9.2.9. Brown Root Disease, *Phellinus noxius*

(i) Symptoms: The characteristic symptoms of the disease are yellowish discolouration of foliage, black encrustation of soil, sand and fungal hyphae cemented to roots, and brown lines in the affected roots. Plants dry up in advanced cases.

(ii) Management

Open up the root system. Trace and prune completely killed and dried roots. Wash partially affected and healthy roots with emisan solution. When the fungicide dries up, apply a thin coating with a wound dressing compound. Refill the soil and drench the base with 3 g copper oxychloride mixed in a litre of water.

9.2.10. Secondary Leaf Fall, *Gloeosporium alborubrum*

The disease occurs during April-October.

(i) Symptoms: Infects tender leaves mostly at leaf tip region where numerous spots coalesce and dry leading to defoliation. Infected leaves crinkle and become distorted.

(ii) Management

Spraying with mancozeb at 2.6 g/litre or carbendazim at 2 g/litre of water gives effective control of the disease.

10 CASHEW NUT

India is the largest producer, importer, processor, exporter and second largest consumer of cashew (*Anacardium occidentale*). In cashew nut, area has increased from 0.176 million hectares in 1961 to 0.843 million hectares in 2005–06. The production in cashew has gone up from 0.079 million tonnes in 1961 to 0.579 million tonnes in 2005–06 (Bijay Kumar, 2007) (Table 10.1). The major cashew growing states include Andhra Pradesh, Karnataka, Kerala, Madhya Pradesh, Maharashtra, Goa, Orissa, Tamil Nadu and West Bengal. India exported cashew kernels worth US$ 553 million during 2006–07. Export of cashew is rising at 27% per annum. These export earnings are exceeded only by coffee and rice among agri-exports.

Table 10.1. State-wise area, production and productivity of cashew nut in India (2005–06)

State	Area in '000 ha	Production in '000 MT	Productivity in MT/ha
Kerala	80.0	67.0	0.837
Tamil Nadu	121.0	56.0	0.479
Karnataka	100.0	45.0	0.450
Andhra Pradesh	170.0	92.0	0.541
Maharashtra	160.0	183.0	1.143
Orissa	120.0	78.0	0.650
West Bengal	10.0	10.0	1.000
Assam	14.0	10.0	0.642
Goa	55.0	27.0	0.490
Gujarat	4.0	4.0	1.000
Meghalaya	6.4	6.4	1.000
Pondicherry	3.0	1.0	0.333
Total	843.4	579.4	

10.1. PESTS

Of the several insect pests reported to infest cashew, the tea mosquito bug and the cashew stem and root borer are of major importance. The production loss from a single pest, the tea mosquito bug alone is estimated to be about 30%. Cashew stem and root borer is yet another pest responsible for killing 5–10% of the productive trees.

10.1.1. Tea Mosquito Bug, *Helopeltis antonnii, Helopeltis schoutedeni* and *H. anacardii*

(i) Damage: Tea mosquito bugs, also known as *Helopeltis* bugs or mirid bugs, are the most important and serious pests of cashew. These bugs are slender, delicate insects, about 7–10 mm long with long legs and antennae, the antenna being nearly twice as long as the body. The females are red and the males are brown to yellowish red. They lay eggs inserted into the soft tissue near the tips of flowering or vegetative shoots. Nymphs (immature bugs) are yellowish in colour. Both adults and nymphs feed on young leaves, young vegetative and flowering shoots, and developing fruits.

Feeding on the stalks of the tender shoots causes elongated green lesions, sometimes accompanied by exudation of gum. Severely damaged shoots die back due to the effect of bug saliva in combination with fungi, which enter the plant tissue through the feeding lesions (Fig 10.1); the subsequent development of numerous auxiliary buds causes a bunched terminal growth known as 'witches' broom'. In case of serious infestations, the trees may appear as if scorched by fire.

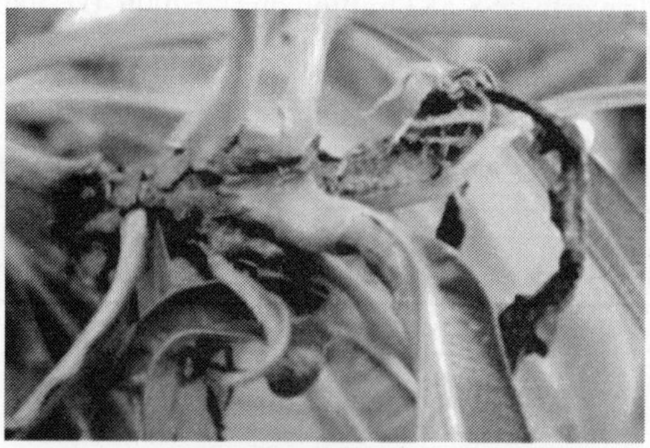

Figure 10.1. Typical damage of tea mosquito bug on a young shoot of cashew by *Helopeltis* bugs. Note necrosis on the shoot (Courtesy: A. M. Varela, ICIPE)

Attacked leaves are deformed and show angular lesions, particularly along the veins (Fig 10.2), which may drop off, so that the leaves appear as if attacked by biting insects.

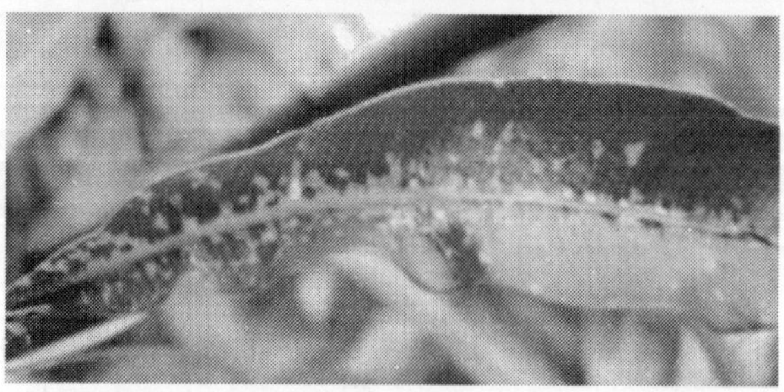

Figure 10.2. Typical leaf damage by *Helopeltis* bugs on a leaf. Note angular spots along the main vein (Courtesy: A. M. Varela, ICIPE).

Bug feeding on developing apples and nuts causes brown sunken spots (Fig 10.3). The growth of trees is seriously retarded and fruit formation of attacked flowering shoots is reduced.

Figure 10.3. Cashew nuts with different levels of *Helopeltis* damage. Left to right: slightly damaged, healthy and severely damaged nuts (Courtesy: A. M. Varela, ICIPE)

(ii) Management

- Monitor the crop regularly. *Helopeltis* attack occurs very suddenly and great vigilance is very important to control this pest, particularly during the rainy season or when water is available leading to flushing (production of young shoots) when *Helopeltis* populations normally build up.
- Conserve natural enemies. Weaver ants build nests on cashew trees providing good protection against this and other bug pests (Fig 10.4).

Figure 10.4. Weaver ants foraging on cashew apples and nuts (Courtesy: A. M. Varela, ICIPE)

- Do not interplant cashew with crops that are host for *Helopeltis* bugs, such as cotton, tea, sweet potato, guava and mango.

- The pest can be easily controlled by spraying 0.05% endosulfan (15 ml in 10 litres of water). Three sprays should be given for complete control of the pest. The first spray of endosulfan should be given in October-November when the new flushes emerge, second of systemic insecticides like monocrotophos and dimethoate in December-January during emergence of inflorescence, and followed by third spray of endosulfan in February at the time of fruit set.

10.1.2. Coconut Bug, *Pseudotheraptus wayi*

(i) Damage: Adult bugs are reddish brown in colour and 12 to 14 mm long. Nymphs are red brown to green brown in colour and have long antennae. Bug feeding causes necrotic bruise-like depressions; a hard lump develops, which can be easily removed when the fruit is peeled. The bug sucks on the developing fruits causing pock marks. The kernels are also affected showing spots, which lower their market value.

(ii) Management

- Conserve natural enemies. Weaver ants nest on cashew trees deterring and feeding on coconut bugs.

10.1.3. Cashew Weevil, *Mecocorynus loripes*

(i) Damage: The cashew weevil is large weevil (Fig 10.5), about 20 mm long, and of a knobbed appearance. It is dark grey-brown in colour. The female weevil lays single eggs in small holes in the bark of the trunk or branches. The larvae are legless grubs, whitish in colour with a brown head. They bore through the bark and move downwards tunneling under the bark while feeding on the sapwood. Brown-black gummy frass is seen on the trunk and main branches. Heavily attacked trees become ringed by damaged sapwood and eventually die. Neglected plantations are likely to be severely attacked. Fully-grown larvae pupate in a chamber about 2 cm below the bark.

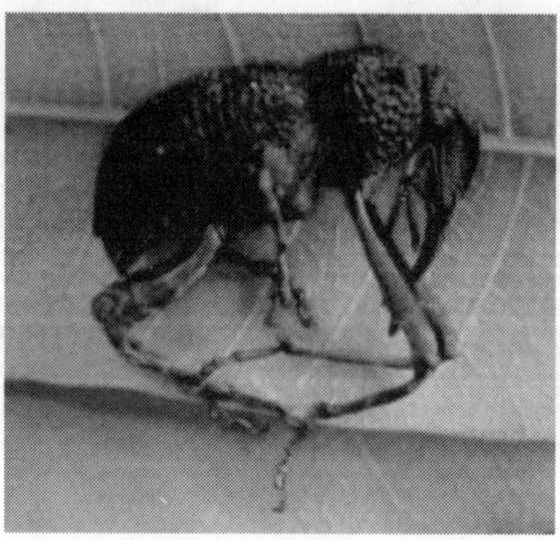

Figure 10.5. Adult cashew weevil (*Mecocorynus loripes*). Real size: about 20 mm long (Courtesy: A. M. Varela, ICIPE)

(ii) Management

- Cut away bark from damaged areas of lightly infested trees and kill the larvae and pupae underneath. Repeat this every month for a further six months if required
- Destroy severely infested trees. First collect and destroy all adult weevils; then fell the tree and remove the bark to expose all larval galleries; kill all larvae and pupa and burn the tree.

10.1.4. Red-Banded Thrips, *Selenothrips rubrocinctus*

(i) Damage: Adults of the red banded thrips are dark brown or blackish. Nymphs are pale yellow with a broad transverse red band on the dorsal side of the abdomen (Fig 10.6). Thrips attack older leaves, flowers and shoots. Attacked leaves drop off leaving bare shoots with few young leaves at the tip. Infestation of flowers causes poor fruit formation. Locally limited infestations may cause considerable damage.

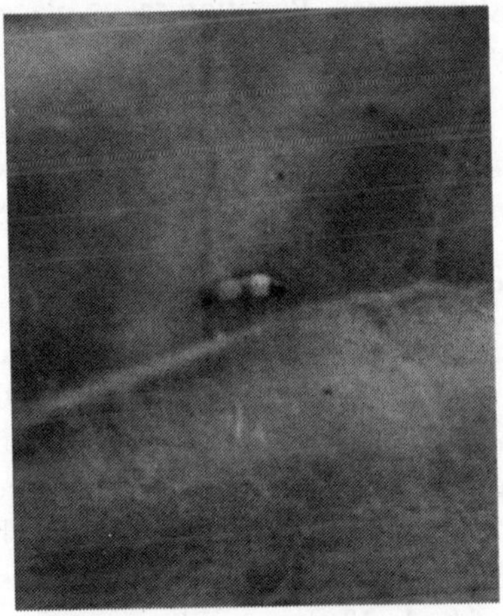

Figure 10.6. Immature stage of the red-banded thrips. Real size: 1mm long. Note typical transverse red band on the abdomen (Courtesy: A. M. Varela, ICIPE)

(ii) Management

- Conserve natural enemies. Anthocorid bugs are important in natural control of thrips.

10.1.5. Mealy Bugs, *Pseudococcus longispinus*

(i) Damage: The long-tailed mealy bug (*Pseudococcus longispinus*) attacks shoots, inflorescences, apples and nuts. Affected parts appear completely white. Trees infested during the flowering stage fail to produce fruits, whereas those infested at the nut swelling stage produce discoloured nuts, which result in a lower grade. However, cutting tests showed no difference in kernel outturn between clean and discoloured nuts. Mealy bugs have been a problem for cashew growers in Tanzania.

(ii) Management

- Conserve natural enemies. Mealy bugs are usually controlled by a wide range of natural enemies. However, use of pesticides may kill these natural enemies leading to mealy bug outbreaks.

10.1.6. Stem Girdler, *Paranaleptes reticulata*

(i) Damage: Adults are long horn beetles, with a body length of 25–35 mm and with antennae longer that the body. The head and the thorax are dark brown; the wing cases are orange with large black blotches giving them a reticulate appearance. Adult beetles girdle branches from 3–8 mm in diameter leaving a V-section cut; only a narrow, central pillar round the pith zone is left, which eventually breaks off. Female beetles lay elongated eggs in transverse slits made in the bark of the girdled branch at points above the girdle. Larvae are yellow in colour and reaches a length of 45 mm when fully grown. They mine in dead wood of the girdled branches. Pupation takes place in the dead wood. The life cycle takes one year. Neglected plantations may be severely damaged.

(ii) Management

Once a year (in November or December) all girdled branches should be collected and burned. Only the dead or dying part of the branch above the girdle needs to be collected.

10.1.7. Stem and Root Borer, *Placaederus ferrugineus*

(i) Damage: This is a serious pest of cashew and is capable of killing the tree outright. The adult is medium sized, reddish brown longicorn beetle, the head and thorax of which are dark brown or almost black. The female lays eggs into the crevices of loose bark in the trunk or the exposed portion of the roots of cashew trees. The grubs that hatch out bore into the bark and feed on the soft tissues (Fig 10.7). The attack could be identified by presence of small holes in collar region, gummosis, extrusion of chewed up fibres and excreta and discolouration of bark. If no control measures are taken up, the leaves start yellowing followed by drying of branches and finally resulting in the death of the tree.

Figure 10.7. Stem borer on cashew nut

(ii) Management

It is very difficult to control the pest in the advanced stage. Therefore, control measures should be taken up at early stage of infestation. For effective control of the pest, swab the trunk of the tree with coal tar and kerosene (1: 2) up to one metre height from the ground. Treat the soil with 500 g BHC 10% to a radius of 45 cm. The trees showing the severe infestation should be removed along with roots and burnt. Heaping the organic waste around the trunk should be avoided.

The grubs are removed by peeling the bark with chisel and hammer and destroyed. The peeled portion of the bark is smeared with Bordeaux paste and the trunk and exposed roots should be swabbed with carbaryl emulsion at 10 g/l of water. Drench the soil around the trunk of the tree with carbaryl at 4 g/l of water. Keep the orchard clean.

10.1.8. Fruit and Nut Borer, *Thylocoptila paurosema*

(i) Damage: It is a pinkish dark actively moving caterpillar which bores into the tender apples and nuts. The entrance of the hole is plugged with fecal matter. Nuts are damaged in case of severe infestation. The caterpillars attack the fruits at all stages and cause the shriveling and premature fall of nuts.

(ii) Management

This pest can be effectively controlled by spraying 0.05% endosulfan.

10.1.9. Twig Borer, *Niphonoclea albata*

(i) Damage: This insect pest is common during the dry season. The adult beetle girdles the small branches causing them to dry up or break and drop to the ground. Its creamish larvae bore into the pith of the branches. As they feed, they move downward until they pupate. All affected twigs and small branches may eventually die.

(ii) Management

Remove or collect all affected twig as well as dried twigs on the ground. Dispose them properly by burning before applying chemical sprays with malathion, carbaryl and/or methyl parathion. Repeat application at 7–10 days interval when necessary.

10.1.10. Leaf Miner, *Acrocercops syngramma*

(i) Damage: Young plants in the nursery and in the orchard are more affected by these pests. Caterpillars of this silvery gray moth mine through the tender leaves, thus, severely damaging them.

(ii) Management

Spray 0.05% phosphamidon as soon as infestation is detected on new leaves.

10.1.11. Saw-Toothed Grain Beetle, *Cryzaephillus surinamensis*

(i) Damage: This pest is known to attack the nuts during storage.

(ii) Management

Nuts should be thoroughly dried and placed in air tight containers. Surface treatment is recommended. For finished products, fumigation is recommended.

10.1.12. Termites, *Neotermes gardneri* and *Trinevitermes biformis*

(i) Damage: Termites attack the roots and the trunk of cashew trees. They burrow on the bark of roots and branches especially of old trees (Fig 10.8). They build their soil mounds or nest on dead parts of the tree trunks and feeds on tree bark. They feed on roots that are remaining underground. Trees succumb when the population is high.

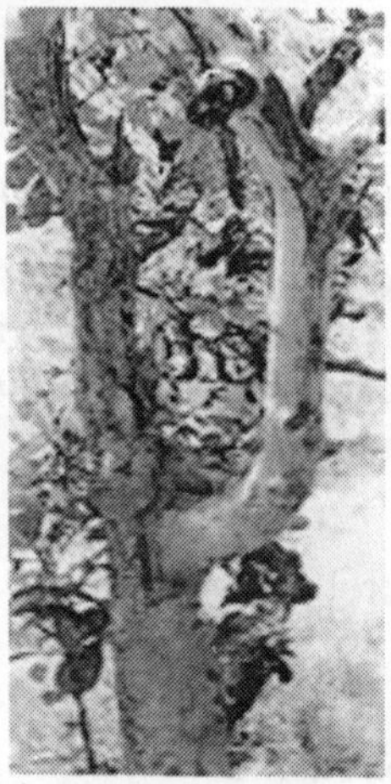

Figure 10.8. Termite infestation on cashew nut tree

(ii) Management

Soil mounds must be destroyed and the tree trunks should be swabbed with coal tar. The queen should be located and killed either mechanically or by spraying with 2% chlordane. Chlordane should not be applied on living parts of the tree because of its long residual effect. Cistin powder could be applied to any part of the tree infested with termites.

10.1.13. Bark Eating Caterpillar, *Indarbela tetraonis*

(i) Damage: Caterpillars bore short distances into the trunk or branches, usually at the forks and takes refuge during the day. During nights, larvae comes out and feeds on bark of the tree moving in the rope like web gallery of wood, silk and excreta (Fig 10.9). Small trees may succumb and large trees fail to put forth a new flush when the infestation is severe.

(ii) Management

Remove the frass and swab the trunk with carbaryl 50% WP emulsion at 10 g/litre of water.

10.1.14. Flower Thrips, *Scirtothrips dorsalis*

(i) Damage: Thrips occur during emergence of new flushes and panicle. Nymphs and adults infest the leaves and suck the sap by lacerating the leaves. Scabs like rough skin appear on floral branches, apples and nuts (Fig 10.10). Shedding of immature nuts and improper filling of kernels are the other symptoms of thrip attack.

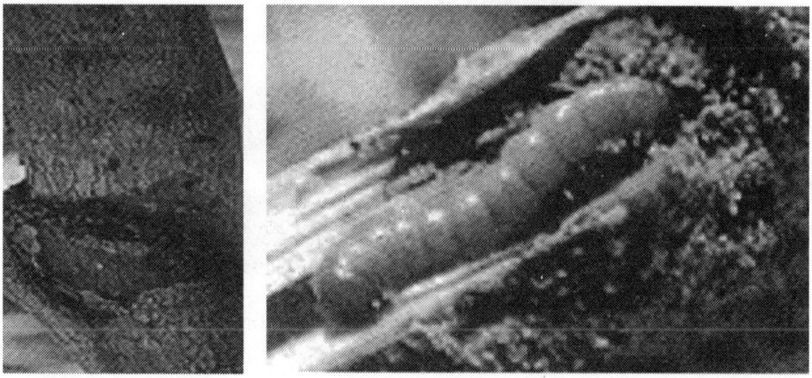

Figure 10.9. Bark eating caterpillar infestation on Cashew nut

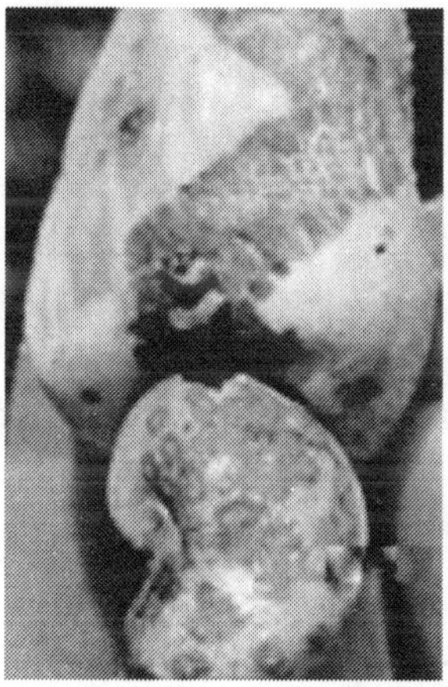

Figure 10.10. Flower thrips infestation on cashew nut

(ii) Management

Spraying 0.03% phosphamindon, dimethoate, methyl-demeton or monocrotophos gives effective control of thrips.

10.1.15. Leaf and Blossom Webber, *Lamida moncusalis*

(i) Damage: Leaf and blossom webber occur during emergence of new flushes and panicle. Caterpillars web the shoot and inflorescence together. They remain inside and feed. Apples and nuts are also covered with webs and their surfaces are scraped (Fig 10.11). Nuts crack and fail to develop. The shoot in the webbed portion dries up.

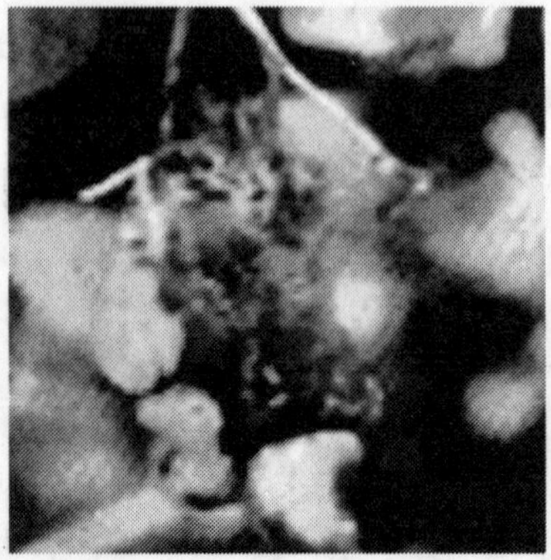

Figure 10.11. Blossom Webber infestation on cashew nut

(ii) Management

Spraying of endosulfan or chlorpyriphos at 2 ml/litre of water gives effective management of the pest.

10.1.16. Slug Caterpillar, *Lamantrida* spp.

(i) Damage: The caterpillar feeds on the leaves causing semi-defoliation.

10.1.17. Vertebrate Pests

(i) Damage: Birds, deer, squirrel, monkey, wild boar, and porcupine damaged cashew in hill region of Karnataka. Seedlings are damaged by rat, wild boar and porcupine. The total loss due to vertebrate pests in cashew at Mudigere, Chikmagalur, in Karnataka was 17% of apple and 21% of nut during 1991–92. A troupe of 8 monkeys on an average, caused 24% nut losses (Chakravarthy, 1993).

Table 10.2. Pest management

Cashew Pests	Symptoms	Control
Stem and root borer, *Plocaederus ferrugineus*	This is a major pest common in old and neglected plantations; it occurs throughout the year, but more prevalent during the rainy season; the adult lays eggs in crevices and cracks of bark on the trunk; the hatched grubs bore into the bark; as a result the vascular tissues are damaged, the ascent of the plant sap is arrested, leaves turn yellow and are shed and finally the tree dies; the grubs pupate in a calcareous cocoon in the soil; the symptoms of infestation include	Remove and destroy severely infested trees; after identifying the infested tree, the bark is peeled, the eggs, pupae and grubs are removed and destroyed; the peeled portion of the bark is smeared with Bordeaux paste; as a prophylactic measure, the orchard should be kept clean and the soil around the trunk should be mixed with 5% carbaryl dust before the onset of monsoon rains; drench the base

Table. 10.2. (Contd.)

Cashew Pests	Symptoms	Control
	the presence of small holes in the collar region, gummosis and extrusion; dark-brown beetles lay eggs under the loose bark of the trunk and roots and white grubs bore into the bark, feed on soft tissues; in severe cases, the beetles may cause death of the trees.	of the tree with 0.05% lindane; fumigate holes with carbon bisulphide or petrol or DDVP.
Leaf and Blossom webber, *Macalla moncusalis*	This is a major pest in Tamil Nadu, appearing regularly throughout the year particularly at the time of emergence of new flush; generally young trees suffer more; the adult moth lays eggs at the growing point; the caterpillar on emergence webs the terminal leaves as well as panicles and feeds inside by scrapping the epidermal layers; the pest can be identified by such webs on the plants; as a result flowers fail to open and there will be no crop; it also feeds on tender, developing nuts and apples by scrapping the epidermal layers, which drop prematurely.	Since the pest is out of reach in the webs, it is better, first to disturb the webs mechanically and then spray with quinalphos 0.5% or endosulfan 0.07%.
Leaf miner, *Acrocerops syngramma*	This pest appears regularly at the time of emergence of new flush; the caterpillars mine into the young and developing leaves and cause blisters on the leaves and feed on the tissues.	The pest can be controlled by spraying phosphomidon 0.05% (0.5 ml/l) or fenitrothion 0.05% (1 ml/l) or endosulfan 0.07% (2 ml/l) concentration at the time of emergence of new flush.
Shoot and inflorescence tip borer, *Chelaria haligramma*	The pest appears throughout the year, particularly at the time of flowering and fruiting; caterpillar feeds on leaves, tender shoots, inflorescence stalks, apples and nuts; at the time of new flush, the caterpillar folds the tender leaves from one margin and feeds inside; then it enters the tender shoots by boring a hole at the tip and feeds inside, causing drying of twigs and particles; at fruiting, the larva enters the fruit at the joint of the apple and nut and feeds inside, causing premature fruit drop.	Fenitrothion 0.05% (1 ml/l) or endosulfan 0.07% (2 ml/l) when sprayed twice, one at the time of new flush and next at the time of fruiting, effectively control the pest.

Table. 10.2. (Contd.)

Cashew Pests	Symptoms	Control
Apple and nut borer, *Nephopteryx* sp.	This pest causes direct loss to the crop; it appears at the time of fruiting; the larvae bore into either apple or nut at the joint and feed inside; as a result the development of kernel is arrested, the nuts get shriveled, apples get hollowed and shriveled and they drop prematurely; the caterpillars reside inside the apple or nut and not amenable for the chemicals; as such preventive measures are more important.	It is advisable to give prophylactic spray at the time of full bloom or fruit set with endosulfan 0.07% or phosphomidon 0.05%.
Tea mosquito bug, *Helopeltis antonii*	Adults and nymphs suck the sap from the shoots, leaves, floral branches and nuts; affected parts dry up (die-back symptoms).	Spray chlorpyrifos, dimethoate or quinalphos at recommended dosage.
Thrips: *Selenothrips rubrocinctus, Rhipiphorothrips cruentatus, Retithrips syriacus, Rhynchothrips raoensis*	The first two are commoner; attack the leaves, shoots and flowers which dry up.	Spray 0.03% phosphamindon, dimethoate, methyl-demeton or monocrotophos.

10.2. DISEASES

As compared to the number of pests attacking cashew and intensity of damage caused by them, the magnitude of the disease problem is rather comparatively low. Among few diseases attacking cashew, die-back and damping-off of seedlings are found to be most destructive.

10.2.1. Dieback or Pink Disease, *Corticium salmonicolor*

(i) Symptoms: This is a fungal disease caused by *Corticium salmonicolor* which usually occurs during the rainy season. The affected branches initially show white or pinkish growth on the bark; a film of silky thread or mycelium develops. Later, the fungus penetrates into the deep tissues and develops a pinkish growth which are the spores resulting in gradual death of the shoots. In advanced stages, the bark splits and peels off. The leaves turn yellow and fall off giving a barren appearance to a tree (Fig 10.12). Affected shoots start drying up from the tip.

(ii) Management

For effective control of the disease, prune all the affected branches below the spot of infection and destroy them. Protect the cut surface by applying Bordeaux paste. The disease can be prevented by giving prophylactic spray of 1% Bordeaux mixture twice a year, first in May-June before the onset of monsoon and second in October.

Figure 10.12. Die-back on cashew nut

10.2.2. Damping-Off

(i) Symptoms: It is a fungal disease caused by *Fusarium* sp., *Pythium* sp., *Phytopthora palmivora* and *Cylindrocladium scoparium* and affects the cashew seedlings in nursery especially when the soil medium gets too wet. The fungi attack either the collar region or the root zone of the tender seedlings that turn pale with water-soaked discoloured areas around the stem resulting in rotting and collapse of seedlings causing severe loss at the nursery stage (Fig 10.13).

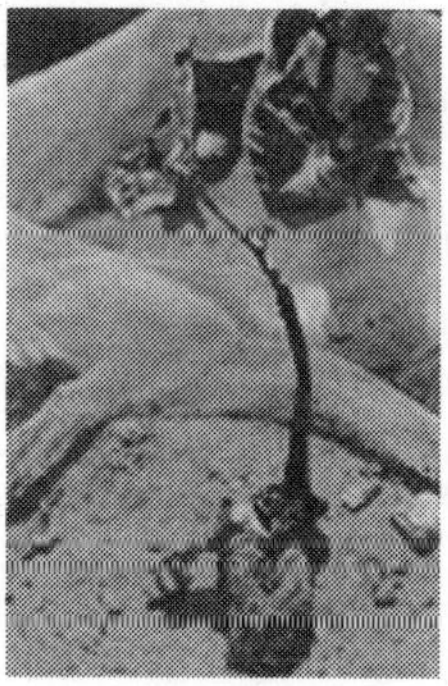

Figure 10.13. Damping-off of cashew seedling

(ii) Management

Seeds for planting should be treated with Arasan before sowing. Soil media for potting should be treated with soil fumigants.

The disease incidence is severe in nursery where drainage conditions are poor. Therefore, adequate drainage should be provided during the rainy season. The disease can be controlled by drenching the bags with 1% Bordeaux mixture or copper oxychloride at 3 g/l of water.

10.2.3. Anthracnose, *Collectorichum gloeosporoides*

(i) Symptoms: The disease attacks young plant tissues and can cause severe crop loss. This disease is caused by the fungus *Collectorichum gloeosporoides* that usually infect tender leaves, shoots, inflorescences, young fruits (apples) and young nuts (Fig 10.14). This disease is most prevalent when there is excessive rainfall coinciding with the appearance of new growth and flowering. Infected parts in their early stage show shiny, water-soaked lesions which later turn reddish-brown. At the lesion site, resinous exudation can be seen. As the disease progresses, the lesions enlarge in size, all affected tender leaves wrinkle, and the young apples and nuts become shriveled. Inflorescences become black. The disease is promoted by warm, damp conditions.

Figure 10.14. Anthracnose on cashew nut

(ii) Management

- Prune dead branches and twigs and remove from the field. They constitute the principal source of infection.
- Though pesticide application on cashew is not practiced, copper based fungicides are known to be effective against anthracnose.
- Spray 1% Bordeaux mixture at new flush and blossom stages.

10.2.4. Powdery Mildew, *Oidium anacardii*

(i) Symptoms: This disease is particularly serious in coastal areas. Infected panicles and leaves are coated with white, powdery fungal growth. In severe attacks the entire panicle may be infected and the fruit and nuts fail to set.

(ii) Management

Pesticide application is not practiced in cashew fields. However, sprays of potassium/ sodium bicarbonate and sulphur provide good control of powdery mildew.

10.2.5. Gummosis, *Diplodia natalensis, Ceratocystis* sp.

(i) Symptoms: The pathogen causes exudation of a brown liquid from longitudinal cracks on the main stem and branches. Later the exudates turns black (Fig 10.15).

Figure 10.15. Gummosis on cashew nut

(ii) Management

The disease can be managed by scraping the wound in the bark and applying Bordeaux paste.

10.2.6. Red Rust, *Cephaleurus virescens*

(i) Symptoms: The pathogen causes circular and slightly raised red rust spots on leaves (Fig 10.16). Spots on upper surface of leaves consist of numerous filaments, which project through cuticle. Release of spores results in fading of spots to cream to white texture.

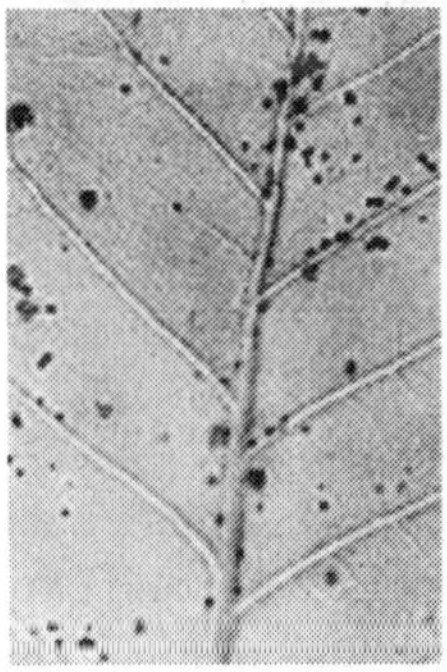

Figure 10.16. Red rust on leaf of cashew nut

(ii) Management

Spraying of copper oxychlodide at 3 g/litre of water gives effective management of the disease.

10.2.7. Blossom and Twig Blight

(i) Management

Spray 1% Bordeaux mixture to seedlings to manage the disease.

11 TAMARIND

Government and policy makers should encourage and promote tamarind (*Tamarindus indicus*) cultivation in arid and semi-arid areas for its economic and nutritional benefits. The major breakthrough in recent years has been the identification of tamarind types with less acidic pulp, commonly referred as sweet tamarinds. In Thailand, more than 50 sweet tamarind cultivars are grown, while in Philippines, 8 selected cultivars are popular among the farmers.

India, the largest producers (Table 11.1) and exporters of tamarind, continue to grow sour types. Tamarind is most suitable for growing in regions that have extended spells of dry weather. Those that farm under adverse climatic conditions will benefit most due to the draught resistant nature of this species. Presently, sweet tamarind plantations are interplanted with annual crops in Thailand. Small scale plantations of sweet tamarind can be raised for the fresh fruit market.

Table 11.1. The major tamarind growing countries and its production (Gunasena and Hughes, 2000)

Countries	Production (in MT)
India	2,50,000 (1964)
Thailand	1,40,000 (1995)
Coasta Rica	221 (1997)
Mexico	37
Puerto Rico	23 (1977)

Due to ease of cultivation, particularly in marginal areas which are often unsuitable for other crops and the potential for domestic use and export, tamarind could be considered a "Tree for the future". Tamarind attains maximum production during tenth year which yields 4 tonnes/ha (40 kg/tree) with a net income of Rs. 38,000 annually.

Another important aspect is its food and nutritional value, which could safeguard food and nutritional security to alleviate poverty and improve health and the standard of living of rural populations. Tamarind is available when other food supplies are low, thus contributing significantly to the nutrition of low income rural households. Tamarind seed protein contains highly favourable amino acids and could complement methionine and cystine in the diet, to alleviate protein malnutrition which is widely prevalent in developing nations.

Tamarind is a valuable timber species widely used for making furniture, tool handles, charcoal, oil mills, rice pounders and fuel wood. The leaves are an important source of food and herbal medicine and the edible pulp of ripe fruits is used as flavouring agent in cooking, soups, jams, chutneys, sauces, juices, etc. The fruit pulp of tamarind is the richest natural source of tartaric acid (8 to 18%) and is the chief acidulant used in the preparation of foods. The major industrial product of tamarind is the seed kernel powder which is an important material used in sizing of textiles, paper and jute.

During 1993, exports reached 11,145 tonnes valued at Rs. 10.8 million (US $ 2,40,300), and were in the form of fresh fruit, pulp, seed powder and paste. The importing countries are mainly Pakistan followed by UAE, Japan and Yemen. Over 75% of the imports are of sweet type, having a premium demand over sour types.

11.1. PESTS

Tamarind trees are liable to be attacked by a large number of insect pests. In India alone, 40 insect pests have been recorded as attacking tamarind, causing severe economic losses (Joseph & Oommen, 1960). The major pests which attack tamarind include shot hole borers, toy beetles, leaf feeding caterpillars, bag worms, mealy bugs and scale insects (Coronel, 1991). Among the worst are the sap suckers, which affect the tender shoots and fresh foliage. The most destructive among these are white flies, thrips, coccids and aphids. The defoliators, which include caterpillars and the chafer beetles, also cause considerable damage. Some of these insects attack the flower buds and the developing fruits and seeds, while others damage the fruits during storage.

11.1.1. Scale Insect, *Aspidiotus tamarindus*

(i) Damage: Some of the most serious pests in India are hard scale insects (*Aonidiella orientalis, Aspidiotus destructor* and *Saissetia oleae*) (Timyan, 1996). The oriental yellow scale (*Aonidiella orientalis*) (Hemiptera: Diaphididae), is a polyphagous pest which attacks tamarind. The eggs are laid under a protected covering such as cracks in the bark of the trees. After hatching, the nymphs crawl in search of suitable succulent parts of the plant. Yellow encrustation over developing fruits can be seen.

(ii) Management

The scale insects can be controlled by removing the affected parts in the initial stages. When the infestation is serious, chemical spraying with 0.1% diazinon or carbosulfan or 0.3% dimethoate or endosulfan 2 ml or carbaryl 3 g/l of water provide effective control (Butani, 1979).

11.1.2. Mealy Bugs

(i) Damage: Mealy bugs (*Nipaecoccus viridis* and *Planococcus lilacinus*) (Hemiptera: Pseudococcidae) also attack tamarind trees. Nymphs and the females suck the sap on the ventral surface of the leaflets, the base of leaf petioles, tender shoots and even the mature shoots. *P. lilacinus* also attacks the roots. The leaflets become chlorotic and defoliation is common. Immature fruit fall is sometimes observed.

(ii) Management

Mealy bugs can be controlled by mechanical removal of infected shoots on the small scale. In serious infestations, chemicals such as 0.1% diazinon or carbosulfan can be sprayed.

P. lilacinus populations living in the soil can be displaced by raking or be killed using 5% aldrin at 500 g/tree (Butani, 1979).

11.1.3. Thrips

(i) Damage: Some thrips are reported to attack tamarind flowers. *Scirtothrips dorsalis* (Thysonoptera: Thripidae) is a polyphagous thrip and the adults live for 10 or 15 days and complete 25 overlapping generations per year. *Ramaswamiahiella subnudula* is another small yellowish thrip which attacks tamarind.

(ii) Management

Thrips can be controlled by spraying with dimethoate at 20–30 ml/10 litres of water or fenthion at 10–15 ml/10 litres of water.

11.1.4. Bruchid Beetle

(i) Damage: The bruchid beetle [*Pachymerus (Coryoborus) gonogra* (Coleoptera: Bruchidae)] is the most serious pest of tamarind in India and Pakistan (Chaturvedi, 1985). The eggs of this insect are laid on the pods or on seeds and the contents are eaten by the larvae. The infestation frequently continues in storage.

Larvae of tamarind weevil (*Caryedon serratus*) are reported as a serious pest which attacks the pods, seeds and fruits in the field and fruits and roasted seeds in storage in India (Butani, 1979), Colombia and Puerto Rico (Timyan, 1996).

The beetle, *Calandra linearis* also attacks fruits and is reported as a common pest in Florida and Hawaii (Anon, 1979).

(ii) Management

Neem oil at 5–20 ml/kg of seed, has been found effective to control bruchid beetles attacking tamarind pods and seeds. The use of neem oil has pronounced adulticidal and ovicidal effects, which are maintained over six months, without any adverse effect on seed viability. This low cost bio-pesticide is very cost effective and efficient in protecting seeds of tamarind from bruchid damage. This method is widely used in the Sahel (Cardet *et al.*, 1998).

The bruchid beetles can also be controlled by fumigation with phosphine from aluminium phosphide tablets (1.5 g/tonnes of pods/seeds) or ethylene bromide or carbon tetrachloride or acrylnitrile (Mittal & Khanna, 1978).

11.1.5. Lac Insect, *Kerria lacca* (Hemiptera: Lacciferidae)

(i) Damage: The lac insect is widely distributed polyphagous pest in India which attacks tamarind stems, twigs and leaves.

(ii) Management

Removing the affected parts in early stages, as commonly practiced by farmers can control the lac insect. In serious infestations, spraying 1% diazinon or carbosulfan is effective.

11.1.6. Leaf Caterpillar, *Achaea janata* (Lepidoptera: Noctuidae)

(i) Damage: Larvae of *Achaea janata* are reported to cause heavy losses when epidemics of the moth infest flowers in tamarind plantations in Tamil Nadu (Ahmed, 1990). In such epidemics over 400–650 larvae can be found on a single tree causing serious growth reduction. It is also considered as a serious pest causing damage to flowers and defoliation of trees in the Salem district of Tamil Nadu, India.

(ii) Management

Spraying 0.5% permethrin or endosulfan 2 ml or carbaryl 3 g/l of water can control this pest.

11.1.7. Aphids, *Toxoptera aurantii* (Hemiptera: Aphididae)

(i) Damage: Aphids are serious pests that attack tamarind. It is a major pest which sucks sap of tender shoots and leaflets causing them to become distorted and covered with molds growing on the secreted honeydew.

(ii) Management

Aphids can be controlled by spraying endosulfan at 1 ml/l of water or dimethoate at 20–40 ml/10 litres of water, repeated every two weeks until the pest is no longer observed.

11.1.8. Storage Beetle, *Pachymeres gonagra*

(i) Damage: The storage beetle bores into fruit on trees and in storage.

(ii) Management

In case of severe infestation, spray endosulfan 2 ml or carbaryl 3 g/litre of water to control the pest.

11.1.9. Vertebrate Pests

Monkeys are reported to devour the fruits greedily while animal browsing causes damage to seedlings unless effectively protected in the early stages (Hocking, 1993).

11.2. DISEASES

11.2.1. Powdery Mildew, *Erysiphe polygoni* and *Oidium* sp.

(i) Symptoms: The powdery mildew disease is of common occurrence in nursery seedlings. the disease causes defoliation and early growth is severely retarded.

(ii) Management

In order to have effective control of powdery mildew, 3–4 sprays of 0.2% wettable sulphur at 15 days interval is the most economical method (Siddaramaiah & Kulkarni, 1982).

11.2.2. Leaf Spot, *Cercospora tamarindi*

(i) Symptoms: In Uttar Pradesh, India, tamarind leaves have been attacked by *Cercospora tamarindi* resulting in severe defoliation (Khan *et al.*, 1988).

(ii) Management

The leaf spot can be easily controlled by spraying with maneb, benomyl or captan.

12 SPICE CROPS

12.1. INDIAN SCENARIO

From time immemorial, India is regarded by the rest of the world as the "magic land of spices". Spices are natural plant products used to improve flavour, aroma, taste and colour of food products, beverages, liquors, pharmaceutical, cosmetic and perfumery products. No other country in the world has such a diverse variety of spice crops as in India. Indian spices are also noted for their excellent aroma, flavour and pungency not so easily matched by any other country. Even in minute quantities, spices are a real delight to all our sensory organs, making the food dishes more palatable. Usage of spices in food processing industries is increasing quite fast in all the countries. Hence, the world demand for spices in recent years is continuously on the rise.

India is the largest producer, consumer and exporter of spices in the world. The annual production of spices is more than 5 million tonnes valued at US$ 1,500 million. India's share of the world spices trade is estimated at 45 to 50% by volume and 25% by value. The present annual production of spices in the country is 5.107 million tonnes from over 2.857 million hectares (Bijay Kumar, 2007) (Table 12.1). The spices contributed 2.76% of total horticulture production in 2005–06. About 8.5% of India's export earnings from agricultural and allied products come from spices. Indian spices flavour foods in over 134 countries. The important spice crops include black pepper, cardamom, ginger, garlic, turmeric and chilli. The major spice-growing states are Andhra Pradesh, Tamil Nadu, Karnataka, Gujarat, Madhya Pradesh, Rajasthan, Uttar Pradesh, West Bengal, Orissa, Kerala, Chhattisgarh and Maharashtra.

Table 12.1. State-wise area, production and productivity of major spices in India (2005–06)

State	Area in '000 ha	Production in '000 MT	Productivity in MT/ha
Andhra Pradesh	326.9	920.5	2.8
Tamil Nadu	130.0	784.0	6.0
Karnataka	258.8	619.8	2.4
Gujarat	426.1	505.6	1.2

Table. 12.1. (Contd.)

State	Area in '000 ha	Production in '000 MT	Productivity in MT/ha
Madhya Pradesh	292.3	345.7	1.2
Rajasthan	347.9	297.1	0.9
Uttar Pradesh	91.0	283.8	3.1
West Bengal	108.0	226.7	2.1
Orissa	145.6	195.3	1.3
Kerala	428.1	191.8	0.4
Chhattisgarh	28.9	168.2	5.8
Maharashtra	112.5	106.0	0.9
Others	162.1	463.2	2.9
Total	**2857.7**	**5107.7**	**1.8**

Since spices form a part of traditional/ethnic medicines, the demand for organically produced spices is growing. Export of organic spices from India has started in right earnest. The country at present exports around 50 tonnes of different varieties of organic spices. Export will get a significant boost in the coming years as more and more farmers switch to organic methods.

12.2. EXPORT OF SPICE PRODUCE

India is the largest producer, consumer and exporter of spices and spice products in the world and produces more than 50 spices. Spice export figures for the year 1996–97 and 2006–07 shows the phenomenal growth of the Indian spice industry. The growth in the spice exports during the ten years is not only a leap, but a quantum leap. The Indian spice export for the year 1996–97 was 2, 25,000 metric tonnes which was worth $350 million; in the year 2006–07 the spice exports touched the figure of 3,75,000 MT which was worth US$ 800 million. Thus, these figures show the exponential growth of the Indian spice industry in a very short time.

Spice exports from India have touched US$ one billion mark during 2007–08. Already 50% of the exports were non-traditional spices like chilli and mint. Export of fenugreek, fennel, coriander and cumin should be promoted. India is the largest producer in the world of ginger, turmeric and black pepper.

India exports around 0.25 million tonnes of spices annually, and account for around 37% of the export volume and 23% of the export value. The major destination of Indian spice exports is USA followed by European Union, Sri Lanka, Japan and Middle East Countries.

12.3. PEST AND DISEASE SCENARIO

See Table 12.2.

Table 12.2. Avoidable yield losses in spice crops due to plant parasitic nematodes

Crop	Nematode(s)	Yield loss (%)	Reference(s)
Black pepper	*R. similis,* *M. incognita*	38.5–64.67	Mohandas & Ramana, 1991
	R. similis	59	Mohandas & Ramana, 1991
	M. incognita	46	Mohandas & Ramana, 1991
Cardamom	*M. incognita*	32–47	Ali, 1986
Ginger	*M. incognita*	29.60–33.35	Ramana *et al.*, 1998 Koshy, 2002 Charles & Kuriyan, 1979
	R. similis	39–73	Sundararaju *et al.*, 1979
Turmeric	*M. incognita*	18.6–25.0	Poornima & Vadivelu, 1998
	R. similis	46–76	Sosamma *et al.*, 1979
Coriander	*M. incognita*	51	Midha & Trivedi, 1991
Cumin	*M. incognita*	34	Midha & Trivedi, 1991
Fennel	*M. incognita*	39	Midha & Trivedi, 1991

Table 12.3. Biological control of plant parasitic nematodes in spice crops

Crop	Nematode	Effective bioagent/s	Reference(s)
Black pepper	*M. incognita*	*P. lilacinus* *P. fluorescens*	Sosamma & Koshy, 1995
		P. penetrans	Eapen *et al.*, 1997
		G. mosseae	Sosamma & Koshy, 1995 Sivaprasad & Sheela, 1999
		G. mosseae, *G. fasciculatum,* *Acaulospora laevis,* *Gigaspora margarita*	Anandraj *et al.*, 1991
Cardamom	*Meloidogyne* spp.	*P. lilacinus,* *Trichoderma* spp.	Eapen & Venugopal, 1995
Turmeric	*M. incognita*	*P. fluorescens*	Seenivasan *et al.*, 2001
Coriander	*M. incognita*	*P. lilacinus*	Midha *et al.*, 2001

13 BLACK PEPPER

Black pepper (*Piper nigrum*), known as the "King of spices", has its origin in the South-Western hills of India. It is a branching and climbing perennial shrub belonging to the family Piperaceae and is grown in the hot and humid parts of the world. Pepper is loved for its spicy factor, piperine. It is used in various culinary seasonings, as a preservative for meat and fish, in medicine and perfumery (pepper oil). In India, it is a common practice to use live trees as standards for climbing pepper. Pepper is the most important spice produced in the country and accounts for 50–60% of the total export earnings. It is being cultivated in 0.223 million hectares producing 0.070 million tonnes of black pepper with an average yield of 1.3 tonnes per hectare.

13.1. PESTS

13.1.1. Pollu Beetle, *Longitarsus nigripennis*

(i) Damage: Adult beetles feed and damage tender leaves and spikes. Grubs bore into and feed on internal tissues. The infested spikes turn black and decay (Fig 13.1).

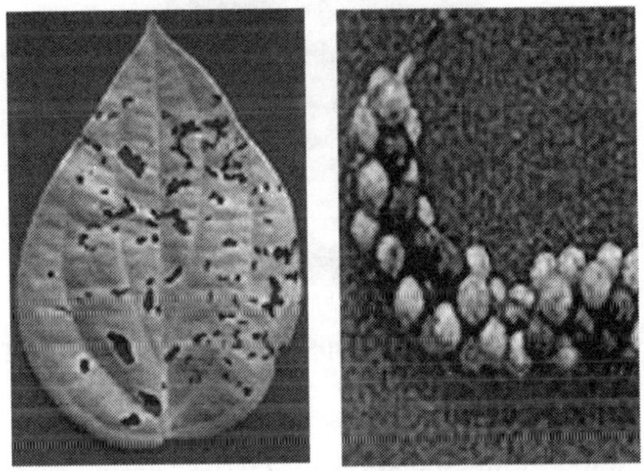

Figure 13.1. Pollu beetle damage on black pepper leaf and spike

(ii) Management

Regulation of shade in the plantation reduces the population of the pest in the field. The pests can be managed by spraying endosulfan or quinalphos each at 0.05% during June-July and September-October. Neemgold (neem-based insecticide) at 0.6%, neem oil at 0.4%

during August, September and October are also found effective for the management of the pest. The underside of leaves (where adults are generally seen) and spikes are to be sprayed thoroughly. An integrated strategy involving regulation of shade and spraying of Neem Gold at 0.6% at 21 days intervals during June-October was found effective against pollu beetle.

13.1.2. Top Shoot Borer, *Cydia hemidoxa*

(i) Damage: Larvae bore into tender terminal shoots and feed on internal tissues, resulting in blackening and decay of affected shoots.

(ii) Management

Spraying monocrotophos or endosulfan each at 0.05% on tender terminal shoots and repeating at monthly intervals during July-October protects the emerging new shoots against the pest.

13.1.3. Leaf Gall Thrips, *Liothrips karnyi*

(i) Damage: Leaf margins curl downwards and inwards resulting in formation of marginal leaf galls. Infested leaves become crinkled and malformed (Fig 13.2).

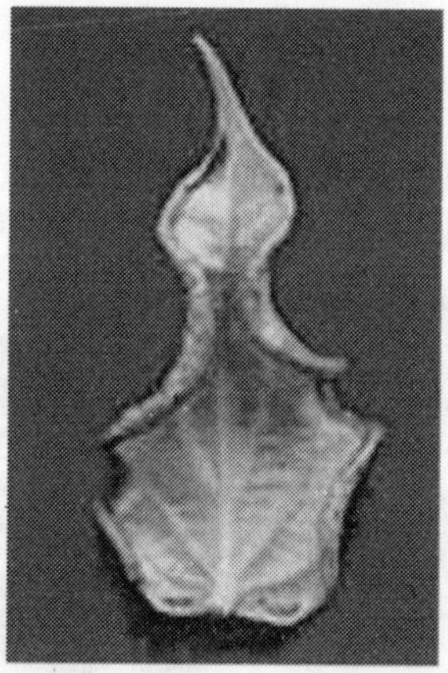

Figure 13.2. Leaf gall thrips damage on black pepper

(ii) Management

Spraying monocrotophos or dimethoate each at 0.05% during emergence of new flushes in the field or cuttings in the nursery gives effective control of the pest.

13.1.4. Scale Insects, *Lepidosaphes piperis, Aspidiotus destructor, Lecanium* sp.

(i) Damage: Scale insects are sedentary and appear as encrustations on stem, leaves and berries (Fig 13.3). Insects feed on plant sap and cause yellowing and wilting of infested portions.

(ii) Management

Clipping off and destroying severely infested branches reduce the pest incidence. Spraying monocrotophos or dimethoate each at 0.05% during early stages of pest infestation and repeating the spray after 21 days is effective in controlling the pest. Scales are controlled by shade management during May-June and application of neem oil (0.6%) during October-November. The pest can be managed by spraying tobacco decoction. Spraying of neem-garlic suspension at 2% was found effective. Spraying of 0.05% dimethoate at fortnightly intervals also gives effective control. *Chilocorus nigrita* is a potential natural enemy against scale insects.

Figure 13.3. Scale infestation on black pepper leaf

13.1.5. Leaf Feeding Caterpillar, *Synegia* sp.

(i) Damage: The pest damages the leaves and spikes of younger vines.

(ii) Management

Spraying 0.05% quinalphos gives effective control of leaf feeding caterpillar.

13.1.6. Mealy Bug, Gall Midge and Aphids

(i) Damage: Mealy bugs, gall midges and aphids infest tender shoots especially in nurseries.

(ii) Management

Spraying 0.05% monocrotophos if the infestation is severe gives effective control of these pests. Mealy bug infestation on roots can be controlled by drenching the soil with 0.075% chlorpyriphos.

13.2. DISEASES

13.2.1. Quick Wilt/Foot Rot, *Phytophthora capsici, P. palmivora*

The disease is more severe in Karnataka and Kerala. The pathogen is responsible for 25–70% loss in yield in Cannanore of Kerala. If management practices are not initiated at early stage, 7.0 to 78.8% loss in yield has been reported (Jahagirdar & Siddaramaiah, 2000). The mortality of pepper vines due to this disease ranged from 5.15 to 61.75% in Cannanore and Kasaragod districts.

(i) Symptoms: The pathogen causes black spots on leaves with fine fibre like projection from advancing margins which rapidly enlarge and cause defoliation (Fig 13.4). Tender leaves and succulent shoot tips of new runner shoots trailing on the soil turn black. When the main stem is affected, the entire vine wilts followed by shedding of spikes with or without black spots.

Figure 13.4. Foot rot infected plants of black pepper

(ii) Epidemiology: In arecanut-black pepper mixed cropping system, high rainfall (1000 mm) and microclimatic conditions like high relative humidity (84–89%), low temperature (22.7–29.6°C) and short sunshine hours (2.8–3.5 hr/day) favoured the disease development (Ramachandran *et al.,* 1988). Soil and infected plant debris in a plantation are the main sources of inoculum. The pathogen spreads through rain splashes. The survival of the pathogen in the infected plant debris was recorded up to 240 days and on alternate hosts up to 140 days. The incidence of quick wilt was maximum during July and minimum during January to May. A significant positive correlation has been established between total rainfall, number of rainy days and relative humidity; and a negative correlation with maximum temperature and sunshine hours.

(iii) Management

Adoption of phytosanitory measures like removal and destruction of dead vines along with root systems from garden as well as use of disease-free planting materials, keeping tillage operations to the minimum to avoid soil disturbance and root damage, providing adequate drainage to reduce water stagnation, providing green legume cover/grass cover to avoid rain splash spread to adjoining vines, application of *Trichoderma* spp. at 75 g/vine along with neem cake at 2 kg/vine during May-June (pre-monsoon) with suitable carrier medium such as coffee husk and well rotten cow dung followed by second round of application during August-September (post-monsoon) gives effective control of the disease. Prophylactic application of Bordeaux paste to collar region during May-June (pre-monsoon) and spraying with 1% Bordeaux mixture or 0.2% copper oxychloride or 0.01% metalaxyl or 0.3% potassium phosphonate helped in reduction of the disease. Spraying of modified panchagavya at 10^1 dilution at 2–3 litres/vine twice as pre-monsoon (May-June) and post-monsoon (August-September) is recommended.

The freshly emerging runner shoots should not be allowed to trail on the ground. They must be either tied back to the standard or pruned off. The branches of support trees should be pruned at the onset of monsoon to avoid build up of humidity and for better penetration of sunlight. Reduced humidity and presence of sunlight reduces the intensity of leaf infection. At the onset of monsoon (May-June), application of *T. harzianum* (10^{10} cfu/g) around the base of the vine is recommended. A second application of *T. harzianum* during August-September is also recommended.

In disease affected gardens, plants treated with *T. harzianum* (IISR 1369 strain with 10^{10} cfu/g) at 50 g/vine along with 10 kg of FYM or 1 kg neem cake during May-June period immediately after the receipt of first early monsoon shower and second dose of inoculum applied during August-September gave effective control of foot rot and slow decline (*Meloidogyne incognita, Radopholus similis*).

Five strains of *Pseudomonas fluorescens* (IISR-11, 8, 51, 6 and 13) gave protection against foot rot (above 55% over control) to the plants, and were also found to be efficient in growth promotion. Maximum disease suppression was obtained in the plants treated with strain IISR-51 (86%).

Integration of *T. harzianum* (IISR 1369) with *Pseudomonas fluorescens* (IISR-41) gave very effective control of foot rot (10% disease incidence as compared to 90% in control) and increased plant height under field conditions. Integration of *T. harzianum* (IISR 1369) with *Pseudomonas fluorescens* (IISR-11 or 6) suppressed the root rot to the extent of 63% over control and improved the vigour of pepper vines.

T. harzianum culture mixed with pre-wetted neem cake or FYM at 1 kg/100 kg and incubated for 2 days. Pre-monsoon application of this mixture at 5 kg/vine below 10 years and 10 kg/vine above 10 years gave good control of wilt disease.

An integrated management strategy involving phytosanitation, cultural practices and application of Bordeaux mixture (1%) and copper oxychloride or potassium phosphonate (Akomin) is effective against foot rot disease. Field trials showed effectiveness of *Trichoderma harzianum* and *T. virens* to manage *Phytophthora* foot rot.

Raising plants in solarized nursery mixture, fortified with *Trichoderma* spp. and endomycorrhizae, has been found effective in suppressing root rot. Endomycorrhizae and *T. harzianum* are effective against *Phytophthora* spp. and *Meloidogyne* spp.

Foliar spray and drenching of root zone with copper oxychloride + metalaxyl (3 g/litre of water) prior to and after monsoon is effective.

Among the several black pepper lines screened for their reaction to *Phytophthora capsici*, the casual agent of foot rot disease, P24, an open pollinated progeny of Perambramundij has been identified as tolerant. Of the various cultivars, Kalluvally showed the minimum disease incidence of 5.1%.

13.2.2. Leaf Rot, *Rhizoctonia solani*

(i) Symptoms: The disease is often serious in nurseries during April-May when warm humid conditions prevail. The fungus infects both leaves and stems. Grayish sunken spots and mycelial threads appear on the leaves. The infected leaves are attached to one another with the mycelial threads. On stem, the infection occurs as dark brown lesions, which spread both upwards and downwards. The new flushes subtending the points of infection gradually droop and dry up.

(ii) Management

Removal and destruction of the infected berries to avoid repeated overlapping of infections is recommended. A prophylactic spray with 1% Bordeaux mixture prevents the disease. Spraying with 1% Bordeaux mixture during June-July, late July and late August was found effective in the management of the disease. Foliar spray with liquid formulation of *Pseudomonas fluorescens* and drenching at both pre- and post-monsoon treatments provided good protection. The treated plants had 35% disease index as compared to 61% in control and the yield was 2.34 kg/vine as compared to 0.78 kg/vine in control. However, this was on par with Bordeaux mixture spray with 29% disease index and 3.65 kg yield/vine.

13.2.3. Basal Wilt, *Sclerotium rolfsii*

(i) Symptoms: The disease is mainly noticed in nurseries during June to September. Grayish lesions appear on stems and leaves. On the leaves, white mycelium is seen at the advancing edges of lesions. The mycelial threads later girdle the stem resulting in drooping of leaves beyond the point of infection and in advanced stages, the rooted cuttings dry up. Small whitish to cream coloured grain like sclerotial bodies appear on the mature lesions.

(ii) Management

The affected cuttings along with defoliated leaves should be removed and destroyed. Later all the cuttings should be sprayed with 1% Bordeaux mixture or 0.2% carbendazim.

13.2.4. Anthracnose/Pollu Disease, *Colletotrichum gloeosporioides*

The early infection of the pathogen caused 77% loss in yield and late infection was responsible for 56% loss in yield.

(i) Symptoms: The pathogen infects the leaves causing angular to irregular yellowish brown to dark brown irregular necrotic leaf spots with a chlorotic halo. Affected berries show brown sunken patches during early stages and further development is affected. In later stages, the discolouration gradually increases and berries show the characteristic cross splitting. Berries turn black and dry.

(ii) Epidemiology: The disease is generally noticed in pure plantations and mixed cropping. Maximum disease was noticed in August-September (28–34%). The pathogen spreads over winters on the plantation itself due to overlapping infections that occurs throughout the year.

(iii) Management

- Removal and destruction of the infected berries helps to prevent repeated overlapping of infections.
- Spraying 1% Bordeaux mixture or 0.2% copper oxychloride during June-July, late July and late August gives effective control of the disease.

13.2.5. Stunt Disease (Little leaf)

(i) Symptoms: Leaves become small and narrow with varying degree of deformation. Leaves appear leathery, puckered and crinkled. Chlorotic spots and streaks on leaves give appearance of mosaic. Vines show shortening of internodes to varying degree.

(ii) Management

Affected plants should be destroyed. Planting material should be collected from the disease-free plants.

13.2.6. Phyllody Disease (Phytoplasma)

(i) Symptoms: Floral buds are transformed into narrow leaf like structures. In advanced stages, leaves become small and chlorotic. Internodes are also shortened. The affected fruiting laterals give a witches' broom appearance. Severely affected plants become unproductive in 2–3 years.

(ii) Management

The infected vines should be collected and destroyed to prevent further spread of the disease.

13.3. NEMATODES

Although several nematodes are reported on black pepper, the two most important pathogens recorded are the burrowing nematode, *Radopholus similis* and the root-knot nematode, *Meloidogyne* spp.

13.3.1. Burrowing Nematode, *Radopholus similis*

Van der Vecht (1950) first reported the association of the burrowing nematode with the "Yellows disease" of black pepper and also proved the pathogenicity of *R. similis*. "Yellows disease" was reported to have wiped out 22 million pepper vines within 20 years in Bangka Island in Indonesia. *R. similis* is found associated with "Slow-wilt" disease of black pepper in Kerala and Karnataka, India (Mohandas & Ramana, 1987; Ramana *et al.*, 1987a).

A population level of 250 nematodes/g of root was constantly recorded with slow wilt affected pepper vines in Kerala (Ramana, 1986). Menon (1949) reported mortality up to

10% of the vines due to the disease. Mohandas and Ramana (1991) reported that *R. similis* was responsible for 59% loss in yield of black pepper.

(i) Distribution: *R. similis* was reported on black pepper from Indonesia (Van der Vecht, 1950), India (Venkitesan, 1972), Malaysia, Thailand (Sher *et al.*, 1969; Reddy, 1977) and Sri Lanka (Gnanapragasam *et al.*, 1985).

(ii) Symptoms: Primary symptoms of slow-wilt (yellows) disease is the appearance of pale yellow drooping leaves on the vines whose number gradually increases and finally the entire foliage becomes yellow. Yellowing is followed by shedding of leaves, cessation of growth and die-back symptoms. Within 3–5 years of the initiation of the primary symptoms, all the leaves are shed and the death of the vine takes place. Young and old plants are susceptible. In bearing vines, shedding of the spikes is another major symptom. At the base of the affected vines, large numbers of shed spikes can be seen. In plantations, barren standards that have lost the vines or standards supporting dead vines are of common occurrence. The foliar symptoms become more pronounced when the soil moisture is depleted.

The symptoms on the roots are the presence of characteristic orange to purple coloured lesions on the thin, white feeding roots which are not clearly seen on old brown roots. The main roots remain devoid of fine feeder roots and extensive root rotting takes place. Subsequently, extensive necrosis of the larger lateral roots takes place (Fig 13.5).

Figure 13.5. Burrowing nematode on black pepper. A-Infected; B-Healthy (Thorne, 1961)

(iii) Life Cycle: The nematode completes its life cycle within 25–30 days at temperature range of 21–31°C. The black pepper isolate of *R. similis* can be cultured axenically on carrot discs (Koshy, 1986).

(iv) Host Range: Crops like coconut and areca nut which are commonly used as live standards are good hosts of *R. similis*. Intercrops like banana, ginger, turmeric and cardamom are also susceptible to *R. similis* infestation.

(v) Histopathology: *R. similis* penetrates the roots within 24 hours of inoculation and the cells around the penetration site become brown (Venkitesan & Setty, 1977). Nematodes do not enter stellar portions of the root and plugging of the xylem vessels with a gum like substance has been reported. Nematodes are found in inter and intracellular positions within the cortex.

(vi) Interaction with Other Pathogens: Yellows disease in Indonesia is caused by a nematode-fungus complex involving *R. similis* and *Fusarium* spp. and possibly other fungi. *R. similis* predisposed black pepper seedlings to attack by a weakly pathogenic isolate of *Fusarium solani* causing severe root damage.

(vii) Management

(a) Cultural Methods: Use of healthy and nematode-free planting material can ensure better establishment, survival and higher yield. Soil solarization of nursery sites or potting mixture and random rotation of nursery sites are reported to be effective in black pepper. Mulching (with Guatemala grass) has an ameliorating effect on the symptoms of slow wilt/pepper yellows (*R. similis*). Planting nematode-free rooted cuttings, uprooting of affected vines and replanting after a period of 9–12 months, use of non-living standards/supports, avoiding *R. similis* susceptible intercrops like banana, ginger, turmeric and exclusion of susceptible trees as standards for trailing the vines reduces the nematode population. Application of organic amendments like neem cake (200 g), green foliage (3–5 kg) or FYM (1 kg per vine) and earthing up in September-October are recommended. Growing marigold as a trap crop and uprooting the trap crop at flowering stage and burning reduce the incidence of nematodes. *R. similis* was not encountered when black pepper was intercropped with coffee (Venkitesan, 1976).

Application of fertilizers at 400 kg N, 180 kg P, 480 kg K, 425 kg Ca and 112 kg Mg in combination with a mulch for effective control of yellows disease.

(b) Chemical Methods: Under Indian conditions, aldicarb/carbofuran/phorate each at 3 g *a.i.*/vine applied in May-June and again in September-October resulted in reduction of the nematode population and remission of foliar yellowing. Among the above three nematicides, phorate proved to be the best (Ramana, 1986; Mohandas & Ramana, 1987a). Aldicarb sulphone at 8kg *a.i.*/ha gave best control of *R. similis* followed by fensulfothion (Venkitesan, 1976).

Various chemicals tried for the management of *R. similis* on black pepper under nursery and field conditions is presented in Table 13.1.

(c) Biological Methods: *Pasteuria penetrans, Paecilomyces lilacinus* and *Pochonia chlamydosporia* are effective in suppressing the root-knot and burrowing nematodes. Five isolates of rhizobacteria (IISR-552, 528, 658, 853 and 859) were good growth promoters (26.7 to 55.6%) having dual nematicidal action (suppressing both *M. incognita* and *R. similis*).

Table 13.1. Effect of various chemicals for the management of *R. similis* in black pepper under nursery and field conditions

Chemical	Dosage	Duration & type of treatment	Effectiveness
(i) Nursery level			
Aldicarb sulfone	4–8 kg *a.i.*/ha	At nursery stage	Eliminated infection
Fensulfothion	4–8 kg *a.i.*/ha	At nursery stage	Eliminated infection
(ii) Field level			
DBCP	40 litres/ha	Field application	Effective in controlling slow wilt disease
Phorate	3 kg *a.i.*/ha	Field application	Effective in controlling slow wilt disease
	3 g *a.i.*/vine	Twice a year	Reduced nematodes population & foliar yellowing

(d) Host Resistance: A local variety at Peringamala, Kerala, India was found not to be invaded by *R. similis*

(e) Integrated Methods: Integrated methods of nematode management that can be suggested are:

- Planting nematode-free rooted cuttings.
- Use of non-living standards and excluding *R. similis* susceptible trees as standards for trailing pepper vines and avoiding infested planting materials of intercrops such as banana, ginger and turmeric.
- Uprooting of affected vines and replanting after 9–12 months with nematode-free rooted cuttings of high yielding tolerant variety and protecting them with phorate at 1 g *a.i.*/vine at planting itself.
- Application of phorate at 3 g *a.i.*/vine in May-June and September-October. If the vine is trailed on a susceptible tree like areca nut, phorate has to be applied at 6 g *a.i.*/standard.
- Application of 5 kg green manure (*Glyricidia, Crotolaria*), 1 kg FYM, along with the recommended dosages of NPK, neem oil cake at 200 g/vine and earthing up to 50 cm radius.
- Avoiding susceptible intercrops such as banana, ginger, turmeric etc. or application of 3 g *a.i.*/banana plant twice a year and 1 g *a.i.*/ginger/turmeric plant at planting.

13.3.2. Root-knot Nematodes, *Meloidogyne incognita, M. javanica*

Butler (1906) reported root-knot nematodes from black pepper in Wynad, Kerala, India. *M. incognita* and *M. javanica* have been reported from India. The root-knot infestation is a serious problem in Government nurseries in Kerala. Up to 91% root-knot infestation was reported from Para, Brazil (Ichinhoe, 1975) and Kerala, India (Ramana *et al.*, 1987a).

An initial inoculum level of 10 J_2 per rooted cutting was found to reduce growth by 15%, while at 1,00,000 J_2 level 50% reduction in growth was observed over one year period (Koshy *et al.*, 1979a). More than 50% death of transplants occurs in field planting of infected cuttings. *M. incognita* was responsible for 46% loss in yield of black pepper (Mohandas & Ramana, 1991).

(i) Distribution: The root-knot nematodes have been reported from India, Brazil, Sarawak, Borneo, Cochin-China, Malaysia, Brunei, Kampuchea, Indonesia, Philippines, Thailand and Vietnam.

(ii) Symptoms: Prominent symptoms of root-knot infestation on black pepper are unthrifty growth and yellowing of leaves. Inter-veinal yellowing of the foliage is also noticeable. The leaves exhibit dense yellowish discolouration of the interveinal areas making the leaf veins prominent with deep green colour. Heavy galling of the root system is also present.

(iii) Host Range: Among commercially used standards, *Oroxylon indicum, Erythrina lithosperma, Ceiba pentandra, Bombax malabaricum* are highly susceptible to root-knot nematodes (Koshy *et al.*, 1977).

(iv) Life Cycle: The eggs are laid in a gelatinous matrix. Its contents undergo a series of determinate cleavage giving rise to the first juvenile stage which moults within the egg and gives rise to second stage juvenile which hatches out. The J_2 juveniles penetrate host plants above the meristematic zone and orient parallel to the stele. It starts feeding on the pericycle cells for several weeks, gradually swelling. A series of 3 moults occur in quick succession. Reproduction may be parthenogenetic or bisexual with a female laying 400–500 eggs. Egg laying is completed within a week. The life cycle under optimum conditions takes 3–4 weeks.

(v) Interaction with Other Pathogens: *M. incognita* and *Fusarium solani* were found associated with black pepper vines. Both the organisms together were found to do more harm than either of them alone. Increased susceptibility of *M. incognita* and *M. javanica* infested pepper to *Phytophthora* infection has been reported.

(vi) Management

(a) Cultural Methods: Live standards for black pepper such as *Artocarpus heterophyllus, A. hirsutus, Ailanthes malabarica, Mesopsis emini, Peltophorum pterocarpum, Swietenia macrophylla, Tamarindus indica, Garuga pinnata* and *Macaranga peltata* are resistant to *M. incognita*.

Growing of non-host cover crops like Siratro (*Macroptilium atropurpureus*) in the interspaces and mulching with Guatemala grass (*Imperata cylindrica*) are recommended to reduce the root-knot nematode (*M. incognita*) population on black pepper in the Amazonian region (Ichinohe, 1980).

Application of neem cake reduces nematode populations followed by increase in yield.

(b) Chemical Methods: Fumigation of nursery potting mixture with methyl bromide is effective in checking the infestation.

Green berry yields can be doubled by four applications of carbofuran incorporated into mound soil at 114 g/vine per application in black pepper fields infested with *M. incognita* and *M. javanica*. Application of aldicarb at 1.25 g *a.i.*/vine or carbofuran at 10 g *a.i.*/vine twice a year, including at planting around cuttings, can reduce populations of *M. incognita* on black pepper. Phenamiphos at 1 g *a.i.*/vine followed by carbofuran and ethoprophos was effective in controlling nematodes.

(c) Biological Methods

Antagonistic Bacteria: : Root-knot nematodes in pepper can be managed by the application of *Bacillus macerans* at 1.2×10^8 cells/vine just before the monsoon period.

Antagonistic Fungi: Application of *Paecilomyces lilacinus* and *Pochonia chlamydosporia* has potential to control nematodes which reduced the rate of infection to 15 and 12%, respectively. Soil application of *P. chlamydosporia* reduced nematode population and increased the yield (5.14 kg/vine) with cost: benefit ratio of 1: 7.12.

Arbuscular Mycorrhizal Fungi: AMF such as *Glomus fasciculatum, G. etunicatum* and *Acaulospora laevis* reduced root-knot nematodes and enhanced the growth of black pepper. Pre-inoculation of *A. laevis* in black pepper *cv.* Panniyur-1 reduced the root-knot index (0.75) followed by *G. mosseae/ Gigaspora margarita* (1.25) compared to control (3.5) (Anandraj *et al.*, 1991).

Effective strains of AMF (*Glomus* sp. and *Gigaspora* sp.) isolated from black pepper rhizosphere were very effective in suppression of root-knot nematodes.

(d) Host Resistance: The root-knot nematode infestation was less in *cvs.* Veliakaniakadan, Karimunda, Kalluvalli, Balankotta, Narayakodi and Padapan. A cultivated germplasm 'Ottaplakal-1' (CLT-P-812) showed resistant reaction to *M. incognita*. Its yield (4.7 kg green/vine) was on par with the popular varieties–Paniyur-1 and Karimunda. It has an average yield of 7526 kg green/ha at fifth year with a potential yield of 17,280 kg green/ha. This variety also has high oleoresin content (13.8%) and was released as 'Pournami' and recommended for cultivation in all black pepper areas having root-knot nematode problem.

(e) Integrated Methods: Black pepper vines combinedly inoculated with *P. lilacinus* and *P. penetrans* had put out 23–112 per cent increase in plant growth over control and were very effective in the management of root-knot nematodes (Sosamma & Koshy, 1995).

Integration of neem cake application at 1 kg/vine along with phorate/carbofuran at 3 g *a.i.*/vine during May-June and again during September-October gave effective control of nematodes infesting black pepper.

Application of aldicarb at 1 g *a.i.*/vine twice a year (May/June and October/November) integrated with fertilizers (N-100 g, P- 40 g, K-140 g/vine) in two equal splits, earthing up to 50 cm radius at the base of the vines and mulching the base of the vines with leaves reduces foliar yellowing by 83% and *M. incognita* population by 33–38% (Venkitesan & Jacob, 1985).

13.3.3. Root-knot nematode, *Meloidogyne incognita* and Foot Rot, *Phytophthora capsicii* Disease Complex

Increased susceptibility of *M. incognita* and *M. javanica* infested cultivars of black pepper to *Phytophthora* infestation has been reported.

(i) Management

(a) Integrated Methods: Integrated management of foot rot (*P. capsicii*) and nematodes (*M. incognita* and *R. similis*) on black pepper was achieved by: (i) mixing AMF and *Trichoderma harzianum* in solarized nursery mixture to raise healthy and robust seedlings, (ii) application of *T. harzianum* and FYM in planting pit, and (iii) field application of neem cake at 1 kg/vine mixed with 50 g of *T. harzianum* during August.

14 CARDAMOM

Cardamom (*Elettaria cardamomum*), also known as the "Queen of Spices" enjoys a unique position in the world market has its origin in the evergreen rain forests of Western Ghats of South India. India enjoyed near monopoly in the field till recently. Guatemala has emerged as a competitor to India. It is a perennial plant with an underground stem (rhizome) and aerial shoots and valued for their fruits (capsules). The fruits are small, trilocular capsules containing 15–20 seeds. India and Guatemala are the main producers and exporters of cardamom. Other cardamom growing countries are Tanzania, Sri Lanka, El Salvador, Vietnam, Laos, Kampuchea and Papua New Guinea. The area under cultivation in India during 2004–05 was 73,237 ha, with an annual production of 11,415 metric tonnes. The export earning during 2006–07 was Rs. 274.7 million by exporting 650 million tonnes of small cardamom. Kerala, Karnataka, and Tamil Nadu produce cardamom in India. While Kerala accounts for 60 per cent of the production, Karnataka contributes 30 per cent, and Tamil Nadu 10 per cent. In Karnataka, cardamom is cultivated mostly in Kodagu, Hassan, and Chikmagalur districts. The plants last for 15 years. They grow in moist, elevated, and shady forest loams. They require a fairly distributed annual rainfall ranging from 2,500 mm to 5,000 mm and are found within an altitude ranging between 600 m and 1,200 m above the mean sea level. It is used for flavouring various food preparations, confectionery, beverages, liquors and medicines.

14.1. PESTS

14.1.1. Rhizome Weevil, *Prodioctes haematicus*

(i) Damage: This is a serious pest in the secondary nursery especially where seedlings are raised continuously year after year. The grubs feed on the rhizome and basal portion of the stem. This results in drying of leaves and breaking of stem at the base.

(ii) Management

Drenching the nursery beds with chlorpyriphos at 0.04% can control the pest.

14.1.2. Shoot Fly, *Formosina flavipes*

(i) Damage: The pest is observed in the nursery during January to May. Dead heart or decay of the central spindle is the external symptom.

(ii) Management

Spraying of quinalphos 0.025% or application of phorate granules at 1 g *a.i.*/m^2 is recommended for control of the pest.

14.1.3. Thrips, *Sciothrips cardamomi*

(i) Damage: This insect is a serious pest of cardamom. Thrips are pronounced during dry months of December-May. Injury is caused by thrips feeding in young leaf sheaths and basal ends of unopened flower bracts. Nymphs and adults feed with their rasping-sucking mouth parts, which lacerate young tissues and suck the juices that ooze from ruptured cells. First the injured area develops a silvery sheen because of air occupying the emptied cell cavities. Due to oxidation, this area becomes a mixture of white, yellow, and brown splotches and streaks. Infestation on the panicle and flower buds results in stunted growth of panicles, shedding of flower buds and warty growth on the surviving capsules (Fig 14.1). The infested capsules are light in weight, inferior in quality and fetch very low price in the market.

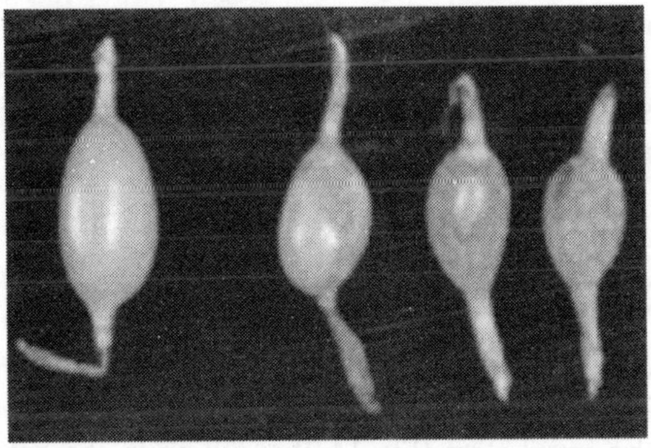

Figure 14.1. Thrips damage on capsules of cardamom

The cardamom thrips is difficult to find, since feeding is confined to young tissues in dark, confined areas such as the sheaths of young leaves, the base of unopened flower bracts, and possibly the spindle of top shoots.

The nymphal stages are sluggish in movement. Adults are also sluggish and rarely take to flight.

This thrips is suspected to be a vector of a mosaic disease that effect cardamom in India.

(ii) Biology: The life cycle (egg to adult) is completed in 25 to 30 days. Reproduction is mostly sexual, though virgin females are able to lay viable eggs during the summer. The white kidney shaped eggs are inserted in leaf sheaths or flower bracts by the sharp ovipositor of the adult. Eggs hatch in about a week. The transparent larvae are approximately 1/25 inch in length. The larval stage consists of three larval stages separated by molts in the course of 15 to 21 days. Pupae resemble the larvae except for the presence of wing pads (incompletely developed wings). There are 2 pupal stages, the prepupa and pupa, neither of which are feeding stages. Pupae become adults in 10 to 15 days. The head and abdomen of this thrips are dark gray brown and the thorax and legs are pale yellowish brown. Females measure approximately 1/4 inch in length; males are slightly smaller.

(iii) Management

General cultural practices involving sanitation, such as the removal of alternative hosts and plant debris within and outside the field and regulation of shade helps in reducing the pest infestation. The removal of dried shoot peels at the base of the stem resulted in a reduction of thrips populations by one-third.

Chemical control is difficult because the thrips remain hidden under the cover of leaf sheaths and bracts. The synthetic pyrethroids gave adequate control. Of the different synthetic pyrethroids, permethrin was the most effective.

Since the pest population is high during dry months from December to May, pesticide application during this period is important. Four sprays with either monocrotophos 0.025% or dimethoate 0.05% or quinalphos 0.025% during March-September gives effective control of thrips. Application of fish oil rosin soap also is beneficial to manage thrips damage. The following EC and Dust formulations are also effective in controlling the thrips:

EC formulations: Quinalphos 0.025%, fenthion 0.03%, phenthoate 0.03%, phosalone 0.05%, monocrotophos 0.025%, fenitrothion 0.05%, formothion 0.03%, dimethoate 0.05%.

Dust formulations: Quinalphos 1.5%, carbaryl 10%, phosalone 4% or phenthoate 4% each at 25 kg/ha.

Cardamom *cv.* Mudigere-1 (Malabar) is tolerant to thrips.

14.1.4. Shoot/Capsule Borer, *Conogethes punctiferalis*

(i) Damage: It is a serious problem to cardamom growers of Kerala, Tamil Nadu and Karnataka. At the early stage of the crop, the caterpillars of this yellow coloured moth bore into the core of the aerial stem (Fig 14.2) resulting in the death of central spindle, which appears as characteristic dead hearts. The caterpillar bores into the stem and feeds on the internal contents. This results in the decay of the central spindle and production of dead heart. Faecal matter of the caterpillar can be seen coming out through the holes.

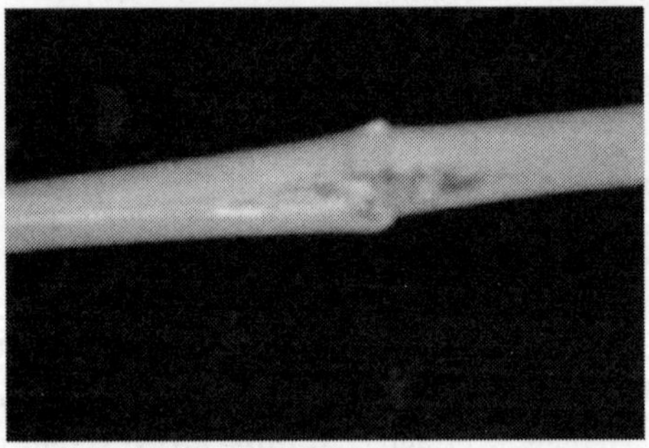

Figure 14.2. Shoot borer infestation on cardamom

At the time of flowering, when the caterpillars attack the panicles and spikes it may lead to flower shedding and drying up of the attacked portions. At a later stage of the crop, the caterpillars bore into the capsules, feed on the seeds and make them hollow. The presence of excreta at the region of attack indicates presence of the caterpillars in the pseudostem, inflorescence and pods.

Pest infestation is pronounced in three seasons *viz.* January-February, June and September-October.

(ii) Management

Late stages of larvae bore into the pseudostem and remain there. Insecticides sprayed at this time may not give adequate control of the pest. For an effective management of the pest, the insecticides have to be targeted on early stages of the larvae, which are usually present within 15–20 days after adult emergence in the field. Spraying fenthion 0.05% or monocrotophos 0.05% is recommended during the months of February-March and September-October. Spraying with quinalphos 0.025%, carbaryl 0.1%, monocrotophos, fenthion or dimethoate at 0.05% or endosulfan or phenthoate at 0.1% is recommended against the shoot borer.

As soon as bore holes of stem borer are noticed, injection of *Bacillus thuringiensis* preparation into the bore hole (0.5 ml in 10 ml water) will kill the larvae so that subsequent resurgence can be reduced.

Cardamom *cv.* Mudigere-1(Malabar) is tolerant to shoot borer.

14.1.5. Leaf Eating Caterpillars, *Eupterote* sp.

(i) Damage: There are 10 species of caterpillars feeding on cardamom leaves. Out of these, seven species are hairy and appear in large numbers during certain seasons causing extensive defoliation.

(ii) Management

For controlling the leaf caterpillars, mechanical collection and destruction of larvae and spraying of any contact insecticide are recommended to reduce the pest damage.

14.1.6. Cardamom White Fly, *Kanakarajiella (Dialeurodes) cardamomi*

(i) Damage: It is a serious pest in cardamom growing tracts of Kerala. The adult is a small soft-bodied insect, about 2 mm long and having two pairs of white wings. The nymphs are elliptical and pale green. The nymphs secrete sticky honeydew, which drops on to lower leaves. On these, black sooty mould develops, which interrupts photosynthesis of the leaves.

(ii) Management

The flies are attracted towards yellow colour. So metal sheets painted yellow and coated with sticky materials, such as castor oil or poly-vinyl butanol would serve as traps. By placing such yellow sticky traps between rows of cardamom plants, population of adults can be monitored and adults trapped to some extent. Nymphs are effectively controlled by spraying the lower surface of leaves with a mixture of neem oil (500 ml) and triton (500 ml) in 100 litres of water. Acephate 0.075% and triazophos 0.04% are equally effective. The spray may be repeated two or three times at 15 days interval.

Nymphs can be controlled by spraying neem oil with soft soap (500 ml neem oil and 500 g soft soap in 100 litres of water).

14.1.7. Root Grub, *Basilepta fulvicorne*

(i) Damage: The grubs of a small, greenish blue beetle cause damage. The grubs are short, stout, pale white in colour and often assume a shape resembling 'C', which feeds on cardamom roots. The symptoms start as yellowing of leaves, which later result in the drying up and death of the plant.

(ii) Management

Collect the beetle with hand nets or sticky traps at the time of mass emergence (March-April and August-September) and destroy. Early stages of the grub which are usually present in soil during May-June and September-October can be controlled either by drenching chlorpyriphos 0.04% at 3–4 litres per clump or by applying phorate at 2–4 g *a.i.*/clump 10–15 cm around the plant.

The population of root rot grubs in cardamom field was reduced significantly by soil application of carbofuran 3G at 100 g/plant followed by spraying imidachloprid at 0.5 ml/litre of water.

14.1.8. Scale, *Aulacaspis* sp.

(i) Damage: This scale insect is found on the lower surface of leaves, leaf sheath, panicles and fruit stalk. As a result of damage, capsules get shriveled, panicles become dry and the leaves become yellow. The pest is mostly seen during summer months.

(ii) Management

Spray monocrotophos or fenthion at 0.05% during the peak season.

14.1.9. Aphids, *Pentalonia nigronervosa*

(i) Damage: Aphids (Fig 14.3) are vectors of Katte virus disease. Colonies of aphids are seen concealed inside leaf sheath of old pseudostem. Aphids suck the sap and devitalize the plant.

Figure 14.3. Aphid on cardamom leaf

(ii) Management

Spraying of 0.05% dimethoate gives effective management of thrips.

14.1.10. Vertebrate Pests

(i) Damage: Wild boar, monkey, rodents and birds are important vertebrate pests of cardamom in south India. Rodents and squirrel constitute rodent pests incurring on an average 10% yield losses in cardamom. Cultivation of cardamom is being severely affected in Koppa, Sringeri and Agumbe in Ckikmagalur district; Thirthahalli in Shimoga district and at Madai, Monnagiri and Galibeedu in Kodagu district of Karnataka due to monkey menace. Damage due to monkeys varied from 2.80 to 17.50% to cardamom suckers.

(ii) Management

Rodent pests can be controlled by clean cultivation, overlapping of panicles from August, timely harvests and mulching which saved 2.56, 0.21, 5.54 and 4.20% capsules, respectively. Locally made wooden snap traps baited with germinating paddy seeds reduced capsule damage from 8.2 to 4.90%. These practices are cheap, practicable and eco-friendly.

Cardamom nurseries can be protected from wild boar damage for 45 days using a mixture of 5 g each of phorate 10G and sand, and suspending the same in punched polybags (10 cm × 10 cm size) along the borders of cardamom nursery (at 3 m distance and 2 m height). Selective shooting would ameliorate the wild boar problem.

Selective debranching of shade trees and watch and ward during early morning (before 7 a.m.) and late evening (after 5 p.m.) hours significantly reduced clump damage from 49 to 3.2% due to monkey menace in cardamom at Mudigere, Ckikmagalur, Karnataka.

Integrated Management: Three cultural practices, *viz.,* dispersion of panicles in an overlapped fashion, clean cultivation (without weeds) and timely harvest together with snap trapping and mulching saved 14.40% cardamom capsules from rodent pests (Srihari & Chakravarthy, 1998). Suspending a mixture of 5 g of phorate 10G granules +5 g sand in punched polybags along borders of cardamom nursery + good drainage + watch and ward reduced damage due to wild boar from 10–15% to 3–5% seedling damage during summer (Chakravarthy, 1993).

14.2. DISEASES

14.2.1. Damping-Off, *Pythium vexans, Rhizoctonia solani*

(i) Symptoms: This disease is caused by *Pythium vexans* and *Rhizoctonia solani*. Infection is observed at the collar region.

(ii) Management

Provide good drainage, and spray and drench the nursery with 1% Bordeaux mixture or 0.2% copper oxychloride.

14.2.2. Nursery Leaf Spot, *Phyllosticta elettariae*

(i) Symptoms: This disease is caused by *Phyllosticta elettariae*. Pale specks appear on the leaf lamina, which dry up and become paper white.

(ii) Management

Spraying the plants with 0.3% copper oxychloride or 0.25% mancozeb at fortnightly intervals is effective in controlling the disease.

14.2.3. Leaf Rot, *Coniothyrium* spp.

The leaf rot is another important nursery disease of cardamom. The disease is common during rainy season.

(i) Symptoms: Water-soaked lesions are formed on the leaves, which later turn to brown colour and cover the entire leaf blade. The infected blade rots and later it spreads to entire shoot. In severe infections, the entire nursery is destroyed.

(ii) Management

Spraying with 1% Bordeaux mixture is advocated for 3–4 times before the onset of monsoon.

14.2.4. Capsule Rot/Azhukal, *Phytophthora* spp.

The disease was first reported in Idukki district of Kerala (Menon *et al.*, 1972). It occurs during July-September period coinciding with south-west monsoon in India. The disease is more serious in Kerala and to some extent in Karnataka. It causes 30% yield loss of cardamom in Idukki district of Kerala (Nambiar & Sarma, 1976).

(i) Symptoms: This is a fungal disease caused by *Phytophthora* spp. occurring during the rainy season. It affects the leaves, tender shoots, panicles and capsules. On the infected leaves, water-soaked lesions appear first and rotting and shedding of leaves along the veins occur there after. The infected capsules become dull greenish brown and decay. This emits a foul smell and subsequently shed. Infection spreads to the panicles also.

(ii) Epidemiology: The disease starts during May, reaches to maximum in August (11.0%) coinciding with south-west monsoon and gradually decreases there after. The disease incidence is directly correlated with the amount of rainfall, relative humidity and soil moisture. The wet and humid conditions and a low temperature prevailing in the plantations are highly favourable for the disease development. The pathogen survives in the form of chlamydospores. The fungus survives for 32 weeks in infected rhizomes in air-dried soil and 42 weeks in moist soil. soil pH between 6 and 7 is more favourable for the survival of the pathogen. Soil and infected plant debris serve as the primary source of inoculum. Colocasia plants are reported as collateral hosts of this pathogen.

(iii) Management

Removal and destruction of the infected parts should be done as a phytosanitary measure just prior to the onset of southwest monsoon. Remove the trash (dried leaves and leaf sheaths) from the basal region of the plant to the extent possible. Movement of planting material as rhizomes from diseased areas to disease-free areas should be avoided. Lopping of branches of shade trees before the onset of monsoon helps in better light penetration and unfavourable climate for fungal infection.

Spray the shoots with 1% Bordeaux mixture with adhesive (rosin soda or any other sticker) by the commencement of the monsoon and continue the spraying operation two or three times up to November-December according to the intensity of the disease and rainfall. Give a copious spray to the panicle with 1% Bordeaux mixture at 3 litres/plant during July-August when the disease intensity is maximum. Spraying with 0.25% metalaxyl MZ or 0.1% fosetyl AL was also found effective in managing the disease.

Trichoderma spp. can be used along with cow dung or neem cake for controlling this disease. Soil inoculation with *T. viride, T. harzianum* and *Bacillus subtilis* resulted in reduced disease incidence (Suseela Bhat *et al.*, 1992a).

Moderate tolerance to this disease was noticed in two cultivars of Mysore and Malabar types (Suseela Bhat *et al.*, 1992b).

14.2.5. Rhizome Rot or Clump Rot, *Pythium vexans, Fusarium oxysporum, Rhizoctonia solani*

(i) Symptoms: The disease occurs during south west monsoon. The symptoms include pale yellow colour on the foliage and premature death of older leaves. Collar portion of the aerial shoots becomes brittle and tiller breaks easily from the rhizome at the bulbous base (Fig 14.4). Water soaked lesions appear first on leaves. The pathogen causes rotting and shredding of leaves along the veins. Rotting develops in the collar region emitting a foul smell. Capsules turn dull greenish brown emitting a foul smell.

Figure 14.4. Clump rot of cardamom

(ii) Management

Soil drenching with 0.2% copper oxychloride (2–3 litres per plant) and repeating this two times at monthly intervals gives effective control of the disease. Drenching the soil with 0.1% Ridomil (Metalaxyl + mancozeb) or applying *Trichoderma viride* (native strain isolated from cardamom plantation) as a prophylactic measure is effective.

As a bio-control measure, inoculate seedlings with native arbuscular mycorrhiza, *Trichoderma* and *Pseudomonas fluorescens* at the time of planting in the nursery and main field, and apply during pre-monsoon period in established plantations.

Solarization of nursery beds before sowing seeds and incorporation of *T. harzianum* has resulted in prevention of damping off and production of pathogen-free healthy seedlings. Soil solarization can be done for sterilizing the nursery mixture. To the sterilized mixture, biocontrol agents such as arbuscular mycorrhizal fungus at 100 g/kg and *T. harzianum* (10^{10} cfu/g) may be added at the time of filling of nursery mixture in polybags. Since the biocontrol agents protect the root system only, the aerial portion may be protected with 1% Bordeaux mixture spray.

Removal and burning of infected panicles and rhizomes from the plantation reduced the disease incidence. Spraying the panicles during pre-monsoon and post-monsoon is recommended. Incorporation of *Trichoderma* spp. multiplied in suitable organic medium (1 kg/clump) prior to the onset of monsoon season is a prophylactic operation. Spraying of 1% Bordeaux mixture at 3 litres/plant with an adhesive by the commencement of monsoon and continuing the spraying operation up to November-December is effective. Integration of *T. harzianum* (IISR 1369) with *Pseudomonas fluorescens* (IISR-11 or 6) suppressed the clump rot to the extent of 36% over control. Soil application of *T. harzianum* or *T. viride* to nursery beds gave disease control ranging from 53 to 84%. Soil application of *T. harzianum* or *T. viride* to the main field resulted in high disease control ranging from 62.3 to 64.8%.

Cardamom *cvs.* ICRI-2 (Mysore) and ICRI-3 (Malbar) are tolerant to capsule rot and rhizome rot disease.

14.2.6. Leaf Blotch Disease, *Phaeodactylium venkatesanum*

(i) Symptoms: The fungus *Phaeodactylium venkatesanum* causes this disease. The disease is characterized by the appearance of large blotches of irregular lesions with alternating shades of light and dark brown necrotic tissues. This is mainly observed on mature leaves. On the lower surface of the lesions ash coloured white superficial growth of the fungus appears during moist weather conditions.

(ii) Management

The fungicides, Bordeaux mixture (1%), mancozeb (0.3%) and carbendazim (0.1%) are effective in controlling the disease.

14.2.7. Leaf Blight or Chenthal, *Colletotrichum gloeosporoides*

(i) Symptoms: The disease occurs during the post-monsoon period. It causes water soaked rectangular lesions on the foliage, which elongate to form parallelly arranged streaks measuring 5 cm. Severely infected plants show a burnt appearance. The lesions are clearly visible even on dried leaves. The incidence of the disease appears to be more severe in areas, which do not have proper shade.

(ii) Management

Providing adequate shade is recommended. Spraying carbendazim or mancozeb or copper oxychloride at 3 g/litre of water at monthly intervals gives effective control of the disease.

14.2.8. Katte or Mosaic

The disease was reported for the first time during 1900 in North Canara district of Karnataka (Mollison, 1900). It is widely distributed in almost all cardamom growing areas. In arecanut-cardamom mixed cropping system, the crop losses due to katte were 10–68, 26–91 and 89–92%, during first, second and third year after infection, respectively (Varma, 1962). In pure crop of cardamom, crop losses of 38, 62 and 68% during first, second and third year after infection, respectively were reported (Venugopal & Naidu, 1984).

(i) Symptoms: This is a virus disease, which is transmitted by the banana aphid, *Pentalonia nigronervosa*. The symptoms consist of discontinuous stripes of light green colour running almost parallel to each other from the mid-rib to the margin of the leaves, which form a mosaic pattern. On young shoots, such stripes are seen on the leaf sheath also. The infected clumps will be smaller in size with fewer tillers. There will be marked dwarfing and stunted growth of the stem. Fruiting is completely absent in younger and early stage infections. Later stage infections results in reduced fruit size. The infected gardens shall be completely destroyed within 2-3 years if management practices are not initiated timely.

(ii) Transmission: The disease is caused by virus belonging to potyvirus group transmitted by the banana aphid, *Pentalonia nigronervosa*. The virus primarily spreads through use of infected rhizomes as planting material. The vector also transmits the virus to large cardamom, *Amomum cannbecarpum*.

(iii) Management

Eradication of the source of inoculum by destroying infected plants and destruction of the vector by insecticide application is effective. Regular application of insecticide against cardamom thrips controls the aphids also. Avoid using katte-infected rhizome for planting. As prophylactic measure, spraying of 0.05% dimethoate or phosphamidon is advocated. A minimum of six sprays need to be taken up to keep the disease under check.

Destruction of plants showing symptoms of the disease should be done promptly once in two months (Naidu & Venugopal, 1982). Removal of all alternate hosts of virus is also recommended.

Waiting Period for Insecticides/Fungicides

Quinalphos	—	30 days
Monocrotophos	—	30 days
Mancozeb	—	30 days

Integrated Pest and Disease Management

- Trashing the clumps of old peels, leaves and suckers during May-June, spraying with Nuvacron (June-July) 35% EC at 1.25 ml/litre up to 1.5 m plant height followed by another spray with Zolone 25% EC at 2 ml/litre 30 to 40 days later, removal of border affected suckers leaving 5 cm from the ground, were found effective in suppressing thrips and other major pests of cardamom.

- Néemark 15G and Nimbecidine were at par with monocrotophos in reducing the thrips damage.
- Root grubs (*Basilepta* sp.) infestation in nursery could be effectively checked by incorporating carbofuran 3G at 15 to 20 kg/ha.
- Damage by rodents and birds could be considerably minimized by adopting trapping, clean cultivation, timely harvest and by baiting with Bromodiolone (0.005%) or Rodafarin cake.
- Mancozeb (0.25%) + Ridomil (0.1%) were found to be the most effective fungicides for controlling damping off and leaf spot diseases in cardamom nursery.
- Clump rot disease could be effectively checked by drenching the affected and neighbouring clumps with 1% Bordeaux mixture once before the onset of monsoon (June) and again after the end of monsoon (August).
- Katte disease spread was found to be effectively checked by removing affected clumps, old and decayed suckers and plant parts by regular surveys and by avoiding planting of clones from affected gardens.
- Use pest and disease tolerant varieties for planting [Cardamom *cvs.* ICRI-2 (Mysore) against capsule rot, ICRI-3 (Malabar) against rhizome rot and Mudigere (Malabar) against shoot borer and thrips].
- Destroy wild hosts of katte disease like *Amomum, Alpinia, Curcuma, Colocasia, Caladium,* etc.
- Destroy capsule rot affected portions and plant debris.
- Use yellow sticky traps to control whiteflies at 15 to 20 traps/ha.
- Apply *Trichoderma viride/T. harzianum* in plant basins to control capsule rot and rhizome rot diseases.
- Spray neem oil at 3–5 ml/litre against whiteflies.

14.3. NEMATODES

The most important nematode problem is the root-knot nematode, *Meloidogyne* spp. The lesion nematode, *Pratylenchus coffeae* and the burrowing nematode, *Radopholus similis* are also known to cause root rotting (D'souza *et al.*, 1970; Kumar *et al.*, 1971; Khan & Nanjappa, 1972; Viswanathan *et al.*, 1974; Sundararaju *et al.*, 1979b).

14.3.1. Root-knot Nematodes, *Meloidogyne* spp.

M. incognita and *M. javanica* have been reported as widely occurring in the cardamom nurseries and plantations in Kerala, Karnataka and Tamil Nadu (Kumar *et al.*, 1971; Koshy *et al.*, 1976; Ali, 1982, 1986). An yield loss of 32–47% due to root-knot infestation has been reported (Ali, 1984, 1986). An initial population of 100 nematodes/plant causes discernible damage to cardamom (Eapen, 1987).

(i) Host Range: Many annual weeds and common shade trees like *Erythrina indica* and *E. lithosperma* present in cardamom plantations are susceptible to root-knot nematodes.

(ii) Symptoms: Heavy infestation of root-knot on mature plants in a plantation causes stunting, reduced tillering, yellowing, premature drying of leaf tips and margins, narrowing of leaf blades, delay in flowering, immature fruit drop and reduction in yield. Infested roots are characterized by excessive branching. Galling of roots is not conspicuous on mature plants.

In primary nursery, the second stage juveniles infest the radical and plumule as a result of which even 50% of the germinating seeds do not emerge. At the two leaf stage itself the infested seedlings show marginal yellowing and drying of leaves and severe galling of roots. Up to 40% of such seedlings do not establish in the secondary nursery. The infested secondary nursery plants exhibit stunting, yellowing, poor tillering, drying of leaf tips and margins and heavy galling of roots (Ali & Koshy, 1982).

(iii) Survival and Spread: The nematodes are spread through infested seedlings and rhizomes used for propagation.

(iv) Interaction with Other Pathogens: In nurseries, presence of *M. incognita* increases incidence of rhizome rot and damping off diseases caused by *Rhizoctonia solani* (Ali, 1986; Eapen, 1987).

In cardamom, *M. incognita* is the predisposing factor for *R. solani* infection, damping off and rhizome rot, prevalent in nurseries (Ali & Venugopal, 1992, 1993).

(v) Management

(a) Physical Methods: Soil solarization for 40–45 days using 400 gauge transparent LDPE was proved ideal pre-sowing treatment for the management of nematodes and other soil-borne diseases in cardamom nurseries. Mean soil temperature during the solarization period increased by 8.7°C in solarized beds and the germination was enhanced by 25.5%, while weed growth was suppressed by 82%. *Pythium vexans* was totally eliminated, while populations of *Rhizoctonia solani, Phyllosticta elettaria* and nematodes were suppressed to varying levels. Solarization was also found to stimulate the growth of cardamom seedlings (Eapen, 1995).

(b) Cultural Methods: It is advisable to change nursery sites every year.

Application of neem cake at 0.5 and 1.0 kg/plant twice a year reduces nematode populations followed by increase in yield (Ali, 1984). Soil application of crushed neem seed can take care of nematode problem.

(c) Chemical Methods: Disinfestation of nursery beds with methyl bromide at 500 g/10 m^2 is effective in controlling root-knot infestation in both primary and secondary nurseries.

Application of aldicarb at 5 kg *a.i.*/ha three times in a year, resulted in increased growth and vigour of seedlings both in primary and secondary nurseries (Koshy *et al.*, 1979; Ali, 1986). Increase in yield of *M. incognita* infested cardamom plants from 47–88% was recorded by the application of aldicarb/carbofuran/phorate at 5 g and 10 g *a.i.*/plant (Ali, 1984).

(d) Biological Methods: AMF such as *Glomus fasciculatum* and *Gigaspora margarita* were recorded to be effective in reducing the root-knot nematode population.

P. lilacinus suppressed *Meloidogyne* spp. population by 58.3–86.9 per cent in cardamom nursery. The number of quality seedlings for transplanting was greater in beds treated with this biocontrol agent (Eapen & Venugopal, 1995).

T. harzianum also suppressed *Meloidogyne* spp. by 58.3–68.9 per cent in cardamom nursery and increased the number of quality seedlings for transplanting (Eapen & Venugopal, 1995).

(e) Integrated Methods: Soil solarization of nursery beds and application of bioagents such as *P. lilacinus* or *Trichoderma* spp. improved growth of cardamom seedlings by suppressing root-knot nematode population (Eapen & Venugopal, 1995). Application of *T. harzianum* multiplied on decomposed coffee husks (7 days old) at the time of sowing at 2.5 kg/bed (4.5 m × 1.0 m) and repeated after 3 months is recommended for the control of root-knot nematodes and damping off in nurseries.

Integration of the following practices can help in the successful management of root-knot nematodes:

- Changing nursery sites frequently.
- Disinfecting nursery beds.
- Introduction of biocontrol agents at nursery level.
- Control of susceptible weeds.
- Exclusion of susceptible shade trees.
- Destruction of infested crop residues.
- Mulching with dead leaves.
- Application of neem cake.

14.3.2. Root-Knot Nematode, *Meloidogyne* sp. and Rhizome Rot, *Rhizoctonia solani* Disease Complex

(i) Symptoms: *M. incognita* was found to predispose cardamom seedlings to *R. solani* infection, which causes damping off and rhizome rot in the primary nursery.

(ii) Management

(a) Chemical Methods: When both fungicide (metalaxyl) and nematicide (carbofuran) were applied for the control of rhizome rot disease and nematodes, the mortality of seedlings was least.

(b) Integrated Methods: *P. lilacinus* in combination with *Trichoderma* spp. suppressed *Meloidogyne* spp. and rhizome rot disease (*R. solani*) complex when incorporated in solarized cardamom nursery beds (Eapen & Venugopal, 1995). Application of carbofuran along with neem cake was found to reduce immature fruit-drop and to increase capsule yield.

Soil solarization alone enhanced the germination by 25.5% and suppressed weed growth by 82.0%. Solarization also enhanced the growth and vigour of cardamom seedlings. The disease complex was suppressed by incorporation of *P. lilacinus*/*T. harzianum* and phorate into the solarized nursery beds. This approach is being adopted on a large scale for the production of nematode-free cardamom seedlings.

15 GINGER

Ginger (*Zingiber officinale*), an herbaceous perennial, belonging to the family Zingiberaceae, is valued for its rhizome. The country of its origin is not definitely known but is presumed to be either from India or China. India is the largest producer (3,05,900 metric tonnes) and exporter of dry ginger in the world (export earning of Rs. 191 million). The other ginger producing countries are Jamaica, Sierra Leone, Nigeria, Southern China, Japan, Taiwan and Australia. In India, It is being cultivated in 85,100 hectares producing 0.306 million tonnes of ginger with an average yield of 3.6 tonnes per hectare.

15.1. PESTS

15.1.1. Shoot Borer, *Dichcrosic punctiferalis*

(i) Damage: A shoot-borer, *Dichocrocis punctiferalis* attacks the crop in India. The larvae bore into the shoots. They can cause them to wilt and die. Caterpillar bores the rhizomes and pseudostem causing dead heart. Extrusion of frass through holes can be observed (Fig 15.1).

Figure 15.1. Shoot borer infestation on ginger

(ii) Management

Spraying 625 litres/ha of 0.07% endosulfan is effective for the management of shoot borer. The shoot borer can also be managed by spraying 0.1% malathion or 0.075%

monocrotophos or 0.05% dimethoate or 0.3% Dipel (*Bacillus thuringiensis* product) at 21 days interval during July-October. The spraying has to be initiated when the first symptom of pest attack is seen on the inner most leaves on the pseudostem.

Pruning and destroying freshly infested pseudostems (at fortnightly intervals) during July-August and spraying 0.1% malathion (at monthly intervals) during September-October is also effective against the pest.

15.1.2. Leaf Folder, *Udaspes folus*

(i) Damage: Larvae of leaf roller cut and fold leaves and feed from within. The adults are medium size butterflies with brownish black wings with white spots. The larvae are dark green in colour.

(ii) Management

Spraying with 0.1% carbaryl or 0.05% dimethoate once in 10 days is effective in management of the pest.

15.1.3. Scale, *Aspidiella bartii*

(i) Damage: The scale insect *Aspidiella bartii* has been found on rhizomes in the Lesser Antilles; they do little damage, but may spoil the appearance of the sample. Scale insects of the genus *Aspidiotus*, including *A. destructor*, have been reported attacking ginger plants in several countries. The scale insects suck the sap from rhizomes. Rhizomes shrivel and dry up (Fig 15.2).

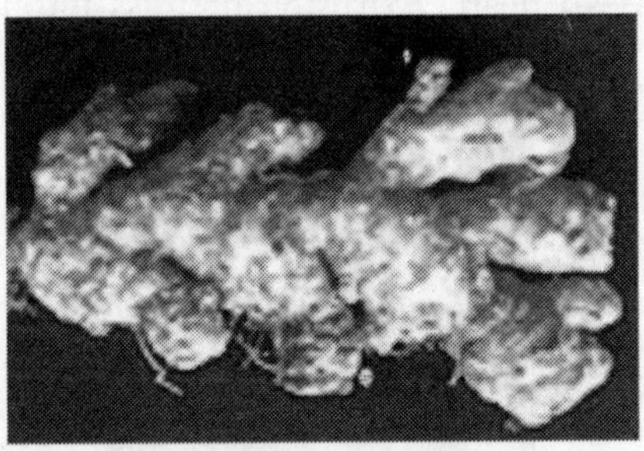

Figure 15.2. Rhizome scale infestation on ginger

(ii) Management

Dipping of rhizomes in 0.1% quinalphos (5 ml/l of water) prior to sowing gives effective control.

Fumigating the rhizomes with carbon bisulphide, after which the scales can be removed in the process of cleaning, may kill them.

15.1.4. Cut Worm

(i) Damage: The primary shoot is eaten just below or just above ground level. The first leaf may also be eaten. Feeding is done at night and larvae can be found in the ground at the base of the shoot during the day. Larvae grow up to 40 mm long, are grey, brown or pink and curl up when disturbed.

(ii) Management

Spray when first damage is evident. Spraying in the late afternoon will give better control.

15.1.5. Pod Borer, *Helicoverpa armigera*

(i) Damage: Young shoots are eaten off at ground level. The caterpillar grows up to 40 mm long and varies in colour from pinkish-brown to light green, with a variable number of dark markings and longitudinal stripes along the body.

(ii) Management

Control with a registered insecticide.

15.1.6. African Black Beetle

(i) Damage: Ginger stems are gnawed approximately 5 cm below ground level, causing plant collapse. Adult beetles are dark brown to black and approximately 12 mm long.

(ii) Management

Spray when first damage is evident. Spraying in the late afternoon will give better control. Insecticide should be applied with sufficient water to wet the soil down to 5 cm.

15.1.7. Stored-Products Pests, *Sitodrepa panacea, Lasioderma serricorne*

(i) Damage: The common stored-products pest attack stored ginger. Paulose (1973) refers to them as the common drug store beetle, *Sitodrepa panacea* and the tobacco borer beetle, *Lasioderma serricorne.*

(ii) Management

Fumigation of the store houses should be carried out regularly to control these pests.

15.1.8. Symphilids

(i) Damage: These feed on the root tips, stimulating the production of many short branching roots, the characteristic "witches broom" effect. This root damage can affect growth and yield.

15.1.9. Scolytid Beetle, *Stephanoderes trinitatis*

(i) Damage: The female burrows into the rhizomes to lay her eggs.

15.1.10. Rhizome Fly, *Chalcidomyia atricornis*

(i) Management

Soil application of phorate at 1 kg *a.i.*/ha gives effective control of rhizome fly.

15.1.11. Stem Borer

(i) Management

Soil application of phorate at 1 kg *a.i.*/ha gives effective control of stem borer.

15.1.12. Mite, *Rhizoglyphus* sp.

(i) Damage: A minute translucent mite of the genus *Rhizoglyphus* sometimes develops in large numbers in stored rhizomes in the Lesser Antilles, and while doing little damage may spoil the appearance of the produce.

(ii) Management

A light dusting of flowers with sulphur will kill the mites and prevent reinfestation.

15.2. DISEASES

15.2.1. Rhizome Rot, *Pythium aphanidermatum*

(i) Symptoms: Soft rot is the most serious disease of ginger in India and in some other countries. It is caused by *Pythium* spp., of which *P. aphanidermatum* is the principal species in India, although *P. butleri*, *P. gracile*, *P. myriotylum*, *P. nigriotilum*, and *P. vexans* have also been recorded. The bases of the aerial shoots become soft, watery and then rot. The affected plants become pale; the tips of the leaves turn yellow, followed by complete yellowing and drying up of the leaves (Fig 15.3). The shoots fall and cease to produce rhizomes. The infection extends to the rhizomes; the inner tissues being reduced to a soft and black, putrefying mass. Losses can be high.

Figure 15.3. Rhizome rot of ginger

(ii) Epidemiology: The disease is favoured by high moisture content of the soil with insufficient drainage. The pathogen perpetuates from year to year through the infected seed material.

(iii) Management

Selection of well drained soils for planting is important for managing the disease, since stagnation of water predisposes the plant to infection. Planting of ginger rhizomes in raised beds instead of flat beds is recommended. It reduces the disease incidence and increases the yield which can fetch Rs. 2,000/- more without incurring any extra expenditure. Disease-free rhizomes should be used for planting. Rhizomes may be treated in a mixture of 250 g of mancozeb (Dithane M-45) and 100 g of carbendazim (Bavistin) in 100 litres of water for 60 min followed by drying in shade for 48 hr before planting. Seed treatment with 0.25% thiram and later spraying with the same chemical also gives effective control of rhizome rot disease. Seed treatment with apron 35 S at 0.07% and mancozeb at 0.3% increased the percentage of germination. Soil drenching with metalaxyl at 0.25% given once before the heavy monsoon was effective in controlling post-emergence rot. Once the disease is located in the field, removal of infected clumps and drenching with mancozeb or copper oxychloride each at 0.3% during August in mid-hills and during September in low-hills checks the spread of the disease. Application of *Trichoderma harzianum* along with neem cake at 1 kg/bed helps in preventing the disease.

When harvesting ginger for seed purposes, do not commence until tops have died down naturally, as the stalk scar is an ideal point for infection to occur. Only harvest when rhizomes are dry and store in new bags in a dry and well-ventilated shed.

Ginger *cvs.* SG 600, Sadhupul, Jamaica, Awacho and B 1 have been reported to exhibit fair amount of resistance.

Integrated disease management involving phytosanitation, soil solarization and seed treatment was found effective in reducing rhizome rot and increasing ginger yield.

15.2.2. *Alternaria* Leaf Spot, *Alternaria* sp.

(i) Symptoms: The pathogen causes oval brown spots with white or grey centre on upper side of leaf. Rhizomes do not develop.

(ii) Management

Disease-free rhizomes should be used for planting. The rhizomes can be treated in a mixture of 250 g of mancozeb (Dithane M-45) and 100 g of carbendazim (Bavistin) in 100 litres of water for 60 min and drying in shade for 48 hr before sowing. Soil drenching of the affected plants with 0.3% copper oxychloride during August in mid-hills and during September in low-hills is also effective. Spraying 0.2% mancozeb or 0.3% copper oxychloride or 1% Bordeaux mixture after the first appearance of leaf spot is effective for the management of the disease.

15.2.3. *Phyllosticta* Leaf-Spot, *Phyllosticta zingiberi*

(i) Symptoms: A leaf-spot disease has been reported from Himachal Pradesh, caused by *Phyllosticta zingiberi*. The same disease was reported in Andhra Pradesh also. But it was not so serious. Oval water-soaked spots occur on leaves, later becoming papery, and may coalesce, causing extensive discolouration and desiccation.

(ii) Management

A 1 per cent Bordeaux mixture spray provides protection against leaf spot disease.

15.2.4. *Colletotrichum* Leaf–Spot, *Colletotrichum zingiberis*

(i) Symptoms: A leaf–spot caused by *Colletotrichum zingiberis* is of common occurrence in different parts of India. The fungus attacks the leaves only, forming round to oval light-yellow spots which may coalesce to form large discoloured areas (Fig 15.4), often drying up in the centre to form a hole. The plants become stunted in growth and the development of the rhizomes is adversely affected.

Figure 15.4. *Colletotrichum* leaf spot on ginger leaf

(ii) Management

Removing and burning the affected parts and prophylactic spraying of 5-5-50 Bordeaux mixture can control the disease.

15.2.5. Yellows/Rhizome Rot, *Fusarium oxysporum* f. sp. *zingiberi*

(i) Symptoms: Rhizome rot is a very common and serious fungal disease which is specific to ginger. Infected plants are stunted and yellow, lower leaves dry out and turn brown. Eventually all above-ground shoots dry out completely. Plant collapse is very slow (up to several weeks) compared with the rapid collapse associated with bacterial wilt infection. Diseased rhizomes show a brown internal discolouration, are normally shriveled in appearance and eventually decay leaving the outer shell intact with fibrous internal tissue remaining. Increased nematode infestations are usually associated with *Fusarium* rhizome rot, accentuating yield losses. *Fusarium* is also responsible for serious loss of planting pieces and poor germination. The disease spreads rapidly in the field during wet weather.

(ii) Management

Disease-free rhizomes should be used for planting. When cutting up seed, discard any pieces showing shriveling or brown discolouration and regularly dip the cutting knife in methylated sprit or a commercial disinfectant solution. Rhizomes may be treated in a mixture of 250 g of mancozeb (Dithane M-45) and 100 g of carbendazim (Bavistin) in 100 litres of water for 60 min followed by drying in shade for 48 hr before planting. Affected plants may be drenched with 0.3% copper oxychloride during August in mid-hills and during September in low-hills. Crop rotation of at least 2–5 years between ginger crops will help to reduce *Fusarium* in the soil. Dipping the seed rhizomes in ethoxyethyl mercury chloride can control it. Avoid areas with heavy nematode infestations.

15.2.6. Leaf Blotch, *Taphrina maculans*

(i) Symptoms: Small oval or irregular brown spots develop on either side of leaves. Leaves become yellow and plants present scorched appearance in severe cases (Fig 15.5).

Figure 15.5. *Taphrina* leaf spot on ginger leaf

(ii) Management

Spraying mancozeb at 2 g/litre of water gives effective management of the disease.

15.2.7. Banded Leaf and Sheath Blight

(i) Symptoms: Large discoloured areas develop alternating with irregular bands. Blotching of leaf and leaf sheath occurs in case of severe infection (Fig 15.6).

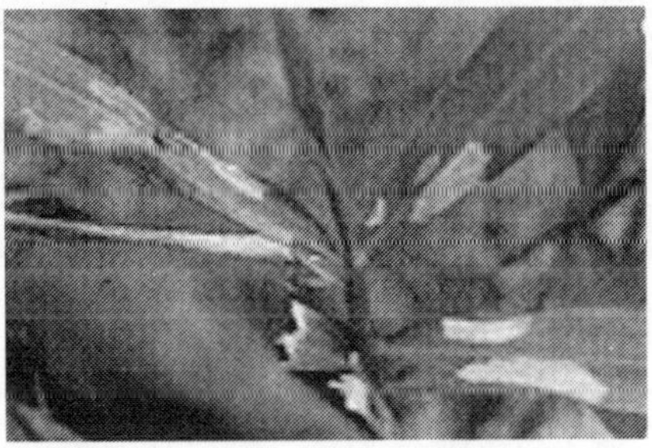

Figure 15.6. Leaf and sheath blight on ginger

(ii) Management

Dipping seed rhizome in a solution containing streptocycline at 2 g in 10 litres of water for 20 min is effective.

15.2.8. Thread Blight, *Pellicularia filamenlosa*

(i) Symptoms: Thread blight is caused by the commonly called shoe string (boot-lace) fungus. A common parasite of forest trees, it normally only occurs in recently cleared forest country and causes yellowing of the stem very similar to *Fusarium* symptoms. When infected plants are examined, black string like threads can be seen adhering to the rhizomes.

(ii) Management

All roots and stumps should be removed before planting newly cleared land to prevent infection.

15.2.9. White Fungus

(i) Symptoms: This disease has caused serious losses on some farms. Infection first occurs beneath the scales on the rhizomes and may progress to produce a deep brown rot over the entire surface. Affected rhizomes become enveloped in a white fungal mycelial growth which may extend up to the basal stem. Mild infections usually produce on above ground symptoms. Severe infections cause the stems to turn yellow and slowly dry out.

(ii) Management

Prepare the ground well in advance of the ginger crop as this fungus will persist on crop residues. The most serious outbreaks of this disease have been encountered in areas previously cropped with sugarcane where a large amount of undecomposed cane trash remained in the soil.

15.2.10. Bacterial Wilt, *Ralstonia solanacearum*

(i) Symptoms: A bacterial wilt of ginger caused by *Ralstonia solanacearum* was recorded from Australia, Hawaii, India, Mauritius and Malaysia. Bacterial wilt is a very serious disease that can spread very rapidly through an area, causing complete destruction. The first symptoms of the disease are yellowing and wilting of the lower leaves, which quickly spread upwards, affecting the whole plant. In advanced stages the base of the pseudostem becomes water-soaked, readily breaking away from the rhizome at ground level and the plants eventually collapse. The vascular tissues become dark brown or black. When an infected stem or rhizome is cut transversely, and a little pressure applied, a milky white exudate flows freely from the cut surface. Diseased rhizomes are much darker than healthy ones. Biotype 4 was found to produce the most severe symptoms.

(ii) Management

Solarization of ginger rhizomes for 2 to 4 hr from 9.00 a.m. to 1.00 p.m. (which raised the rhizome temperature to 55°C during summer) just before planting, eliminated the seed-borne bacterial infection from ginger rhizomes. The seed rhizomes may be treated with 200 ppm streptocycline for 30 min and shade dried before planting. Once the disease is noticed in the field, all beds should be drenched with 1% Bordeaux mixture or 0.2% copper oxychloride.

Only take planting material from areas known to be free of this disease. Infested areas should not be replanted with ginger, but planted with a perennial grass and fenced off to stop animals and people walking through and spreading the disease. All crop residues where the disease is known to occur should be destroyed, by burning if possible. Implements and boots should be washed down with a commercial agricultural disinfectant.

15.2.11. Bacterial Soft Rot, *Erwinia* sp.

(i) Symptoms: Normally bacterial soft rot is only a storage rot. The bacteria are present in most soils but field infection usually only occurs in waterlogged areas. *Erwinia* sp. has caused severe losses of stored rhizomes on some farms, but is not considered a serious storage problem where precautions are taken. Softening of the tissue is accompanied by production of a strong odour and the rhizome eventually collapses completely. Bacterial soft rot differs from other rhizome rots in that putrid odour is produced in this soft rot.

(ii) Management

Make sure the area is well drained. When harvesting ginger for seed purposes, do not commence harvesting until tops have died down naturally, as the stalk scar is an ideal point for infection to occur. Only harvest when it is dry and store in new bags or clean bins in a dry and well-ventilated shed.

15.2.12. Big Bud (Bunchy Top)

(i) Symptoms: The mycoplasma-like organism which causes this disease also causes big bud on a wide range of plants and is similar in characteristics to a virus. The disease is spread by leaf hoppers.

(ii) Management

All diseased plants should be detected and rouged out.

Integrated Disease Management

Selection of healthy planting material from disease-free gardens (soft rot, *Pythium aphanidermatum, P. myriotylum*; bacterial wilt, *Ralstonia solanacearum*; and nematodes, *Meloidogyne* spp., *Pratylenchus* spp.) is essential. Hot water treatment of the seed rhizomes at 51°C for 10 min eliminated nematodes and surface bound pathogens also. Strict phytosanitation, provision of better drainage and cultivating ginger in raised beds are some of the important cultural operations to be adopted to reduce crop losses due to the biotic factors.

As a seed disinfection procedure, a novel technique called rhizome solarization has been devised at the Indian Institute of Spices Research, Calicut. The disinfected rhizomes when treated with *T. harzianum* and rhizobacterial strain consortia as seed treatment and soil application resulted in higher yields and growth promotion and soft rot suppression and minimal bacterial wilt. In addition, in combination with *Glomus* spp., the disease was absent probably through growth mediation. *T. harzianum* in combination with *Pseudomonas fluorescens* showed a synergistic effect in reducing the soft rot infection.

Application of *T. harzianum, T. viride, T. hamatum* and *T. virens* as seed treatment and soil application along with neem cake at 100 kg/ha gave protection and increased yield. Application of *T. viride* along with saw dust/neem cake was found to be highly effective in suppression of rhizome rot. There was synergistic effect in protection where soil solarization followed by biocontrol was adopted. *Trichoderma* spp. and *T. viriens* were found to be highly effective in suppressing *Fusarium* yellows and *Pratylenchus coffeae*.

- Disease-free, healthy seed rhizomes should be used for planting.
- Rhizomes may be treated in a mixture of 250 g of mancozeb (Dithane M-45) and 100 g of carbendazim (Bavistin) in 100 litres of water for 60 min followed by drying in shade for 48 hr before planting.
- Adequate drainage should be provided in the field.
- The disease-affected clumps should be removed and the soil around the affected plants drenched with 0.3% copper oxychloride during August in mid-hills and during September in low-hills.
- Spraying with 0.3% copper oxychloride at 10 days interval with the appearance of leaf spot disease.

15.3. NEMATODES

Although a large number of nematode species have been reported from ginger, the root-knot nematode, *Meloidogyne* spp., the burrowing nematode, *Radopholus similis* and the lesion nematode, *Pratylenchus coffeae* are the most important.

15.3.1. Root-knot Nematodes, *Meloidogyne* spp.

M. incognita, M. javanica, M. arenaria and *M. hapla* have been reported to be associated with ginger from various countries. Under pot conditions, an initial inoculum level of 10,000 nematodes per plant over a period of six months caused 74% reduction in rhizome weight. A population level of one juvenile/30 g of soil was found to cause significant reduction in yield (Sukumaran & Sundararaju, 1986). Kaur (1987) estimated 41–59% yield loss in ginger when the crop was raised using apparently healthy rhizomes in nematode infested fields.

(i) Symptoms: Heavily infested plants exhibit stunting and chlorotic leaves with marginal necrosis. The root-knot nematodes cause galling and rotting of roots and underground rhizomes. Infested rhizomes show brown, water-soaked lesions in the outer tissues, particularly in the angles between shoots. The J_2 of *M. incognita* invade the rhizome through the axils of leaf sheaths in the shoot apex. In fibrous roots, penetration occurs in the area of differentiation and in fleshy roots. The entire length of root is invaded. In the rhizomes and fleshy roots extensive internal lesions develop.

(ii) Life Cycle: In fleshy and fibrous roots the nematode develops to maturity in 24 days, while in rhizomes it takes 43 days at a soil temperature of 30°C (Huang, 1966).

(iii) Spread: Infested rhizomes help in long distance dissemination of nematodes.

(iv) Histopathology: In all infested tissues, except rhizome meristems, abnormal xylem and hyperplastic parenchyma are observed. Giant cells are found in all infected tissues. Infection sites in differentiated rhizomes and fleshy roots, wound cork and lignified wall thickening of endodermis and pericycle are formed (Huang, 1966; Shah & Raju, 1977).

(v) Interaction with Other Pathogens: Bacterial wilt of ginger, caused by *R. solanacearum* is influenced by *M. incognita* (Samuel & Mathew, 1983).

M. incognita had synergistic effect in rhizome rot/soft rot (*Pythium aphanidermatum)* of ginger, particularly when nematodes were inoculated to the plants prior to fungal inoculation. The damage to the crop was more and the onset of the disease was early when *M. incognita* was inoculated 40 days prior to fungal inoculation (Ramana *et al.*, 1998).

(vi) Management

(a) Physical Methods: Summer ploughing at an interval of 15 days and use of hot water treated ginger rhizomes (at 50°C for 10 min) as seed proved promising in reducing the nematode population and in increasing the yield up to 65%.

(b) Cultural Methods: Ginger rhizomes used for planting should be collected from nematode-free fields.

Organic Amendments: Application of well decomposed FYM/compost at 25–30 t/ha or neem cake at 2 t/ha and mulching with green leaves at 10-12 t/ha at planting and repeating the mulching during the growth period helps in reducing nematode multiplication (Kaur, 1987).

Two applications of neem cake at 1 t/ha as basal and 45 DAP significantly reduced nematode population and increased the yield.

Crop Rotation: Crop rotation with cereals and millets should be practiced, preferably with rice at least once in 3 years. For ginger cultivation in Fiji, a ginger-taro-fallow rotation has been recommended. Intercropping of ginger with maize, capsicum and mulching with FYM are effective in reducing the nematode population.

(c) Chemical Methods: Kaur (1987) found that presowing dip of ginger rhizomes in 0.04% phenamiphos (Nemacur) or 0.18% Mancozeb for 30 min helps in reduction of nematode population resulting thereby in higher crop yields. In Himachal Pradesh, application of phenamiphos at 3 kg *a.i.*/ha resulted in 70–144% increase in yield of ginger in fields infected with *M. incognita* (Kaur, 1987).

Two applications of carbofuran at 1 kg *a.i.*/ha as basal and 45 DAP significantly reduced nematode population and increased the yield.

(d) Integrated Methods: *P. lilacinus* along with *T. harzianum*/*P. chlamydosporia* was effective in suppression of root-knot nematodes under field conditions.

Application of neem cake at 1 t/ha at planting, followed by carbofuran at 1 kg *a.i.*/ha at 45 DAP gave effective control of nematodes associated with ginger.

The following integrated nematode management measures are suggested:

(a) Production of nematode-free planting material by:

- Selecting an area where ginger has not been grown in the previous season and has no history of severe nematode infestation.

- Preparation of land and fumigation of soil with DD or EDB at 330 litres/ha in August (at a depth of 20 cm in rows 30 cm apart). The time interval between fumigation and planting should be at least two weeks.
- Selection of nematode-free planting material and treatment in hot water at 40°C for 20 min. It is followed by cooling the rhizomes before cutting and dipping in benomyl. Seed should be planted within one week of hot water treatment.
- Growing under saw dust mulch. If saw dust is not available, phenamiphos granules should be sprinkled over the soil between the plants at 11 kg/ha in mid-November and again in mid-January.

(b) Fumigation of land two or more weeks before planting.

15.3.2. Burrowing Nematode, *Radopholus similis*

R. similis infestation in ginger was first reported in India by Charles and Kuriyan (1979). The coconut isolate of *R. similis* in Kerala reproduced well on ginger (Koshy & Sosamma, 1975, 1977). *R. similis* has been reported from more than 50% of the total area in Fiji causing yield reduction up to 40%. An initial inoculum level of 10,000 nematodes/plant causes 74% reduction in rhizome weight (Sundararaju *et al.*, 1979b).

(i) Symptoms: *R. similis* infestation causes stunting, reduced vigour and tillering of infested plants and they tend to mature and dry out faster than healthy plants. The top most leaves become chlorotic with scorched tips. Lesions on the rhizomes results in small, shallow, sunken, water-soaked areas (Sundararaju *et al.*, 1979a). Intra-cellular migration of nematodes lead to large infection channels or galleries within the rhizomes.

(ii) Spread: The perpetuation and dissemination of the nematode takes place through infested rhizomes used for planting.

(iii) Management

The methods suggested for the control of root-knot nematodes were found effective in reducing the burrowing nematode infestation also.

15.3.3. Lesion Nematode, *Pratylenchus coffeae*

(i) Symptoms: The association of *P. coffeae* on ginger has been reported from Kerala (Charles & Kurian, 1979) and Himachal Pradesh. An initial inoculum level of even 10 nematodes proved highly pathogenic to 15-day old ginger plants in sterilized soil (Kaur, 1987).

(ii) Management

(a) Chemical Methods: In Himachal Pradesh, application of fenamiphos at 3 kg *a.i.*/ha resulted in 70–144% increase in yield of ginger in fields infected with *P. coffeae* (Kaur, 1987).

16 TURMERIC

Turmeric (*Curcuma longa*) is best known as an important spice crop. It is believed to have its origin in South-East Asia. Turmeric is valued for its rhizomes, often processed and used commercially. It is mainly cultivated in India and to a small extent in China, Indonesia, Peru and Jamaica. It is an important source of yellow dye and is also used as a colouring matter in food industries, confectionery and drugs. In India, it is cultivated in 1,49,400 ha with annual production of 5,29,000 tonnes during 2004–05. India's export earning from turmeric was Rs. 92 million annually.

16.1. PESTS

16.1.1. Rhizome Fly, *Mimegralla cocruleifrons*

(i) Management

Soil application of phorate or carbofuran at 1 kg *a.i.*/ha or spraying twice with 0.05% monocrotophos gives effective control of the pest.

Disease-free, healthy rhizomes should be selected for planting. Proper drainage should be provided.

16.1.2. Rhizome Scale, *Aspidiella hartii*

(i) Damage: In Africa and the South Pacific, the rhizome scale infests rhizomes in the field (at later stages) and in storage. Adult female scales are circular (about 1 mm in diameter) and light brown to grey and appear as encrustations on the rhizomes (Fig 16.1). They feed on sap and when the rhizomes are severely infested, they become shriveled and desiccated affecting their germination.

Figure 16.1. Infestation of scale on rhizome of turmeric

(ii) Management

Dipping the seed rhizomes in 0.5% malathion or quinalphos for 30 min prior to storage and at the time of planting is effective for the management of scale. Rhizome treatment with 0.075% quinalphos or phosalone, monocrotophos for 20–30 min was more effective for obtaining a higher recovery of rhizomes, higher number of sprouts and lesser incidence of rhizome scale.

Two Hymenopteran parasites, *Physcus* sp. and *Adelencyrtus moderatus*, attack this pest.

16.1.3. Thrips, *Panchaetothrips indicus*

(i) Damage: The thrips infest turmeric leaves causing them to roll, turn pale and gradually dry up. The pest infestation is more common during post-monsoon period especially in drier regions of the country.

(ii) Management

Spraying 0.05% dimethoate or 0.1% phosphamidon is effective for the management of thrips.

16.1.4. Leaf Eating Caterpillar, *Udaspes folus*

(i) Damage: In India the hesperiid caterpillar *Udaspes folus* ('grass demon') is also a serious foliage feeding insect. Larva folds the leaves, lives inside and feeds on the leaves.

(ii) Management

Spraying 0.2% carbaryl or dimethoate or phosphamidon gives protection against leaf eating caterpillar.

16.1.5. Leaf Feeding Beetles, *Lema* spp.

(i) Damage: Adults and larvae of leaf feeding beetles feed on leaves with parallel feeding marks on them.

(ii) Management

The pest can be controlled by dusting with 4% endosulfan dust or spraying 0.1% malathion.

16.1.6. Shoot Borer, *Conogethes punctiferalis*

(i) Damage: In Asia, caterpillars of the shoot borer *Conogethes punctiferalis* (synonym: *Dichocrocis punctiferalis*) attack the central shoot and killing it resulting in dead heart. Extrusion of frass through holes can be seen (Fig 16.2).

(ii) Management

Pruning and destroying freshly infested pseudostems (at fortnightly intervals) during July-August and spraying 0.1% malathion or 0.05% endosulfan (at monthly intervals) during September-October resulted in lower incidence of the pest.

Figure 16.2. Shoot borer infestation on turmeric

Spraying 0.1% malathion or 0.075% monocrotophos or 0.3% Dipel (*Bacillus thuringiensis* product) at 21 days interval during July to October is effective in controlling the pest infestation. The spraying has to be initiated when the first symptom of pest attack is seen on the inner most leaves on the pseudostem.

16.1.7. Lace Wing Bug, *Stephanitis typicus*

(i) Damage: In Asia the tinged bug, *Stephanitis typicus*, causes leaf discolouration by sucking the sap resulting in drying up of leaves. The pest infestation is more common during the post-monsoon period especially in drier regions of the country. Nymphs and adults suck sap in colonies, from under the leaf surface. Leaves turn yellow and black resinous excretions in the form of tiny spots may be seen between the damaged areas (Fig 16.3).

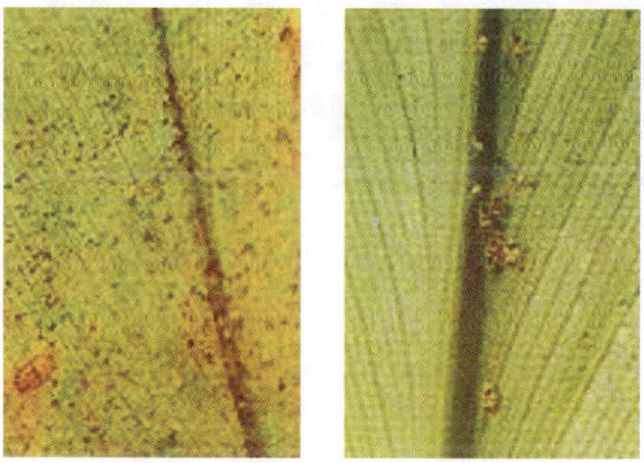

Figure 16.3. Lace wing bug infestation on turmeric leaves

(ii) Management

Spraying dimethoate or phosphamidon each at 0.05% is effective in managing the pest.

16.1.8. Root Grubs

(i) Damage: Root grubs occasionally feed on tender rhizomes, roots and base of pseudostems causing yellowing and wilting of shoots.

(ii) Management

The pest can be controlled by soil drenching with 0.075% chlorpyriphos.

16.1.9. Mite

(i) Management

The mites can be controlled by foliar spraying with 0.05% methyl parathion.

16.2. DISEASES

16.2.1. Rhizome Rot, *Pythium aphanidermatum*

(i) Symptoms: Rhizome rot caused by *Pythium aphanidermatum* shows progressive drying-up of the leaves of infected plants. The base of the aerial shoots shows water-soaked soft lesions. As the disease progresses infection gradually passes to the rhizomes, which begin to rot and become soft. The bright orange colour of the rhizomes changes into brown. The disease may be confined to a few isolated plants or may occur in patches (Fig 16.4). In severe attacks the yield is considerably reduced.

Figure 16.4. Rhizome rot of turmeric

(ii) Management

Use disease-free rhizomes for seed purpose. One of the effective control methods is eradication and burning of infected plants. In serious cases it may be advisable to disinfect the soil with an authorized fungicide (*e.g.* metalaxyl and mancozeb). Rotting of rhizomes can be minimized by drenching the soil with 0.3% chestnut compound or 0.3% mancozeb at 15 days interval soon after the disease is noticed. Treat the rhizomes by dipping in a mixture of 250 g mancozeb (Dithane M-45) and 100 g carbendazim (Bavistin) in 100 litres of water for 1 hr before storage as well as before sowing. Treatment of rhizomes with hot water at 51°C for 10 min and soil application of *Trichoderma harzianum* mixed with neem cake resulted in minimum incidence of rhizome rot and maximum yield.

Some cultivars (*e.g.* 'Suvarna' in India) have shown field tolerance to rhizome rot disease.

16.2.2. Wilt, *Fusarium solani*

(i) Symptoms: The pathogen is responsible for progressive drying up of leaves, which first proceeds along the margin and later the entire leaf dries up. The root system is adversely affected and is reduced to a few decaying brown roots. Infection gradually spreads to the rhizome which begins to rot and becomes soft. The bright orange colour of the rhizome is changed into different shades of brown.

(ii) Management

Drenching the soil at the root region with copper oxychloride 2.5 g/l of water or metalaxyl 1 g/l of water is effective against the disease.

16.2.3. Leaf Spot or Blotch, *Taphrina maculans*

(i) Symptoms: Leaf spot or leaf blotch, caused by *Taphrina maculans*, is characterized by the appearance of large number of spots on both surfaces of the leaves, 1–2 mm in diameter, coalescing freely. The spots first appear as pale yellow discolorations which gradually turn to reddish brown. Infected leaves are distorted, have a reddish brown appearance and soon become yellow (Fig 16.5).

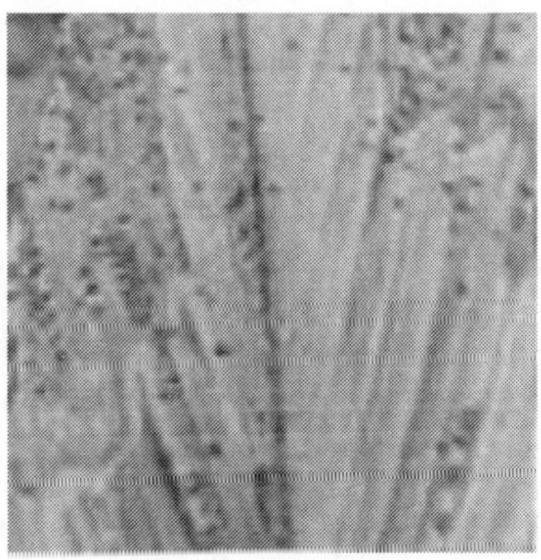

Figure 16.5. *Taphrina* leaf spot of turmeric

(ii) Management

Spraying the crop with 0.25% mancozeb or 0.3% zineb or 1% Bordeaux mixture at 10 days interval gives effective control of the disease.

16.2.4. Leaf Spot, *Colletotrichum capsici*

(i) Symptoms: Another leaf spot disease is caused by *Colletotrichum capsici*, provoking spots of variable size, enlarging to 4–5 cm × 3 cm and frequently coalescing over most of the leaf, which then dries up. Mature spots have a grayish white centre with dark brown margin surrounded by a yellow halo (Fig 16.6). The central portion of the spot becomes thin and papery. Small spots may coalesce to form irregular necrotic patches. In very severe infestation, most leaves dry up, presenting a scorched appearance, resulting in yield losses of more than 50%. Excess shade and intercropping favours the disease development.

Figure 16.6. *Colletotrichum* leaf spot on turmeric

(ii) Management

The disease can be checked by spraying Bordeaux mixture once before the symptoms appear. Captan and zineb, applied at monthly intervals, control the disease adequately. Planting materials should be selected from disease-free areas and treated with an authorized fungicide before planting. Spraying mancozeb or copper oxychloride at 3 g/l of water at 15 days interval gives effective control of the disease.

IDM

Selection of healthy planting material from disease-free gardens (soft rot, *Pythium aphanidermatum, P. myriotylum*; bacterial wilt, *Ralstonia solanacearum*; and nematodes, *Meloidogyne* spp., *Pratylenchus* spp.) is essential. Hot water treatment of the seed rhizomes at 51°C for 10 min eliminated nematodes and surface bound pathogens also. Strict phytosanitation, provision of better drainage and cultivating turmeric in raised beds are some of the important cultural operations to be adopted to reduce crop losses due to the biotic factors.

As a seed disinfection procedure, a novel technique called rhizome solarization has been devised at the Indian Institute of Spices Research, Calicut. The disinfected rhizomes when treated with *T. harzianum* and rhizobacterial strain consortia as seed treatment and soil application resulted in higher yields and growth promotion and soft rot suppression and minimal bacterial wilt. In addition, in combination with *Glomus* spp., the disease was absent probably through growth mediation. *T. harzianum* in combination with *Pseudomonas fluorescens* showed a synergistic effect in reducing the soft rot infection.

Application of *T. harzianum, T. viride, T. hamatum* and *T. virens* as seed treatment and soil application along with neem cake at 100 kg/ha gave protection and increased yield. Application of *T. viride* along with saw dust/neem cake was found to be highly effective in suppression of rhizome rot. There was synergistic effect in protection where soil solarization followed by biocontrol was adopted. *Trichoderma* spp. and *T. viriens* were found to be highly effective in suppressing *Fusarium* yellows and *Pratylenchus coffeae.*

16.3. NEMATODES

Of the many plant parasitic nematodes reported in association with turmeric, the root-knot nematode, *Meloidogyne* spp., the burrowing nematode, *Radopholus similis* and the lesion nematode, *Pratylenchus coffeae* are important.

16.3.1. Root-knot Nematodes, *Meloidogyne* spp.

M. incognita and *M. javanica* have been reported on turmeric of which *M. incognita* is more important. In pot experiments, 100,000 nematodes/plant resulted in 76.6% reduction in the rhizome weight after 6 months (Sukumaran *et al.*, 1986).

(i) Symptoms: Turmeric plants infested with *M. incognita* exhibit stunting, yellowing, reduced tillering and marginal and tip drying of leaves. Galling and rotting of roots can also be noticed. Infested rhizomes have brown, water soaked areas in the outer tissues and lose their bright yellow colour (Mani *et al.*, 1987). High populations of *M. incognita* in field cause stunting, yellowing and withering of plants in large patches. Premature death of plants takes place leaving a poor crop stand.

(ii) Management

(a) Physical Methods: Use of nematode-free rhizomes for fresh planting, washing seed material to free of soil and drying in shade before planting helps to reduce the inoculum. Hot water treatment of rhizomes at 50–55°C for 10 min or 45°C for 30 min eliminates nematode infection. Soil solarization of beds for 40 days during summer reduces nematode population in soil.

Rabbing was most effective for the management of root-knot nematodes on turmeric which gave maximum plant height and rhizome yield with lowest root-knot index (Patel *et al.*, 2001).

Summer ploughing at an interval of 15 days and use of hot water treated turmeric rhizomes as seed proved promising in reducing the nematode population and in increasing the yield up to 65%.

(b) Cultural Methods: Crop rotation with cereals and millets should be practiced, preferably with rice at least once in 3 years. Deep summer ploughing with furrow turner ploughs should be practiced. Application of well decomposed FYM/compost at 25–30 t/ha or neem cake at 2 t/ha and mulching with green leaves at 10–12 t/ha at planting and repeating the mulching during the growth period helps in reducing nematode multiplication (Kaur, 1987).

(c) Chemical Methods: Application of phenamiphos at 2.5 kg *a.i.*/ha one day before planting gave yield increase of 59–187% (Patel *et al.*, 1982). Aldicarb and carbofuran at 1 kg *a.i.*/ha increased yield by 71% and 68%, respectively over control with a cost: benefit ratio of 1 : 6 in aldicarb and 1 : 2 in carbofuran treatments (Gunasekharan *et al.*, 1987). Carbofuran at 4 kg *a.i.*/ha applied in rows to a four-month old turmeric crop has resulted in 81.6% reduction in root-knot nematode population (Mani *et al.*, 1987).

(d) Biological Methods: Complete suppression of nematodes could be achieved by the application of *Fusarium oxysporum* (isolate IISR-11) at 50 g/bed of 3 sq. m. at the time of planting. *Paecilomyces lilacinus, Pochonia chlamydosporia* and *Aspergillus nidulans* suppressed root-knot nematode population.

Turmeric rhizome treatment with *P. fluorescens* at 10 g/kg increased plant growth characters (58 and 31% increase in pseudo stem height and number of tillers over control) and decreased nematode population (40%) and gall indices (25%) (Seenivasan *et al.*, 2001).

(e) Host Resistance: Turmeric *cvs.* Kodur, Chayapasupu, Duggirala, Guntur-1, Guntur-9, Rajampet, Sugandham and Appalapadu are resistant to *M. incognita* (Gunasekharan *et al.*, 1987).

(f) Integrated Methods: Soil application of *T. harzianum* + neem cake at 1 t/ha is recommended.

16.3.2. Burrowing Nematode, *Radopholus similis*

Koshy and Sosamma (1975) reported turmeric as a host of *R. similis*. An initial inoculum level of 10 nematodes/plant resulted in 35% reduction in rhizome weight after four months. A reduction of 65 and 76% in rhizome weight was observed after 4 and 8 months, respectively with an inoculum level of 100,000 nematodes (Sosamma *et al.*, 1979).

(i) Symptoms: Drying of leaf tips and margins is conspicuous. Infested rhizomes have shallow, water-soaked brownish areas on the surface. The golden yellow colour of healthy rhizomes change to yolk-yellow colour on infestation. Infested plants age and dry faster than healthy plants. Rotting of roots takes place and the decayed roots remain devoid of cortex and stellar portions (Sosamma *et al.*, 1979).

(ii) Survival and Spread: The nematodes are spread through infested planting material. Populations of *R. similis* from coconut are known to infest turmeric and the use of turmeric as an intercrop in *R. similis* infested coconut and areca nut based farming systems should be avoided.

(iii) Management

(a) Cultural Methods: Avoiding use of turmeric as an intercrop in *R. similis* infested coconut and areca nut based farming systems and use of only clean, nematode-free rhizomes for fresh planting helps in reducing the nematode population.

16.3.3. Lesion Nematode, *Pratylenchus coffeae*

(i) Symptoms: *P. coffeae* has been reported to cause discolouration and rotting of mature rhizomes of wild turmeric, *Curcuma aromatica*. Affected rhizomes show dark brown necrotic lesions internally. The rhizomes in advanced stages of infection become red to dark brown in colour. Rhizomes become less turgid and wrinkled with dry rot symptoms. Fingers are more affected than mother rhizomes (Sarma *et al*., 1974).

17 CHILLI

Chilli (*Capsicum annuum*) is valued for its diverse commercial uses. India is a major producer, exporter and consumer of chilli. Indian chillies reach over 90 countries in the world. Bangladesh, Bahrain, Canada, Sri Lanka, Saudi Arabia, USA and UAE are the leading importing countries. In India, chilli is grown in almost all states. It is being cultivated in 0.834 million hectares producing 0.847 million tonnes of chillies with an average yield of 1.0 tonne per hectare. Andhra Pradesh has been the largest chilli-growing state followed by Karnataka and Maharashtra. Productivity of chilli is highest in Andhra Pradesh, followed by Arunachal Pradesh and Punjab. The productivity of chilli in Andhra Pradesh and Punjab is higher since the crop is raised under irrigated conditions compared with Maharashtra and Karnataka. Chilli is rich in vitamins A and C.

17.1. PESTS

Chilli thrips (*Scirtothrips dorsalis),* yellow mite (*Polyphagotarsonemus latus)* and fruit borers, *Helicoverpa armigera* and *Spodoptera litura* are the major pests of these crops. The thrips and mite damage often occur at the same leaf and is called churda murda complex.

17.1.1. Thrips, *Scirtothrips dorsalis*

(i) Damage: Both adults and nymphs suck the sap from young leaves. Affected leaves curl upwards along the margin, and get crinkled and reduced in size (Fig 17.1). When the incidence is severe, leaves drop and cause heavy reduction in yield. The pest is serious during dry monsoon periods and summer months.

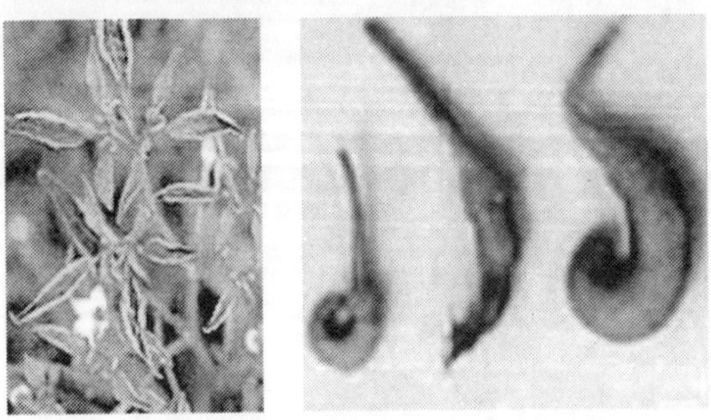

Figure 17.1. Thrips on leaves and fruits of chilli

(ii) Management

- Spray acephate 75 SP at 1.5 g/l or fipronil at 1 ml/l or lamda cyhalothrin 5 EC at 0.75 ml/l or imidachloprid 200 SL at 0.3 ml/l alternately at fortnightly interval. Never repeat the same chemical again and again.
- Spray of difenthiuron at 225 g *a.i.*/ha and monocrotophos at 150 g *a.i.*/ha effectively checked the population of thrips on chilli with higher yield.
- Mix acephate 0.5 g, 2 ml of pongamia oil and 1 ml sticker and make an emulsion (add a little water and shake thoroughly in a bottle) and make the volume to 1 litre and spray.
- Spraying of Vertimec at 0.3 to 0.4 ml/l was found effective against thrips and mites for 45 days.
- Soil application of aldicarb at 2.8 kg *a.i.*/ha 15 days after transplanting effectively checked thrips and mites and resulted in minimum leaf curl incidence with higher yield of marketable green chilli fruits (Patel *et al.*, 1983).
- Soil application of carbofuran at 0.5 kg *a.i.*/ha 15 days after transplanting followed by spray of 0.1% triazophos commencing from 40 days after transplanting at 10 days interval effectively controlled thrips and increased marketable green chilli fruits (Patel *et al.*, 1997).
- Chilli *cvs.* NS 1101, NS 1701(Namdhari seeds), NP 46A, Pusa Jwala, Arka Lohit and K 2 (Kovilpatti) are tolerant to thrips.
- *Chrysoperla carnea* can be released at 2 grubs/plant.

Pongamia oil 2 ml when mixed with 0.5 g acephate +2 ml sticker gave excellent control of thrips, *Scirtothrips dorsalis* in chillies, even during summer (Krishna Murthy & Krishna Kumar, 2002). The thrips damage rating was reduced to 20.67 in the above treatment as compared to 45.67 in control. Similarly, the yield was also increased in treated plots over control (Table 17.1).

Table 17.1. Thrips damage and yield under different management programmes in chilli (Summer, 2002)

Treatments	Doses	Rating of thrips damage/5 leaves	Yield kg/plot
Acepahte +Pongamia oil	0.5 g + 2 ml	20.67	1.69
Dimethoate + neem oil	2 ml + 2 ml	30.00	1.26
Pongamia oil	20 ml	45.67	1.05
Neem oil	20 ml	40.67	3.50
Dimethoate	2 ml	25.00	0.90
Neem seed powder	4%	40.00	0.05
Acephate	0.75 g	10.67	1.35
Control	—	45.67	0.02
CD 5%	—	**7.33**	**0.10**

17.1.2. Pod Borer, *Helicoverpa armigera*

(i) Damage: Tomato fruit borer, *H. armigera* is also a serious pest in chilli and capsicum. It affects the marketable value of the chilli crop to a great extent, if proper care is not taken (Fig 17.2).

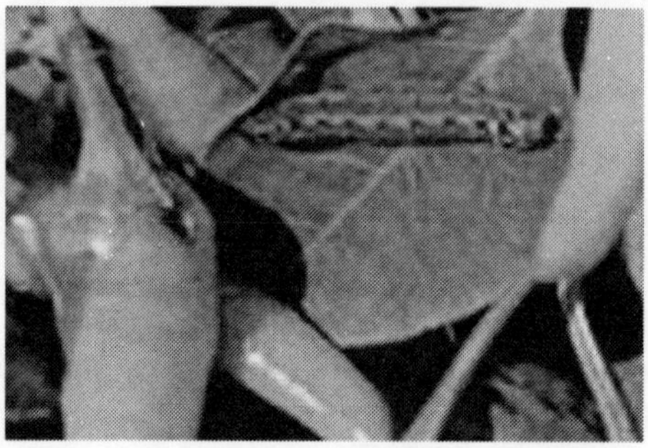

Figure 17.2. Pod borer, *Helicoverpa armigera* damage on chilli fruit

(ii) Management

- Two rounds of sprays of *Ha* NPV at 250 LE/ha (6×10^{12} PIB/ha) at 10 days interval were found effective in reducing the loss caused by borers.
- Inundative release of *Trichogramma* sp. -6 releases of *Trichogramma brasiliensis, T. pretiosum or T. chelonis* at 50,000/ha at weekly intervals.
- Use marigold as trap crop (one row of marigold for every 18 rows of chilli) for managing *H. armigera*.
- Installation of pheromone traps at 5/ha.
- Hand collection of larvae on main and trap crops.
- Spraying with 5% NSKE or *Bt* at 500 g/ha.
- Spraying 0.07% endosulfan or monocrotophos or 0.15% carbaryl or 0.05% chlorpyriphos or quinalphos is effective for the management of the pest.

17.1.3. Aphid, *Myzus persicae*

(i) Damage: The aphid also becomes a serious pest on chilli occasionally. The infestation may be noticed in a few pockets of a chilli crop. Aphids suck plant sap and cause reduction in plant vigour. Small green insects are clobbered underneath the leaf surface (Fig 17.3). Sooty mold appears on leaves due to secretion of honeydew by aphids. Feeding of ants on aphids may be observed.

(ii) Management

Sprays of systemic insecticide or 1% neem or pongamia soap gives good control of this pest.

Parasitoids, *Aphidius platensis* and *Aphelinus* sp. produced 80–98.8% parasitism in an insecticide-free field (Mani & Krishnamoorthy, 1994). The predators, *Chrysoperla carnea* and *Mallada boninensis*, very effectively controlled aphids on chilli when releases were made at 1: 5 ratio of predator: prey (Mani & Krishnamoorthy, 1997a).

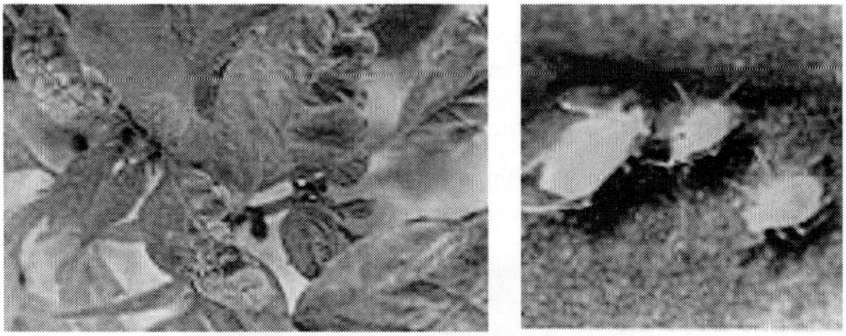

Figure 17.3. Aphid infestation on chilli leaf

17.1.4. Tobacco Worm, *Spodoptera litura*

(i) Damage: Feeding holes on the leaves can be observed due to tobacco worm infestation. Larval entry holes can also be observed near the calyx. The pest infestation causes dropping of young fruits. Mature pods with seed damage, fecal pellets with seed and watery cavity can be seen.

(ii) Management

Collect and destroy egg masses and immature larvae of *S. litura*. Use poison baiting (10 kg rice flour + 1 kg of jaggery + 250 g of methomyl 40 SP) for *S. litura*. Repeat the baiting 2–3 times, if necessary. For *S. exigua,* spray indoxacarb 14.5 SC at 0.75 ml/l or spinosad at 0.75 ml/l or thiodicarb 75 WP at 0.75 g/l. Spraying 0.07% endosulfan or monocrotophos or 0.15% carbaryl or 0.05% chlorpyriphos or quinalphos is effective for the management of the pest.

Two rounds of sprays of *Sl* NPV at 250 LE/ha (6×10^9 PIB/ha) at 10 days interval were found effective in reducing the loss caused by *S. litura*. It can also be controlled by installation of pheromone traps at 5/ha (Lures should be replaced once in 15 days), hand collection of egg masses and early instar larvae, release of *Telenomus remus* at 50,000/ha, installation of bird perches at 50/ha and spraying with 4% NSKE.

17.1.5. Leaf Curl Complex

(i) Symptoms: Chilli leaf curl complex is mainly caused by yellow mite and thrips infestation. It is a major problem on the chilli crop. Sometimes, the Gemini virus also causes enation leaf curl symptom. Reduction in size of leaves, shortening of veins, puckering, sometimes mottling and stunting of plants are common symptoms of the disease. Veins are prominently visible on the lower side of the leaves. The infected plants remain stunted and bear no fruits.

(ii) Management

Management of the viruses as well as thrips and mites involves periodical spraying of insecticides and acaricides. Alternate sprays of dicofol at 2.5 ml/litre of water followed by 0.25% wettable sulphur may be taken up to prevent the development of resistance. Two sprays of systemic insecticide are essential for the control of thrips and mite vectors.

17.1.6. White Grub, *Holotrichia consanguinea*

(i) Damage: White grubs cause wilting of plants. The plants can be easily pulled out. C-shaped white grubs are found near root region (Fig 17.4) and cause damage to roots.

Figure 17.4. White grubs on capsicum

(ii) Management

Only well rotten farm yard manure should used to prevent root grubs. Collection of beetles from plants during night hours and killing them in kerosene water reduced the infestation. Application of neem cake at 250 kg/ha along with FYM is effective against the pest. In severe cases, application of phorate at 2.5 kg *a.i.*/ha is effective. Fungal and bacterial pathogens, *Metarrhizium anisopliae* and *Bacillus popillae* var. *holotrichiae* are also effective in reducing the infestation.

17.1.7. Yellow Mite, *Polyphagotarsonemus latus*

(i) Damage: These infest terminal young leaves like thrips. Mite affected leaves curl downwards and appear twisted (Fig 17.5). Elongation of petioles is another symptom of white mite infestation. Damage is high in hot and humid seasons.

(ii) Management

- Apply wettable sulphur 80 WP at 3 g/l or any acaricide (directing the spray on the ventral surface of leaves).
- Spraying of pongamia oil mixed with insecticides given under thrips management will also reduce mites.
- Spray neem seed powder extract 4% at 10 days interval is effective when the pest incidence is low. As and when the pest incidence increases give synthetic acaricide spray.

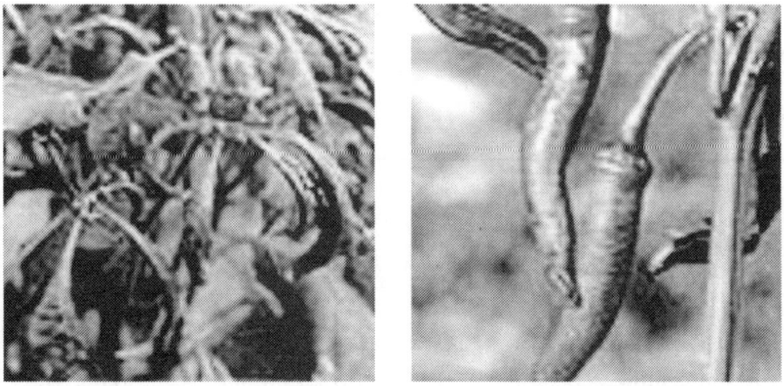

Figure 17.5. Yellow mites on leaves and fruits of chilli

- Spraying of Vertimec at 0.3 to 0.4 ml/l was found effective against thrips and mites for 45 days.
- *Chrysoperla carnea* can be released at 2 grubs/plant.

IPM

The following IPM can be adopted for managing chilli insect pests:

- In nursery, spray a systemic insecticide like imidacloprid (0.3 ml/l) or acephate (0.75 ml/l).
- Dip the roots in a systemic insecticide like imidacloprid (0.3 ml/l) or acephate (0.75 ml/l) for 5 min just before transplant.
- Apply 250 kg/ha neem cake to soil while preparing land.
- Spray Acephate 75 SP at 1.5 g/l or fipronil at 1 ml/l or lamda cyhalothrin 5 EC at 0.75 ml/l or imidacloprid 200 SL at 0.3 ml/l alternately at fortnightly interval. Never repeat the same chemical again and again.
- Mix acephate 0.5 g, 2 ml of pongamia oil and 1 ml sticker and make an emulsion (add a little water and shake thoroughly in a bottle) and make the volume to 1 litre and spray.
- In endemic areas where *Helicoverpa* is the major problem, use marigold as trap crop and take 3–4 NPV sprays at flowering
- In endemic area of *Spodoptera*, give 3–4 sprays of species specific NPV (either *S. litura* or *S. exigua).*
- Monitor for *Spodoptera* eggs and leaf damage. Mechanically collect and destroy eggs and gregarious larvae.
- Give poison baits to grown up *Spodoptera* larvae.

Pesticide Residues

The persistence of terminal pesticide residues and waiting periods for chilli are presented in Table 17.2. (Awasthi, 2001).

Table 17.2. Terminal pesticide residues and their waiting periods for chilli

Pesticide schedule	Dosage Concentration (%)	Terminal residues (ppm/days)	Waiting period (days)
Carbaryl spray at fruiting stage	0.10	0.14/10	2
Monocrotophos spray at fruiting stage	0.05	0.21/7	7
Quinalphos spray at fruiting stage	0.05	0.20/15	15
Cypermethrin spray at fruiting stage	0.01	0.06/10	3
Fenvalerate spray at fruiting stage	0.01	0.02/15	5
Carbofuran soil application at planting	1.0 kg *a.i.*/ha	0.05/20	60

17.2. DISEASES

17.2.1. Damping-Off, *Pythium aphanidermatum*

(i) Symptoms: The pathogen is responsible for death of seedlings in nursery beds.

(ii) Management

Soil solarization using polythene sheet for 30 days during peak summer was most effective and resulted in least occurrence of damping off (1.28%) when compared to control (30.64%). Seed treatment with *Trichoderma viride* resulted in lowest incidence of damping off (4.7%).

Delivery of *T. harzianum* through solid matrix priming reduced the incidence of damping off in nursery. SMP is a process in which moistened seeds are mixed with an organic carrier and the moisture content is brought to a level just below that required for seed sprouting. SMP can further enhance the efficacy of the antagonist in treated seeds. *Trichoderma* spp. grow on seed surface during priming process and increase in numbers and other microbes may not easily dislodge it.

Integration of soil solarization, application of neem cake and *T. viride* gives effective control of the disease.

17.2.2. Phomopsis Blight and Soft Rot, *Phomopsis vexans, Diaporthe vexans*

(i) Symptoms: Infected seeds may cause damping-off in the nursery. Leaves of seedlings show clearly defined circular, light brown spots mostly on lower leaves. Centre of the spots is slightly lighter in colour. Old spots are having numerous black dot-like pycnidia. These spots are papery, often crack and sometimes with shot holes. Almost similar symptoms are observed on the leaves of transplanted crop in the main field. Sometimes irregular big lesions also are observed on leaves, particularly in cool and humid weather. Stem lesions start from basal part of nodal portion as grey dry rot with constriction. This leads to drying of few twigs or partial wilting of the portion of the plant. Pale to light brown sunken spots develop on the old fruits. Individual spots expand and coalesce to cover the entire fruit which leads to fruit rot.

(ii) Management

Deep summer ploughing, seed treatment in hot water at 50°C for 30 min, crop rotation with non-solanaceous crops and collecting and destruction of crop debris shall keep the disease under check. Seeds should be collected from healthy fruits and disease-free field. Seed treatment with 0.25% carbendazim is essential to eliminate any seed inoculum. At least 3-year crop rotation is required with non-host crops to manage the disease. Protective sprays of 0.1% carbendazim at 10 days interval check the disease initiation in the field. Use tolerant varieties like Ramnagar Giant, Pusa Hybrid-5, KS-24 and Green Long.

17.2.3. Wilt, *Fusarium oxysporum*

(i) Symptoms: Symptoms include leaf chlorosis, vascular discoloration, and wilting of chilli pepper plants. High temperature and high moisture are conducive to symptom development.

(ii) Management

Reducing saturated conditions in soil is an important step in managing the disease. If economically justified, soil fumigants and solarization may be used to reduce pathogen populations in soil.

Seedling root-dip treatment and soil application of non-pathogenic *Fusarium* (Fo 52) significantly suppressed the wilt incidence. The combination of Fo 52 with *P. fluorescens* gave maximum inhibition of wilt.

17.2.4. *Phytophthora* Root Rot, Fruit Rot and Leaf Blight, *Phytophthora capsici*

(i) Symptoms: *Phytophthora* root rot is caused by the fungal organism *Phytophthora capsici*. Infection of the root and lower portion of the stem (Fig 17.6) leads to plant wilting, which is the most conspicuous symptom of the disease. In furrow-irrigated fields, there is a row-delimited pattern of wilted plants (Fig 17.6). Over time, wilting is displayed over the entire field. *Phytophthora capsici* produces specialized swimming cells, known as zoospores, which enable spread of the pathogen from plant to plant.

Figure 17. 6. Wilting of plant and a field severely affected by *Phytophthora* root rot (Courtesy: D. Lindsey)

This is a disease of rainy season, characterized by small, water-soaked spots appearing on fruits leading to complete rot of fruits. On leaves, water-soaked bleached spots appear, resulting in blighting. It causes damage to plants and affect fruit yield.

(ii) Management

Management of *P. capsici* consists of reducing soil water saturation; reducing populations of the pathogen through crop rotation, soil solarization, and application of organic amendments; and restricting pathogen dispersal through mulching and judicious use of fungicides such as metalaxyl and mefenoxam (Ristaino & Johnston, 1999).

Use of disease-free seeds for raising seedlings, collection and destruction of rotten fruits and soil application of neem cake + *Trichoderma viride* reduce the disease incidence.

Treat seeds with chlorothalanil or captan or metalaxyl at 2 g/litre. Spray metalaxyl or mancozeb at 0.2% just before onset of monsoon and 1% Bordeaux mixture or copper oxychloride at 0.3% at 8–10 days interval. Spray of cymoxanil, propineb mancozeb or chlorothalanil at 0.2% is also helpful to fruits.

17.2.5. Anthracnose or Dieback, *Colletotrichum dematium*

(i) Symptoms: This is characterized by small grey spots on leaves. It causes necrosis of twigs, lower branches or the entire top of the plant may wither away. Black dots (acervuli) are formed all over the necrotic surface of the affected twig. Later dark brown to black sunken spots having pink spore masses appear on fruits (Fig 17.7). It may cause losses up to 25% under suitable weather conditions. Fruit rot extends to the seed cavity making it internally seed-borne.

(ii) Management

Disease-free seeds should be collected from healthy fruits. Avoid apical damage during transplanting and also in the main field particularly at flowering stage. Well drained and porus soil reduces the disease incidence. Pick rotten fruits and destroy.

Figure 17.7. Anthracnose on chilli fruit

Treat the seed with 0.25% carbendazim or thiram 75 WP at 2 g/kg seed. Seedlings should be sprayed with 0.1% carbendazim before transplanting. Spray propineb or mancozeb at 0.25% or carbendazim or thiophanate methyl at 0.1% or 0.3% copper oxychloride at fortnightly intervals in the main field. Collect diseased plants and burn. To control post-harvest infection, spraying with benomyl/thiophanate methyl at 1 g/litre helps in reducing the disease incidence.

If it is a seed crop, spraying with metalaxyl mancozeb or cymoxanil mancozeb or Al-Fosetyl at 0.2% at fortnightly intervals gives very good control.

Cv. Arka Lohit is tolerant to anthracnose disease.

17.2.6. Leaf Blight, *Alternaria alternata, Cercospora capsici*

(i) Symptoms: Symptoms of *Cercospora* leaf blight are observed as circular spots with light grey to white grey centres with dark brown margins (Fig 17.8). In severe cases, small spots coalesce and lead to defoliation. *Alternaria* leaf spots are covered with a dark grayish brown to black spores. In moist condition, sporulation of both the pathogens was observed on the spots. Disease leads to severe defoliation and affects flowering and fruiting in infected plants.

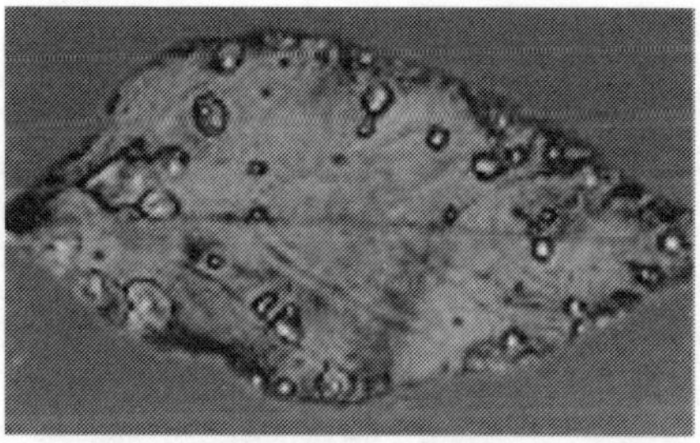

Figure 17.8. Cercospora leaf spot on chilli leaf

(ii) Management

Selection of disease-free seed is important to check the primary infection. Plant vigour should be maintained by spraying 0.2% tricel after fourth picking of green chillies. Foliar spray of 0.2% chlorothalanil is a must to protect the crop from the disease.

17.2.7. Powdery Mildew, *Levillula taurica*

(i) Symptoms: Powdery mildew infection causes whitish powdery patches both on upper and lower leaf surface (Fig 17.9).

Figure 17.9. Powdery mildew on chilli leaf

(ii) Management

Baycor at 0.2%, Devisulphur at 0.3% and Bayleton at 0.05% have been reported to be very effective for the management of this disease (Keshwal & Choubay, 1983).

17.2.8. *Verticillium* Wilt, *Verticillium dahliae* and *V. albo-atrum*

(i) Symptoms: Verticillium wilt of chilli is caused by *Verticillium dahliae* and *V. albo-atrum*. Visual symptoms of the disease include stunting, defoliation, and wilting, with discoloration of the vascular system. In the field, the disease first appears in a single plant or a small group of wilting plants (foci) scattered throughout the field (Fig 17.10), depending on the distribution of the pathogen in soil. The disease gradually spreads from these foci, and this expansion may cover the entire area of plantings. Infected plants may exhibit physiological changes such as reduction in photosynthesis, increased plant transpiration, and respiration.

Figure 17.10. Early stage of *Verticillium* wilt, and scattered plants of chilli affected by *Verticillium* wilt (Courtesy: D. Lindsey)

(ii) Management

Application of chloropicrin and metham-sodium fumigants, and use of organic amendments (Lazarovits *et al.*, 2000), may decrease *Verticillium* wilt through reduction of pathogen populations in soil.

17.2.9. *Rhizoctonia* Root Rot, *Rhizoctonia solani*

(i) Symptoms: *Rhizoctonia* root rot caused by *Rhizoctonia solani* generally affects seedlings. But, *R. solani* can also infect mature plants and induce root rot (Fig 17.11), which leads to wilting and death of chilli plants. *Rhizoctonia solani* persists in soils and organic debris.

(ii) Management

Management of this fungal pathogen relies on using high-quality seed treated with fungicides such as thiram or captan, and avoiding saturated soil conditions. In fields with high pathogen populations, control measures may include the use of soil fumigants and soil solarisation.

Figure 17.11. Chilli plant infected by *Rhizoctonia solani*, with a reddish-brown discoloration at the soil line level of the stem and (Courtesy: P. Bosland)

17.2.10. Wilt/Collar Rot, *Sclerotium rolfsii*

(i) Management

Soil application of Bioderma (*Trichoderma viride*), avoiding excess irrigation and maintaining wider spacing shall reduce the disease incidence.

17.2.11. Bacterial Wilt, *Ralstonia solanacearum*

(i) Management

Cvs. Kandhari, Pungent Pride, Cherry Red, Vattal, Dark Purple, Long Red and Pant C-1 were reported resistant. Manjari and Ujwala developed by Kerala Agricultural University, Vellanikkara, are also highly resistant to bacterial wilt.

17.2.12. Bacterial Leaf Spot, *Xanthomonas campestris* pv. *vesicatoria*

Bacterial leaf spot, also called as 'Scab spot' was first observed in the USA on chillies in 1912. At present, it is prevalent in many countries (Patel, 1972) including India where it is present in almost all the chilli growing areas. Loss in chilli yield due to disease has been reported up to 16.6% (Shekhawat & Chakravarti, 1977a).

(i) Symptoms: Small brown spots appear on leaves, petiole, stem and fruit of chillies which are distinctly raised (Fig 17.12). These spots turn into greasy or black spots.

(ii) Epidemiology: Pathogen is seed-borne (Kore & Dhaliwal1981) and also survives on stray chilli plants (Shekhawat & Chakravarti, 1978b). The bacterium survives for 14 and 10 months in seed at 8 and 10–38°C, respectively. It survives for 150 days in plant debris. Disease spreads to healthy plants through wind splashed rains and to new area through infected seed and planting material.

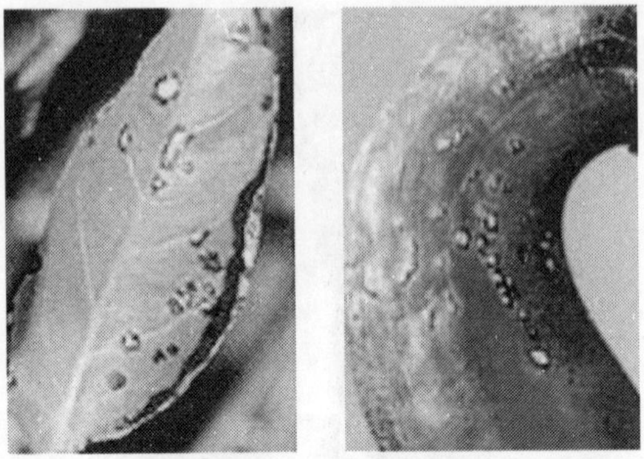

Figure 17.12. Bacterial leaf spot on chilli leaf and fruit

Maximum disease development has been found during July-October (Shekhawat & Chakravarti, 1976) and chilli plants at the age of 42–49 days are found most susceptible (Kore & Dhaliwal, 1981).

(iii) Management

Hot water treatment of seeds at 50°C for 25 min is reported to be effective. Seed treatment with mercuric chloride (1: 1000) is also recommended for the control of the disease. Spraying of 3: 3: 50 Bordeaux mixture or 0.2% zineb or 20 ppm streptocycline or 100 ppm agrimycin + copper oxychloride or copper oxychloride + streptocycline or 200 ppm paushamycin + copper oxychloride is beneficial to check the disease (Raju & Rao, 1984). Chilli *cv.* Jwala is reported to be field resistant.

17.2.13. Chilli Mosaic Virus

(i) Symptoms: The important symptoms of this disease are mosaic, mottling, distortion and shortening of leaves (Fig 17.13). The infected plants show stunted growth and in case of early infection, they do not bear fruits. However, in late infection, few fruits are formed but they are small, deformed and unfit for marketing. Sometimes, yellow or green patches or streaks are produced on the fruits.

Figure 17.13. TMV on chilli leaves

The mosaic disease is caused by many viruses in chilli/capsicum. The major ones are tobacco mosaic virus (TMV), potato virus Y (PVY), potato virus X (PVX), cucumber mosaic virus (CMV), tobacco etch virus (TEV) and pepper veinal mottle virus (PVMV).

(ii) Transmission: All the viruses are mechanically sap transmitted. TMV and PVX have no insect vectors, but other viruses *viz.*, CMV, PVMV and PVY are transmitted by different species of aphids, *Aphis craccivora, A. gossypii* and *Myzus persicae*. These aphids transmit the viruses in non-persistent manner.

(iii) Management

- Raise nursery under nylon net.
- Remove all the infected plants from the field.
- Soil application of carbofuran at 1.5 kg *a.i.*/ha at the time of sowing in the nursery followed by 1–2 foliar sprays with 0.05% nuvacron.
- Spraying of 0.05% nuvacron or 0.02% metasystox at 10 days interval to control aphid vectors.
- Spraying of 1.5 to 2.8% mineral oil at 5 to 6 days interval has been found effective in reducing the disease incidence.
- Use tolerant *cvs.* NS 1101 and 1701.

17.2.14. Leaf Curl Virus

The disease causes heavy losses in yield and quality of fruits. If the plants are infected within 20–25 days after transplanting, the loss in yield goes up to 80–90%. However, in case of later infections, the loss is comparatively less (Singh *et al.*, 1979b).

(i) Symptoms: Symptoms include abaxial and adaxial curling of the leaves accompanied by puckering and blistering of interveinal areas and thickening and swelling of the veins. In advanced stages, maxillary buds are stimulated to produce clusters of leaves which are reduced in size (Fig 17.14). The whole plant assumes bushy appearance with stunted growth. Flowers and fruits are few and small, being curled at the end. The virus has a wide host range. However, important hosts are tobacco, bell pepper and several weed hosts.

Figure 17.14. Leaf curl virus on chilli leaves

(ii) Transmission: The virus is transmitted by the whitefly, *Bemisia tabaci*.

(iii) Epidemiology: The disease spread is more during summer months which favour whitefly multiplication and disease spread.

(iv) Management

- Rouging of infected plants from the field.
- Removal of all the weeds from and around the field.
- Soil application of carbofuran at 1.5 kg *a.i.*/ha in the nursery and at the time of transplanting followed by 2–3 sprays of dimethoate or monocrotophos both at 0.05% at 10 days interval prevents whitefly multiplication and spread of the disease (Singh *et al.*, 1979b).

Integrated Management of Insect-borne Viruses

Use of nylon net cover in nursery beds, soil application of carbofuran at 1 kg *a.i.*/ha in combination with foliar application of 0.1% imidachloprid at 10 days interval effectively reduced the vector population as well as the disease incidence.

At Coimbatore, soil application of carbofuran at 1 kg *a.i.*/ha + covering the nursery bed with nylon net of 400 mesh followed by 2 sprays of 2% neem oil at 15 and 35 DAP recorded lesser virus incidence and highest yield with cost: benefit ratio of 1: 3.3.

At Sabour, use of nylon net and soil application of carbofuran at 1 kg *a.i.*/ha in seed bed and application of carbofuran at 1.5 kg *a.i.*/ha in the main field a week after transplanting + 3 sprays of nuvacron at 1 ml/litre was most effective in managing the disease with cost: benefit ratio of 1: 5.64.

IDM

- Crop rotation with French beans reduces bacterial wilt.
- Crop rotation with cereals, sesame and mustard, so also intercropping with marigold, onion, garlic reduce nematode infestation.
- Soil solarization of nursery beds for 2–3 weeks using transparent polythene sheets (100–160 gauge) kills soil borne pathogens and nematodes.
- Raising nursery in raised seed beds (10–15 cm) provides good drainage and thereby avoid damping-off.
- Seed treatment with *Trichoderma viride/T. harzianum/Pseudomonas fluorescens* at 20 g/kg of seeds to prevent damping-off and nematodes.
- Deep summer ploughing helps in exposing pathogens and their resting stages to sun's heat.
- Collection and destruction of diseased plants/plant parts from the field from time to time.
- Soil application of neem cake at 250 kg/ha as basal dose to control nematodes.
- Soil application of FYM enriched with *Trichoderma viride* at 1 MT/ha.

Table 17.3. Pest and disease management

Chemical	Time of application (days after planting)
Monocrotophos 1.5 ml + wettable sulphur 3 g + teepol 0.5 ml/litre of water	15
Monocrotophos 1.5 ml + wettable sulphur 3 g + teepol 0.5 ml/litre of water	25
Cypermethrin 0.5 ml + wettable sulphur 3 g + teepol 0.5 ml/litre of water	40
Endosulfan 2 ml + carbendazim 1 g + teepol 0.5 ml/litre of water	70
Endosulfan 2 ml + metalaxyl and mancozeb (ridomil) 2 g + teepol 0.5 ml/litre of water	80
Monocrotophos 1.5 ml + metalaxyl and mancozeb (ridomil) 2 g + teepol 0.5 ml/litre of water	90
Monocrotophos 1.5 ml + metalaxyl and mancozeb (ridomil) 2 g + teepol 0.5 ml/litre of water	110

Repeat sprays at 130 days after planting, if necessary.

17.3. NEMATODES

The root-knot nematode is the major limiting factor in successful production of chilli crop.

17.3.1. Root-knot Nematodes, *Meloidogyne* spp.

Root-knot nematodes induce much smaller galls on the roots of chillies. *M. incognita* was responsible for 24.54 to 28.00% loss in fruit yield of chillies (Singh *et al.*, 2003).

(i) Symptoms: Root galls on chilli are frequently small.

(ii) Management

(a) Cultural Methods: Decrease in *M. incognita* population on chillies occurs when the field was left fallow or following crop rotation with marigold, spinach and bottle gourd (Khan *et al.*, 1975). Growing of chilli along with marigold gave highest degree of reduction in root-knot nematode population on chilli roots, followed by onion, garlic and asparagus (Trivedi & Tiagi, 1984).

Row application of fly ash at 0.6 kg/sq.m. enhanced the fruit yield by 21.3% and inhibited the reproduction of *M. incognita* (Khan & Ghadipur, 2004).

Botanicals: Soil application of decaffeinated tea leaves effectively reduced root galling on chilli plant (Trivedi & Tiagi, 1984).

(b) Chemical Methods: Aldicarb and carbofuran at 3 per cent were effective against root-knot nematodes infecting chilli (Kandasamy & Sivakumar, 1981).

Dichlofenthion was found effective against *M. incognita* in chillies (Alam *et al.,* 1973, Saxena *et al.,* 1974).

Aldicarb, carbofuran, ethoprophos and phenamiphos each at 1 to 2 kg *a.i.* per ha were found effective in reducing the root-knot nematode population and in increasing fruit yields of chillies (Handa & Mathur, 1981).

(c) Biological Methods: Pandey and Trivedi (1992) reported a significant reduction in number of galls, eggs/egg mass, hatching and final soil population of *M. incognita* infecting chillies in the presence of *P. lilacinus*.

Avermectins effectively reduced root-knot nematode infection in chilli under field conditions (Parvatha Reddy & Nagesh, 2002).

(d) Host Resistance: Chilli *cvs.* Pusa Jwala, Wonder Hot, Teja, Utkal Abha, Utkal Rashmi were found resistant to root-knot nematodes (Patnaik *et al.*, 2004).

17.3.2. Root-knot Nematode, *Meloidogyne javanica* and *Fusarium oxysporum* Wilt Disease Complex

(i) Management

(a) Integrated Methods: *Pseudomonas aeruginosa* and *P. lilacinus* when used together significantly reduced infection of the disease complex on chilli (Perveen *et al.*, 1998).

18 MINTS

Mints (*Mentha* spp.) are a group of aromatic herbs belonging to the family Lamiaceae, which are considered to be the most important cash crops in Indo-Gangetic plains. Cultivation of mints has been done in a large scale in many tropical and sub-tropical countries of the world including India, China, Brazil, Japan and USA. Indian farmer grow it as a bonus crop on an area of more than 1,50,000 ha in Central and Northern parts of Indo-Gangetic plains. The crop fit well in the traditional food based cropping system with other crops like paddy, wheat, potato, mustard, maize, okra, carrot, onion, spinach, pigeon pea, cowpea, etc. Similarly it can be grown as an intercrop with sugarcane and some legumes. In India, the highest area under commercial cultivation of mint has been in Uttar Pradesh (Terai region), but during last few years it has spread through out the country, especially in North and North-Western states including Madhya Pradesh.

Among different oil yielding species of mint, Japanese mint (*M. arvensis*) occupies the highest area under mint cultivation followed by Peppermint (*Mentha spicata*), Spear mint (*M. piperata*), Bergamot mint (*M. citrata*) and Scotch spearmint (*M. cardiaca*).

Japanese mint essential oil is used as a source of menthol, menthyl acetate, menthone and terpenes which finds wide use in medicine, perfumes, food and cosmetics. The oil of peppermint, which contains less menthol but has a sweeter aroma and taste, is used in tooth paste, chewing gums, candies, high grade liquors, medicines and other pharmaceutical preparations. Spearmint oil having carvone and limonene, is mainly used for flavouring; whereas, Bergamot mint oil is mainly used in perfumery and flavouring due to higher content of linayl acetate and linalool.

18.1. PESTS

18.1.1. Semilooper, *Thysanoplusia orichalcea*

(i) Damage: The pest infests the crop round the year. It has overlapping generations and the life cycle is completed in 27 to 38 days (Prem Sagar & Ramji, 1991). Eggs are found on lower surface of leaves. Larvae devour the leaves and cause severe defoliation.

(ii) Management

Spraying of methyl parathion or quinalphos each at 0.05% or fenvalerate at 0.01% provides god control. NSKE 4% or 1% neem oil reduces damage by semilooper to foliage.

18.1.2. Leaf Webber, *Syngamia abruptalis*

(i) Damage: Infestation of leaf webber is severe during May-June and August-October. Larvae web leaves together and feed on them resulting in skeletonization and drying of damaged leaves. Pupation takes place inside the web. The life cycle is completed in 31 to 43 days in different months.

(ii) Management

Clipping and destruction of webs reduce webber population. Spraying of 0.005% cypermethrin followed by 0.05% quinalphos or 0.1% dichlorvos or 0.07% endosulfan gives successful control of leaf webber.

18.1.3. Aphid, *Aphis affinis*

(i) Damage: This pest is active during March-April. About 15 generations are reported on this crop (Prem Sagar & Singh, 1981). Colonies of the aphid are found on tender shoots and lower side of leaves. They suck the sap resulting in discolouration, deformation and falling of leaves. Affected shoots become stunted in growth and wither.

(ii) Management

Spraying of 4% PSKE or 1% neem or pongamia oil reduces aphid infestation. Dimethoate or oxydemeton methyl each at 0.05% also provides good control of the aphids.

18.1.4. Whitefly

(i) Damage: The initial adult population of whitefly was found building up from March and was maximum in the month of May.

(ii) Management

Among genotypes evaluated for their resistance potential to whitefly population, significantly low number of flies was observed in *cv.* 'Kushal'.

18.1.5. Leaf Roller, *Syngamia abruptalis*

(i) Damage: The caterpillar folds the leaf in the form of a roll and feed inside the leaf tissue. The edges of the leaves are held together with silk-like filaments. Leaves get rolled in August-September.

(ii) Management

Spraying of 0.1% phosphamidon or 0.07% endosulfan or thiodan at 1.5 ml/litre of water gives effective control of leaf roller. Two-three sprayings are sufficient to control the pest.

18.1.6. Red Pumpkin Beetle, *Aulocophora favicollis*

(i) Damage: Leaves and buds are eaten away

(ii) Management

Spraying 0.05% malathion twice at 20 days interval is effective for the management of beetles.

18.1.7. Hairy Caterpillar, *Diacrisia obliqua*

(i) Damage: The caterpillar starts eating the under-surface of the leaves. And the leaves are damaged

(ii) Management

Spraying of folidol or malathion each at 0.05% twice at 20 days interval is effective for the control of the pest.

18.1.8. Cutworms, *Agrotis flammatra*

(i) Damage: Pest damages the young plants at the collar region during spring season.

(ii) Management

Soil treatment with Phorate 10G before planting the crop is recommended.

18.2. DISEASES

18.2.1. Powdery Mildew, *Erysiphe cichoracearum*

This disease was first reported by Ganguli and Pandotra (1962) on Japanese mint. It causes heavy toll in April-May leading to reduction in yield by 20% due to defoliation.

(i) Symptoms: The important symptoms of the disease are; small chlorotic spots appearing on the surface and brownish discolouration on the corresponding lower surface prior to the appearance of powdery phase. Later, circular white powdery patches appear mostly on the upper portions of the surviving leaves (Fig 18.1). Heavy defoliation occurs in severe cases having a banded appearance on stem and branches. Later, these white patches turn to purplish black and produce large number of perithecia.

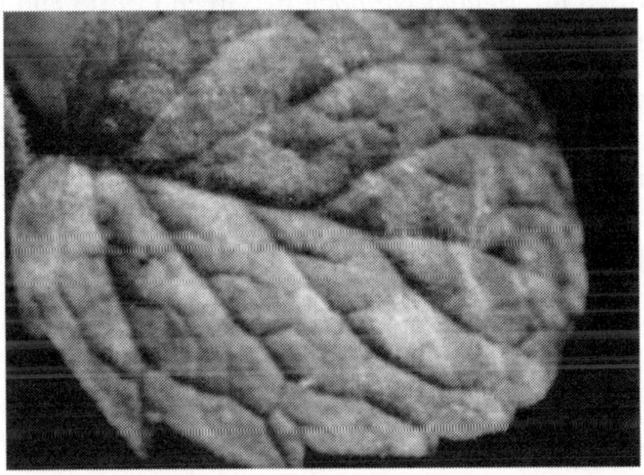

Figure 18.1. Powdery mildew on mint leaf

(ii) Management

To control powdery mildew, remove infected plants and do not water from above.

18.2.2. Wilt, *Verticillium dahliae*

(i) Symptoms: Another fungal wilt disease *Verticillium dahliae* is a major problem on mint crops. The typical symptoms of the disease are dwarfing, unilateral development of the branches, etiolation of leaves, wilting (Fig 18.2) and ultimate death of the plants. The pathogen is soil-borne and spread through soil, roots and stolons used as seed material. None of the commercial varieties are resistant.

Figure 18.2. Verticillium wilt of mentha

(ii) Management

The disease can be controlled using crop rotation and fungicides.

18.2.3. Stolon Rot, *Macrophomina phaseoli* (*Rhizoctonia bataticola*)

The disease was reported on *Mentha cordiaca* by Green (1961) and on *M. arvensis* and *M. spicata* by Hussain and Janardhan (1965).

(i) Symptoms: The important primary symptoms are yellowing of leaves and stunted growth of the plants. The wilting of the plants occurs in advance stages leading to the death of above-ground portion. Under-ground stolons exhibit pinkish brown lesions in the earlier stages and gradually turn into dark brown to black patches. These patches increase in size and finally result in the soft decay of a portion or the entire affected stolon.

(ii) Management

Avoid excess irrigation. Uproot and burn the affected plants. Apply *Trichoderma* sp. to soil.

18.2.4. *Sclorotium* Rot, *Sclorotium rolfsii*

Ganguly and Pandotra (1964) reported this disease from India.

(i) Symptoms: The characteristic symptoms of the disease are appearance of white mycelial strands around infected regions of the stem, which gradually become pale and the plant starts wilting. In later stages, sclerotial bodies develop and start girdling the stem which results in toppling of the plant at the collar region and finally the death of the infected plant.

(ii) Management

Pandotra and Shastri (1968) reported that *Mentha spicata* var. *luciniata* and *M. arvensis* var. *piperascens* were able to tolerate the disease even after infection.

18.2.5. Rust, *Puccinia menthae*

The most significant disease problem encountered with mint is a rust fungus which, if left unchecked, will totally defoliate the plants. Scotch spearmint appears to be more sensitive to rust attack than the native spearmint, but both are prone to significant oil loss if the rust is left unchecked. Ganguly and Pandotra (1962) reported this disease from India. Although this disease is known to cause severe losses in other countries where this crop is grown extensively, but under Jammu conditions, this disease appears almost at the end of the crop season and causes comparatively very little or no damage to the standing crop. This disease mainly causes defoliation which ultimately reduces the yield of mint oil.

(i) Symptoms: Brown, circular pustules on the undersides of leaves are noticed followed by leaf yellowing and defoliation. A yellow spore stage (uredospores) is in reddish blisters on young shoots in spring, and a black spore stage (teliospores) is on stems and regrowth in fall (Fig 18.3). Telial stage is characterized by swelling on the upper portion of the stem. Swollen shoots with elongated, chlorotic internodes are associated with systemic infection.

Figure 18.3. Lesions of rust on the underside of mint leaf

Puccinia menthae overwinters on mint stubble and on wild and escaped mint. It completes its complex life cycle on one host (autoecious) and is macrocyclic (five spore stages), of which the aecial stage can be systemic in plants.

(ii) Management

- Rabbing (flaming) in April 1 to May 15 is absolutely necessary for rust control.
- Early harvest (before leaves drop due to rust) may limit rust damage.
- Spray chlorothalonil or pycroclostrobin or propiconazole. Begin applications when

plants are 5 to 10 cm high or when conditions become favourable for disease. Make second application 10 to 14 days after the first application. Do not apply within 90 days of harvest.

- The disease can be controlled using crop rotation.
- Removalanddestructionofdiseaseaffectedplants.

18.2.6. Leaf Blight, *Alternaria* sp.

(i) Symptoms: The main symptoms of the disease are appearance of the round to oval or slightly irregular dark brown spots on the upper surface of the leaves with concentric zones surrounding the pale yellow margin. Later, these spots coalesce, enlarge and occupy the entire leaf or a portion of the leaf which leads to the dropping of leaves.

(ii) Management

To control leaf blight, remove infected plants and try reducing soil moisture.

18.2.7. Leaf Spot, *Curvularia lunata*

Thakur *et al.* (1974) reported this disease on Japanese mint.

(i) Symptoms: The typical symptoms appear in the form of small dirty brownish spots, scattered all over the leaves in initial stages. These spots are very minute, which increase in their size later on and form big spherical or irregular patches. In the advance stage, the spots coalesce and form large dark brown patches of irregular size. The symptoms of the disease appear at all the stages of the crop growth of leaves, but the bigger spots generally appear on the older and lower leaves. The remaining portion of the infected leaves become chlorotic and finally wither away.

(ii) Management

To control leaf spot, remove infected plants and try reducing soil moisture.

18.2.8. Begomo Virus

(i) Symptoms: Typical symptoms were foliar rugosity, light yellow mosaic, upward curling, crinkling and retarded growth leading to drastic reduction in the herb's yield (Fig 18.4). In 2007, the disease incidence was recorded to be in the range of 50 to 60%. The disease also appeared in the Terai region of Uttaranchal province. From the naturally infected plants, the disease was transmitted to healthy seedlings of *M. spicata cv.* Viridis by whiteflies (*Bemisia tabaci*), but not by aphids or mechanical inoculation. Viruliferous whiteflies did not transmit the infection to other mint species (*M. arvensis; M. citrata* or *M. piperita*) in glasshouse tests.

(ii) Management

Out of four samplings in field from March to May, the Begomo virus infection was found minimum in variety Kushal as compared to others. The maximum infestation of Begomo virus was found in cultivar Shivalik.

Figure 18.4. Symptoms of Begomo virus on mentha leaves

18.3. NEMATODES

Among the recognized species of plant parasitic nematodes, root-knot and root lesion nematodes have been considered most important.

18.3.1. Root-knot Nematodes, *Meloidogyne incognita, M. javanica, M. hapla*

Root-knot disease of Menthol/Japanese mint, Spearmint, Scotch Spearmint, Peppermint and Bergamot mint caused by *M. incognita* and *M. javanica* was observed for the first time in Lucknow and Terai region of Uttar Pradesh which reduce the herbage yield and oil content (Haseeb & Pandey, 1989b). *M. incognita* dominates over *M. javanica* in mixed infection. The quality of mint oil was also adversely affected due to nematode infection. The root-knot nematode, *M. incognita* caused 40.2% loss in herbage yield and 46.6% loss in oil yield in menthol mint (Pandey, 2001).

(i) Symptoms: Initial damage occurs at 1 juvenile/g of soil with marked reduction in chlorophyll content, rate of photosynthesis and oil yield (Pandey, 1988). Symptoms of the disease include occasional yellowing of leaves which in a month time spread to a large portion of the foliage. Growth ceases soon after yellowing. Leaves turn yellow and thin, scorching easily and eventually turning brown. Initially symptoms appear in patches as the reduced plant growth with smaller leaf size and temporary wilting under slight stress of water during hot sun. As the crop grow, especially after first harvest, symptoms become more severe and appears as stunted growth with yellowing of leaves, while veins remain green.

The below-ground symptoms are the large numbers of small sized galls with large egg masses on the roots. Under severe infection, initiation of lateral roots and rootlets on suckers is checked. As a result, uptake of nutrients is inhibited, which in turn produces deficiency symptoms on the aerial portion of the plants (Haseeb & Pandey, 1989b).

(ii) Interaction with Other Pathogens: The severity of root-knot disease of Japanese mint was increased under field conditions in the presence of *Rhizoctonia solani*. The *M. incognita-R. solani* disease complex caused severe reduction in growth and oil yield at lower inoculum densities, and at higher inoculum levels, young plants could not survive.

(iii) Management

(a) Physical Methods: Hot water treatment of *M. arvensis* suckers at 48°C for 30 min gave complete eradication of *M. incognita* without any adverse effect to suckers and roots.

(b) Cultural Methods

Nematode-free Planting Material: The nematodes which damage mints are generally endoparasitic (*M. incognita* and *P. thornei*) in nature and they can easily spread through suckers (propagating materials) from infested planting sites to uninfested planting sites. Therefore, it is essential and utmost important to use nematode-free planting material to check the spread of nematodes (Pandey *et al.*, 2003).

Crop Rotation: Cultivation of some non-host crops in rotation with Japanese mint, water logging, rice cultivation during rainy season caused reduction in the population of root-knot nematodes. Crop rotation sequences of Menthol mint with maize, potato, mustard, pigeon pea, paddy and wheat were effective in reducing the root-knot nematode population and in increasing the yield (Table 18.1).

Table 18.1. Effective crop rotation sequences for the management of root-knot nematode in Menthol mint

Crop rotation sequence	Root-knot index	Net benefit (in rupees)
Maize–potato–menthol mint	Mild	83,000
Paddy–potato–menthol mint	Mild	80,000
Maize–mustard–menthol mint	Mild	76,000
Pigeon pea–menthol mint	Mild	72,000
Paddy–menthol mint	Mild	75,000
Paddy–wheat–menthol mint	Mild	68,000

Organic Amendments: The severity of the nematode was reduced to a considerable extent in clayey soils. The high organic matter content in the soil suppressed the nematodes and enhanced the growth and oil yield. Soil application of neem/castor/mahua oil cakes at 1.0 g N/kg of soil was found effective against *M. incognita* and resulted in increase of oil yield and also plant growth parameters (Pandey *et al.*, 1992). Incorporation of neem cake 15 days before transplanting of *M. spicata* has been found the best approach for managing the population of *M. hapla* (Khan & Khanna, 1992).

The application of hydro-distillation wastes of *M. arvensis, M. piperita, Cymbopogon martini* (palmarosa) and *C. winterianus* (citronella) in the soil (1 month before transplantation) was found effective in reducing the nematode population under field conditions.

Application of different oil seed-cakes improved the growth of Japanese mint under field conditions coupled with increased oil yield and reduced root-knot and lesion nematode population. However, best results were achieved when the soil was treated with neem cake. The use of vermicompost and different distillation wastes were found to enhance the growth and yield of different mint species and reduced nematode population significantly.

Maximum reduction in *M. incognita* population in Menthol mint was recorded in neem cake treated soil followed by mustard cake and *Trichoderma harzianum*. Higher herbage yield (21 t/ha) was observed in the soil treated with oil cakes in comparison to control (14.66 t/ha). The crop produced had significantly higher oil content (2.7%) and yield (151 kg/ha) in neem cake treatment followed by mustard cake (2.7% oil and 143.6 kg/ha yield), *T. harzianum* (2.5% oil and 116.4 kg/ha yield) and *Glomus aggregatum* (2.4% oil and 112 kg/ha yield). The untreated crop gave 1.8% oil and 70.4 kg/ha yield (Pandey *et al.*, 1998) (Table 18.2).

Table 18.2. Effect of bio-organics and a pesticide on herb and oil yield of menthol mint and root-knot indices of *Meloidogyne incognita*

Treatment	Herbage yield (t/ha)	% oil content	Oil yield (kg/ha)	Root-knot indices
Untreated control	1.466	1.8	70.4	3.66
Carbofuran	1.880 (+28.4)	2.3	114.2 (+62.2)	2.33 (–34.7)
Mustard cake	1.998 (+36.3)	2.7	143.6 (+104.0)	2.00 (–45.4)
Neem cake	2.100 (+43.2)	2.7	151.0 (+114.5)	2.00 (–45.4)
Glomus aggregatum	1.758 (+19.9)	2.4	112.2 (+59.4)	2.66 (–27.3)
Trichoderma harzianum	1.750 (+19.4)	2.5	116.4 (+65.3)	2.66 (–27.3)
CD (P = 0.05)	**0.284**	**0.5**	**40.4**	**0.1**

(c) Chemical Methods: Application of carbofuran at 3 kg *a.i.*/ha increased plant growth and oil yield of *M. arvensis* and reduced the root-knot nematode population.

(d) Biological Methods: Application of arbuscular mycorrhizal fungi such as *G. aggregatum, G. fasciculatum* and *G. mosseae* improved plant growth, enhanced herbage and oil yield and effectively inhibited root-knot nematode infection (Pandey *et al.*, 1997) (Table 18.3).

(e) Host Resistance: The lowest root-knot (*M. incognita*) infection level was found on Kalka. Spear mint *cvs.* Neera and Arka were resistant to *M. incognita* (Pandey, 2003). Peppermint *cv.* Kukrail; Spearmint *cvs.* Arka, Neera and Neer-Kalka; and Bergamot mint *cv.* Kiran were found resistant to *M. incognita* (Pandey & Patra, 2001). The genotypes; MAS 35, MAS-216 and cultivar *Shivalik* were found moderately resistant to *M. incognita.*

(f) Integrated Methods: Application of carbofuran and neem cake in combination improved the growth of Japanese mint and oil yield and reduced root-knot nematode population very effectively.

Table 18.3. Effect of arbuscular mycorrhizal fungi on the productivity of mint infected with *Meloidogyne incognita*

Treatment	Fresh herbage weight (g)	% Oil yield	% Mycorrhizal colonization	Root-knot index
Untreated-Inoculated	285.0	0.38	—	3.6
Glomus aggregatum (Ga)	322.0	0.40	42.3	2.3
G. fasciculatum (Gf)	333.0	0.41	58.2	2.0
G. mosseae (Gm)	388.0	0.48	63.5	1.6
Ga + Gf + Gm	360.0	0.46	78.5	1.3

The maximum reduction in *M. incognita* population was recorded in beds treated with mustard cake along with *T. harzianum* and produced significantly higher herbage and oil yields (Pandey, 2005a).

18.3.2. Lesion Nematodes, *Pratylenchus minyus, P. thornei, P. penetrans*

Pathogenicity of *P. penetrans* on *M. piperita* has been demonstrated and reported to cause yield losses up to 34%. *M. arvensis* favours the high population build up of the nematode followed by *M. cardiaca, M. piperita* and *M. citrata.*

(i) Symptoms: High population of *P. thornei* was found to be associated with different mints and caused wilting and chlorosis of the plants (Haseeb, 1992). The lesion nematode causes light brown to dark coloured lesions on the suckers and roots. In severe cases whole underground portion becomes black and rotting of cortical portion takes place, first due to the nematode infection and later because of the invasion of secondary pathogens on necrotic cells. The other symptoms include stunting of the plant, reduction in growth of root/suckers and herbage yield within 4–5 months after nematode infection (Bergeson & Green, 1979; Rhoades, 1984).

Reduction in foliage and root/sucker growth occurs up to 39 and 66%, respectively (Bergeson, 1963).

(ii) Interaction with Other Pathogens: *P. penetrans* and *P. minyus* increased the incidence and severity of *Verticillium albo-atrum* and *V. dahliae* f. sp. *menthae* wilt on mentha.

Bergeson (1963) reported that when *P. penetrans* infected *M. piperita* plants were inoculated with *V. albo-atrum,* diagnostic symptoms of wilt appeared approximately two weeks earlier than the plants inoculated with the fungus in the absence of nematode.

Presence of *P. miniyus* increased both incidence and severity of wilt (*V. dahliae* f. sp. *menthae*) disease, reduced the period for the development of wilt by 2 to 3 weeks, increased the reproduction of nematode and reduced the dry weight of peppermint plants up to 68% when both the pathogens were present (Faulkner & Skotland, 1965).

(iii) Management

(a) Cultural Methods: Application of neem and mustard cakes at 1.5 t/ha reduced the lesion nematode population (*P. thornei*), improved the growth and oil yield of *M. piperita, M. citrata* and *M. spicata* (Shukla & Haseeb, 1996).

(b) Chemical Methods: Mint plants treated with phenamiphos at 5.6 kg per ha gave maximum reduction of *P. scribneri* population (Rhoades, 1984). Ingham *et al.* (1988) found that ethoprophos (Mocap) at 6 kg/ha was highly effective in reducing the population of *P. penetrans* both in soil and root tissue of *M. piperita*. However, oxamyl (Vydate) at 1 kg/ha reduced the population of *Pratylenchus* spp. on the crop significantly.

(c) Host Resistance: Menthol mint *cvs.* Kosi and Kalka were resistant to *P. thornei* and had least *M. incognita* infection. Bergamot mint *cv.* Kiran and Spear mint *cv.* Neer-Kalka were having least *P. thornei* infection.

18.3.3. Reniform Nematode, *Rotylenchulus reniformis*

Haseeb (1992) reported *R. reniformis* on *M. arvensis, M. cardiaca, M. citrata, M. piperita* and *M. spicata. M. piperita* and *M. spicata* are good hosts of *R. reniformis*. The reniform nematode was found to reproduce well on *M. arvensis* and to cause significant reduction in herb and oil yields.

(i) Symptoms: *R. reniformis* was observed to cause stunted growth of mint plants, withering of branches with chlorotic leaves at Hessaraghatta near Bangalore (Khan & Parvatha Reddy, 1992).

19 CINNAMON

Cinnamon, *Cinnamomum cassia* is a thin inner bark of a cinnamon tree. The bark contains about 1% essential oil. The active principles are eugenol, cineole and cinnamaldehyde. It is used in stick form in fruit preserves. It is used ground for cakes, cookies and puddings. It is also used in spicing sauces, pickles, pulav and biryani.

19.1. PESTS

19.1.1. Psyllid, *Trioza cinnamomi* (Homoptera: Triozidae)

(i) Damage: A jumping plant louse, which is principally associated with the foliage of cinnamon leaves causing leaf galls. Rajapakse and Kulasekera (1982) reported that *T. cinnamomi* is the most important insect pest of cinnamon in Sri Lanka. Mani (1973) also reported that an unknown phyllid also causes formation of galls on cinnamon in India.

The feeding by the psyllid causes abnormal cell development and the formation of galls. The developing galls enclose the feeding psyllid.

(ii) Management

Regular pruning could control the *Trioza* infested gall-bearing shoots. Natural enemies often limit the psyllids. Quinalphos (0.05%) or dimethoate (0.06%) are also recommended to control psyllid infested leaves and shoots in India. The above insecticides should be sprayed thrice at 15 days interval during the vegetative bud stage to control the homopteran psyllid. If required, repeat the spray at 10–12 days interval.

19.1.2. Cinnamon Butterfly, *Chilasa clytia* (Lepidoptera: Papillionidae)

(i) Damage: The cinnamon butterfly is the most serious pest of cinnamon especially in younger plantations and in the nursery and is generally seen during the post monsoon period. The adults are large sized butterflies and occur in two forms. One of the forms has blackish brown wings with white spots on outer margins; the other form has black wings with bluish white markings. Fully-grown larvae are pale yellow with dark stripes on the sides and measure about 2.5 cm in length (Fig 19.1). The larvae feed on tender and slightly mature leaves; in severe cases of infestation, the entire plant is defoliated and only midribs of leaves with portions of veins are justify behind.

They may become serious insect pests in young plantations which are abundant from December to June. The butterfly lays eggs singly on the upper and lower surfaces of young leaves. The eggs are spherical, waxy looking and orange-yellow in colour. The larvae are velvety black or dark green, feed on tender and slightly mature leaves. In cases of infestation, the entire plant is defoliated and only midribs of leaves with portions of veins are left behind. The larval period (5 instars) lasts for 11–17 days. The fully-grown larvae are pale yellow with dark stripes on the sides and measures about 2.5 cm in length. The caterpillars are said to resemble bird droppings (defensive mimicry). The pupae are cylindrical, remarkably well camouflaged and fixed to a branch so as to resemble the end of a broken twig. The pupal period lasts for 11–13 days and the total life cycle 24–36 days. The moth is seen in large numbers in monsoon months and spends the dry months as pupae.

Figure 19.1. Larva and adult of Cinnamon butterfly

(ii) Management

Eggs are heavily parasitized by the egg parasitoid *Telenomus remus*. Butani (1983) reported that hand removal of larvae and pupae in light infestations and spraying quinalphos 0.05% on tender and partly mature leaves in severe infestations can control the pest.

19.1.3. Common Blue Bottle, *Graphium sarpedon* (Lepidoptera: Papililionidae)

(i) Damage: The blue bottles are common and seen all the year around in cinnamon growing areas. The eggs are laid on the under surface of the tender leaves and are completely round and smooth and yellow when first laid. Egg period lasts 5–6 days. The first instar caterpillar is very spiny and of smoke colour. They are very sluggish and first keep to the underside of the leaves feeding on them and later seem to favour the mid rib on the upper side of the cinnamon leaves. As they grew older, they become green and quite well concealed by their green camouflage. The larval stage (5 instars) lasts 29–31 days. They pupate on the underside of the leaves, stalk or small branch and the pupal period lasts for 19–20 days. The life cycle is about 59–60 days. The larvae of the Common Jay, *Graphium doson* is also reported to be feeding on cinnamon tender leaves.

19.1.4. Leaf Miner, *Conopomorpha civica*

Infestation by the leaf miner is more common during the monsoon period and generally nursery seedlings are seriously affected. The adult is a minute silvery gray moth. The larvae are pale gray initially and later become pink measuring about 10 mm in length.

(i) Symptoms

- Larvae feed on the tissues between the upper and lower epidermis of tender leaves resulting in linear mines that end in 'blister' like patches.
- The infested leaves become crinkled and the mined areas dry up leading to the formation of large holes on the leaves.

(ii) Management

Spraying 0.05% quinalphos during emergence of new flushes is effective in preventing the pest infestation.

19.1.5. Mite, *Eriophyes boisi* (Acarina: Eriophyiidae)

(i) Damage: A mite *Eriophyes* sp. has also been found to cause morphologically different galls in cinnamon leaves (Mani, 1973). Perera *et al.* (1985) reported that the mite, *Eriophyes dociersi* is also an important pest causing leaf galls in Sri Lanka, and the number of new galls found on young and medium sized leaves were greater than on old leaves. They also reported that number of new galls was found to increase rapidly after rainfall with the flushing of new leaves. These mite-infested galls can be found on the upper leaf surface and they are pinkish in color at the beginning and become green on maturity. The females of mites lay eggs singly or in clusters on the upper surface or lower surface on newly emerged tender leaves. The feeding of the nymphs on the sap stimulates the development of pale green to yellowish galls on the leaf surface.

The feeding by the mite causes abnormal cell development and the formation of galls. The developing galls enclose the feeding mite.

(ii) Management

The numbers of effective miticides for controlling eriophyid mites are limited. Two miticides that may be effective include abamectin and bifenthrin, but these pesticides need to be applied before the mite enter the leaves and cause injury.

Regular pruning could control the *Eriophyes* infested gall-bearing shoots. Application of 30 ml monocrotophos/25 litres of water at 4 days interval resulted in lowest number of galls (Rajapakse & Ratnasekera, 1997).

19.2. DISEASES

19.2.1. Leaf Spot/Blight and Dieback, *Colletotrichum gloeosporoides*

The leaf spots/blight caused by a fungus *Colletotrichum gloeosporioides* (Anandaraj & Devasahayam, 2004; Kumara, 1999) is found in almost all cinnamon growing areas in Sri Lanka mainly affecting the foliage. It appears in all stages of the plant growth from seedling stage to maturity and the entire foliage can be affected in severe infections. Kumara (1999) reported that 18% foliar damage due to this disease in Matara, Sri Lanka with no significant correlation with cinnamon yield and the disease severity. Considerably higher disease incidence was observed in lands with high shades, high planting densities, poor weeding, and improper pruning Kumara (1999a, b).

(i) Symptoms: The symptoms in young seedlings include, small brown specks on leaf laminae which, later coalesce to form irregular patches. Bhat *et al.* (1988) reported that seedlings of *Cinnamomum zeylanicum* in India affected by a leaf spot disease leading to partial drying of the plants and those seedlings yielded *C. gloeosporioides (Glomerella cingulata).* Symptoms were reproduced 7 days after inoculation of healthy seedlings.

Small, brownish, leaf spots on older leaves can be seen, especially in shaded areas. These small specks coalesce to form large necrotic blotches giving scorching appearance. Spots later become papery with dark brown margins. In other cases the central portion is shed, forming shot hole appearance. Lesions may extend from tip of the leaf or from the margins. In severe infections, lesions may be larger than the half of the leaf. In some seedlings, the infection spreads to the stem causing a dieback (Karunakaran & Nair, 1980).

In India, reddish elongated spots were observed arising from the margin and resulting in defoliation (Prakasam, 1991). This also caused shot hole symptoms in later stages. However, the fungus causing this disease has been identified as *Phytophthora capsici* (Prakasam, 1991). Pu and Chang (1999) also reported brown to black spots on *C. verum* leaves in Thaiwan. These spots later coalesced and the infected leaves were shed. The pathogen was identified as *C. gloeosporioides* (Syn. *Glomerella cingulata).*

(ii) Management

Pruning the affected branches and spraying 1% Bordeaux mixture are recommended for the control of the disease.

Copper based fungicides are being used by growers in Sri Lanka to control this disease. For other crops for this type of diseases caused by *C. gloeosporioides,* a range of fungicides such as chlorothalonil, carbendazim, mancozeb or thiphanate-methyl have been recommended at 7–10 days intervals. Reducing shade especially during nursery stages may reduce the disease severity.

19.2.2. Grey Leaf Spots/Blight, *Pestalotia* spp.

Pestalotia cinnamomi causing grey blight is also considered as one of the commonest disease of cinnamon (Anon, 1996). In India, the disease was reported to be caused by *P. palmarum* on *C. verum* causing foliar damage up to 90% (Karunakaran *et al.*, 1993). *Pestalotia furierea* causing similar leaf spot symptoms were reported from Dominican Republic and also from Pakistan.

(i) Symptoms: Small yellow brown spots appear on the cinnamon leaves, at first. Later, these spots turn grey with a border and spread to the lamina. Dark acervuli are produced in older lesions appearing as black dots in the centre. This disease can cause severe damage and defoliation

(ii) Management

Pruning the die-back affected branches and spraying with 1% Bordeaux mixture are recommended for the control of the disease.

19.2.3. Stripe Canker, *Phytophthora cinnamomi*

Phytophthora cinnamomi causing stripe canker attacks shoots and young stems of cinnamon. Rands (1922) first reported *P. cinnamomi* causing severe losses to forest trees and avocado, as causal agent of stripe canker of *C. verum*. The fungus also affect *C. campora, C. culitlawan* and *C. sintok*. The fungus *Phytophthora cinnamomi* from pineapple was also reported pathogenic on cinnamon, but with reduced virulence (Anandaraj & Devasahayam, 2004).

(i) Symptoms: Stripe canker is found on the trunks and branches, particularly of young trees of *C. verum* or more generally on *C. burmannii* in Indonesia (Mehrlich, 1934). Vertical stripes are seen on the stems with amber colour exudates at the advancing margins which harden later. Vertical stripes of dead bark are most numerous near ground level. The disease is prevalent on ill-drained soils and causes up to 42% damage. Fungus produces chlamydospores and characterized by non-papillate sporangia which can be obtained by incubating a non-sterile percolate of field soil.

(ii) Management

Improving soil drainage conditions helps to keep the disease incidence at a lower level. Phytosanitation and wound dressing with tar have also been recommended as control measures for this disease (Anandaraj & Devasahayam, 2004).

19.2.4. Pink Disease, *Corticium salmonicolor*

Pink disease caused by *Corticium salmonicolor* (syn. *C. javanicum)* has been reported on cinnamon and considered as an important disease in Sri Lanka, India and Indonesia (Weiss, 2002).

(i) Symptoms: Formation of pale pinkish white encrusted areas on stems or branches is the visible symptoms at the beginning. Later infection spreads destroying bark and finally leading to the death of the smaller shoots. Pathogen also reported to attack mango, jackfruit, custard apple and other fruit trees often grown in the vicinity of cinnamon plantations.

(ii) Management

Burning of affected pruning's and other plant parts need to be practiced to reduce the disease incidence.

19.2.5. Seedling Blight, *Diplodia* sp.

(i) Symptoms: The disease occurs on the seedlings in the nursery. The pathogen causes light brown patches on the stem, which girdles the stem resulting in mortality of seedlings.

(ii) Management

Pruning the die-back affected branches and spraying with 1% Bordeaux mixture are recommended for the control of the disease.

19.2.6. Brown Root Rot, *Phellinus lamaensis*

This disease is caused by *Phellinus lamaensis*, damages cinnamon plants to some extent. Chang (1992) reported brown root disease caused by fungus *P. noxius* on *C. camphora.*

(i) Symptoms: Affected roots turn brown. Wilting and death of the aerial portions can be observed in these plants. The fungus was found to affect adjacent shade and ornamental trees such as *Dilonix regia*, *Annona* sp. and *Prunus* sp. (Da Graca *et al.*, 1980).

20 OTHER SPICE CROPS

20.1. VANILLA, *VANILLA PLANIFOLIA, V. ANDAMANICA*

There are three main commercial preparations *viz.,* whole pod, powder and extract of natural vanilla. Vanilla flavoring in food may be achieved by adding vanilla extract or by cooking vanilla pods in the liquid preparation. A stronger aroma may be attained if the pods are split in two, exposing more of the pod's surface area to the liquid. In this case, the pods' seeds are mixed into the preparation. Natural vanilla gives a brown or yellow colour to preparations, depending on the concentration. Good quality vanilla has a strong aromatic flavour, but food with small amounts of low quality vanilla or artificial vanilla-like flavorings are far more common, since true vanilla is much more expensive.

A major use of vanilla is in flavouring ice cream. The most common flavour of ice cream is vanilla. Although vanilla is a prized flavoring agent on its own, it is also used to enhance the flavour of other substances, to which its own flavour is often complementary, such as chocolate, custard, caramel, coffee etc.

The cosmetics industry uses vanilla to make perfume. The essential oils of vanilla and vanillin are sometimes used in aromatherapy.

20.1.1. Pests

Very few serious insect pests have been recorded on vanilla in India.

20.1.1.1. Leaf Feeding Beetles and Caterpillars

(i) Damage: A few species of leaf feeding caterpillars and beetles feed on leaves and tender stems.

(ii) Management

The pests can be controlled by spraying 0.05% quinalphos.

20.1.1.2. Sucking Bug, *Halymorpha* sp.

(i) Damage: The adults and nymphs of the sucking bug infest tender shoot tips and emerging inflorescence resulting in their drying and rotting.

(ii) Management

Spraying endosulfan or quinalphos each at 0.05% on tender shoot tips and emerging inflorescence is effective for the management of the pest.

20.1.1.3. Snails and Slugs

(i) Damage: Snails and slugs feed and damage tender shoot tips and leaves especially in moist and shaded areas in the plantations.

(ii) Management

Hand picking and poison baiting with snailkill helps in preventing the pest infestation.

20.1.2. Diseases

20.1.2.1. Wilt, *Fusarium oxysporum* f. sp. *vanille*

(i) Symptoms: *Fusarium* wilt is the most serious disease of vanilla. The disease is more prevalent in younger plantation especially during the monsoon season. The infection starts at leaf axil and spreads to the inter nodal region resulting in rotting and drying of the stem above the point of infection. The fungus also causes leaf rot on the plant.

(ii) Management

Phytosanitation such as removal and destruction of dead and diseased tissues and provision of adequate drainage in the plantation is important. Spraying and drenching with 0.1% carbendazim is effective against the disease. Addition of organic amendments, which encourages growth of antagonistic organisms, also reduces the intensity of the disease in the plantation.

Soil application of compost mounds enriched with *T. harzianum* twice resulted in substantial reduction in wilt. Soil application of rhizobacterial strain consortia (IISR-147 and IISR-148) with *T. harzianum* was effective in disease suppression of both *Phyophthora meadii* and *F. o.* f. sp. *vanillae.*

20.1.2.2. *Phytophthora* Rot, *Phytophthora meadii*

(i) Symptoms: The pathogen causes rotting of beans, leaves and stems. In severe cases, all the beans in a bunch are completely rotten. The disease is more severe during the monsoon especially in shaded plantations and poorly drained soils.

(ii) Management

Phytosanitation, spraying of 1% Bordeaux mixture and soil drenching with 0.2% copper oxychloride is effective against rot caused by *Phytophthora meadii.* Soil application of compost mounds enriched with *T. harzianum* twice resulted in substantial reduction in *Phytophthora* rot. Soil application of rhizobacterial strain consortia (IISR-147 and IISR-148) with *T. harzianum* was effective in disease suppression of both *P. meadii* and *F. o.* f. sp. *vanillae.*

20.1.2.3. Root Rot, *Sclerotium rolfsii*

(i) Symptoms: Root rot is generally seen during the onset of monsoon rains. Initially, rotting of root tips occurs and later the rotting extends to the whole root system followed by yellowing and wilting of vines.

The disease also affects leaves and beans. The infected beans develop reddish brown sunken lesions initially and later the entire bean rots. In severe cases, all the beans in the bunch are covered with a thick white feathery fungal mat.

(ii) Management

Soil drenching with 0.1% carbendazim is effective for the management of the disease.

20.1.2.4. Shoot Tip Rot, *Colletotrichum vanillae*

(i) Symptoms: The disease is characterized by rotting of shoot tips and is more common during monsoon season.

(ii) Management

Soil drenching with 0.1% carbendazim is effective for the management of the disease.

20.1.2.5. Dry Rot, *Fusarium* sp, *Rhizoctonia* sp.

(i) Symptoms: Dry rot affects all parts of the plant. Shriveling of stems, roots and leaves followed by yellowing are the common symptoms of the disease. The disease is more severe during summer. Intermittent drying of stems and leaves occurs at advanced stages of the disease.

(ii) Management

For control of dry rot disease, remove diseased plants and weeds along with surrounding soil where the disease is observed, mulch the base of stem with dry leaves before and after monsoon, avoid injury to roots during cultivation and drench soil around the base of the stem with 0.5% Bordeaux mixture.

20.2. NUTMEG

Nutmeg (*Myristica fragrans*) is the dry hard wrinkled seed or pit of the nutmeg fruit. Mace is the orange red fleshy covering of the nutmeg kernel. They contain 7–14% essential oil. This essential oil contains highly toxic compound known as myristicin, used in small amounts to flavour puddings and fruit pie.

Myristicin present in nutmeg seeds could lead to delirium and deep stupor. Smaller amounts of nutmeg may cause vomiting and colic. Nutmeg has antimicrobial property. The volatile components of these spices are believed to be responsible for this property.

20.2.1. Pests

20.2.1.1. Black Scale, *Saissetia nigra*

(i) Damage: The black scale infests tender stems (pencil thick) and leaves especially in the nursery and sometimes young plants in the field. The scales are clustered together and are black, oval and dome-shaped. They feed on plant sap and severe infestation causes the shoots to wilt and dry up. The infested shoots invariably develop sooty mould cover.

(ii) Management

Spraying quinalphos at 2 ml/litre of water gives effective management of the black scale insects.

20.2.1.2. White Scale, *Pseudaulcapsis cockerelli*

(i) Damage: The white scale is grayish white, flat and shaped like a fish scale and occurs clustered together on the lower surface of the leaves especially in nursery seedlings. The pest infestation results in yellow streak and spots on the affected leaves. In severe infestations, the leaves wilt and dry.

(ii) Management

Spraying quinalphos at 2 ml/litre of water gives effective control of the white scale insects.

20.2.1.3. Shield Scale, *Protupulvinaria mangiferae*

(i) Damage: The shield scale is creamy brown and oval and occurs on tender leaves (Fig 20.1) and stems especially in nursery seedlings. The pest infestation result in wilting of leaves and shoots.

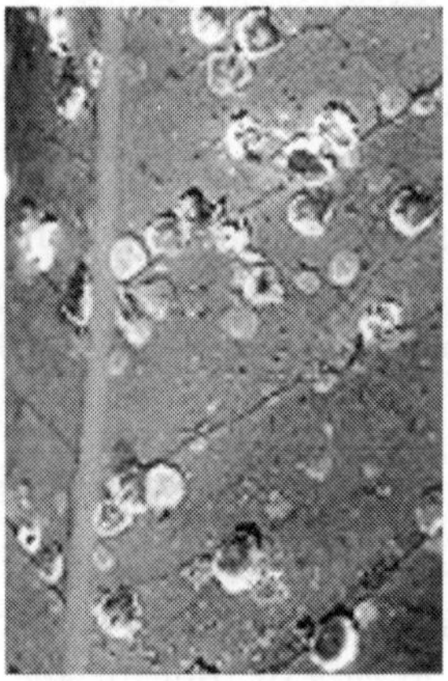

Figure 20.1. Shield scale infestation on nutmeg leaf

(ii) Management

Spraying monocrotophos at 0.05% (1.25 ml/litre of water) is effective against shield scale insects.

20.2.1.4. Stem Borer and Mealy Bug

(i) Management

Stem borer and mealy bug infestation can be controlled by spraying with solution of 0.5 ml dimethoate in 1 litre of water.

20.2.2. Diseases

20.2.2.1. Dieback, *Diplodia natalensis*

(i) Symptoms: The disease is characterized by drying up of mature and immature branches from the tip downwards.

(ii) Management

The infected branches should be cut and removed and the cut ends pasted with 1% Bordeaux mixture or 0.2% copper oxychloride.

20.2.2.2. Thread Blight, *Marasmius pulchermia, M. equicrinus*

(i) Symptoms: Two types of blights are noticed in nutmeg. The first is a white thread blight caused by *Marsimus pulchermia* wherein white hyphae aggregate to form fungal thread that traverses along the stem underneath the leaves in a fan shaped or irregular manner causing blight in the affected portions. The dried up leaves with mycelium form a major source of inoculum for the spread of the disease.

The second type of blight is called horse hair blight caused by *M. equicrinus.* Fine black silky thread of the fungus form an irregular, loose network on the stems and leaves. These strands cause blight of stems and leaves. However, these threads hold up the detached, dried leaves on the tree, giving the appearance of a bird's nest, when viewed from a distance.

Both the diseases are severe under heavy shade.

(ii) Management

These diseases can be managed by adopting phytosanitation and shade regulation. Prune the affected twigs and dried hanging leaves. In severely affected gardens, spraying 1% Bordeaux mixture or 0.2% copper oxychloride may be undertaken in addition to cultural practices.

20.2.2.3. Fruit Rot, *Phytophthora* sp., *Diplodia natalensis*

(i) Symptoms: Immature fruit split, fruit rot and fruit drop are serious in a majority of nutmeg gardens in Kerala. Immature fruit splitting and shedding are noticed in some trees without any apparent infection. In case of fruit rot, the infection starts from the pedicel as dark lesions and gradually spread to the fruit, causing brown discolouraton of the rind resulting in rotting. In advanced stages, the mace also rots emitting a foul smell. The fallen fruits become enveloped with the growth of the organism.

(ii) Management

Spraying 1% Bordeaux mixture or 0.2% copper oxychloride when the fruits are half mature helps to reduce the disease incidence.

20.2.2.4. Shot-Hole, *Colletotrichum gloeosporoides*

(i) Symptoms: Sunken spots surrounded by a yellow halo are the initial symptoms. Subsequently the central portion of the necrotic region drops off resulting in shot hole symptoms. Dieback symptoms are also observed in some of the mature branches. On young seedlings drying of the leaves and subsequent defoliation are seen. In advanced stages, the necrotic spots become brittle and fall off resulting in shot holes.

(ii) Management

The disease can be controlled by spraying 1% Bordeaux mixture or 0.2% copper oxychloride two or three times during rainy season.

20.2.2.5. Wilt, *Rosellinia* sp.

(i) Symptoms: This is the most important disease of nutmeg in Grenada. Although this disease is associated with several fungi, *Rosellinia* sp. has featured most in the association but its causal role in the disease has not been demonstrated (Muller *et al.*, 1990). Thus the cause of nutmeg wilt is still unknown. It is generally assumed that the agent is a pathogen but there may well be other predisposing factors. The symptom associated with nutmeg wilt is the gradual wilting of a plant (Fig 20.2) and dieback of the roots.

Figure 20.2. Wilt of nutmeg

20.3. CORIANDER, *CORIANDRUM SATIVUM*

Coriander (*Coriandrum sativum*) is used as a common flavouring substance. The stems, leaves and fruits have a pleasant aroma. The whole young plant is used in preparing chutney. Its leaves are used for flavouring curries, sauces and soups. Dry seeds are extensively used in preparation of curry powder, pickling spices, sauces and seasonings. The seeds contain 0.5–1% essential oil which has an active principle, an isomer of geraniol. This plant contains vitamin A. In medicines, its seeds are used as a carminative, refrigerant, diuretic, flatulence, vomiting and intestinal disorders. The decoction is used in indigestion, cold and other diseases. The oil from seeds is used in sweet meats, cocoa eatables, liquor and for agreeable smell and taste. Coriander seeds contain thalides which increase the levels of anticancer protective enzymes. The oil is also useful as an insecticide.

It is being cultivated in 0.282 million hectares producing 0.172 million tonnes of coriander seeds with an average yield of 0.6 tonnes per hectare. In India, coriander is cultivated in Andhra Pradesh, Madhya Pradesh, Karnataka, Tamil Nadu and Uttar Pradesh.

20.3.1. Pests

20.3.1.1. Aphid

(i) **Damage:** Aphid infestation affects the yield of the crop both in quality and quantity.

(ii) Management

Spraying dimethoate or monocrotophos or chlorpyriphos each at 0.1% or 0.07% endosulfan at 10 days interval is effective for the management of the pest. Pollination in coriander is brought about by honeybees. Hence, care should be taken in choosing the insecticide that does not damage honeybees or at least causes minimum damage. Endosulfan has been considered relatively safe which can be sprayed in evening hours as the activity of the honeybees remain low.

20.3.1.2. Caterpillars, *Helicoverpa armigera, Spodoptera litura*

(i) Damage: These are cosmopolitan caterpillars damage coriander by defoliation and consumption of tender buds.

(ii) Management

Spraying methyl parathion or monocrotophos or chlorpyriphos each at 0.1% at 10 days interval gives effective control of the above pests.

20.3.1.3. Cut Worm

(i) Management

Application of 4% endosulfan dust at 20–25 kg/ha or 0.1% chlorpyriphos at the time of ploughing is effective for the management of cut worms.

20.3.1.4. Mite, *Tetranychus* sp.

(i) Damage: Frequently mites attack coriander at the seed formation stage. The infestation is more severe on the young inflorescence as well as leaves. They suck the sap of leaves. The leaves of whole plant become whitish yellow and appear sickly. If the infestation is not checked timely, the yield is reduced and grains become shriveled.

(ii) Management

Mites can be effectively controlled by spraying the crop with 0.02% ethion. The control will be more effective if it is used as prophylactic measure or at least applied at the initial stage of the pest attack. It can also be controlled by spraying 0.07% dicofol or 0.03% phosphamidon.

20.3.2. Diseases

20.3.2.1. Wilt, *Fusarium* sp.

(i) Symptoms: Of late, coriander wilt is becoming a serious disease which causes significant yield loss. The leaves of infected plants will turn yellow and dries up.

(ii) Management

Follow crop rotation. Deep ploughing during summer should be taken up. Seed treatment and soil application with *T. harzianum* ware effective for the management of the disease.

Seed treatment with *P. fluorescens* at 10 g/kg gave the lowest Fusarial wilt incidence of 14.4% (36.6% in control) with an yield of 603 kg/ha (423 kg/ha in control). *T. viride* as seed treatment (4 g/kg) and soil application (5 kg/ha) reduced the Fusarial wilt incidence by 58% with an yield of 575 kg/ha compared to 433 kg/ha in control.

20.3.2.2. Powdery Mildew, *Erysiphe polygoni*

(i) Symptoms: It is the most serious disease and prevalent whenever the crop is grown. Powdery mildew causes about 50% loss in yield as it attacks flowers and seeds. The affected leaves and shoots appear flowery. Infection results in yellowing and shedding of leaves resulting in reduction of yield. The disease is common during dry weather and low temperatures (January-February). Heavily infected flowers fail to form seeds and dry up.

(ii) Management

The disease can be effectively controlled by soil application of wettable sulphur twice (once at the time of flower initiation and repeated after 15 days). The disease can be controlled by spraying with 0.1% karathane or 0.1% carbendazim at 15 days interval. The disease can also be controlled by three sprayings of 0.05% dinocap (ICBR Rs. 30.10) or 3 dustings of sulphur at 25 kg/ha (ICBR Rs. 28.32). The first application is carried after 45 days of sowing followed by 2 subsequent applications at 15 days interval.

20.3.2.3. Leaf Spot, *Alternaria alternata*

(i) Symptoms: Once this disease becomes evident, seed losses are inevitable. This disease is seed-borne and the best control is to use clean seed lines. This disease is usually transferred from crop to crop.

(ii) Management

Preventative control of this fungal disease is important. However, it has become common practice to use 2–3 kg/ha of mancozeb product at 10% flowering to control *Alternaria alternata*. Fortnightly to three weekly treatments are applied during the flowering period to protect new growth, especially where irrigation is used. Coriander *cv.* Gujarat Coriander 1 is tolerant to blight.

20.3.2.4. Bacterial Leaf Spot, *Pseudomonas syringae* pv. *coriandricola*

(i) Symptoms: This is a seed borne, bacterial disease against which there is no commercially available preventative treatment. This disease remains dormant in the plant until 'stress' is applied to growing plants (*e.g.* frost, hail damage). The pathogen causes leaf spots (Fig 20.3) and can cause total crop loss.

Figure 20.3. Bacterial leaf spot on coriander

(ii) Management

Growers should use only seed that has been tested for freedom from the disease or seed with infection levels below 0.2%. Once the disease is evident in irrigated crops, it is suggested to stop watering for 2–4 weeks in an effort to reduce the spread caused by water droplets.

20.3.3. Nematodes

The root-knot nematode is an important limiting factor in successful production of coriander crop.

20.3.3.1. Root-knot Nematode, *Meloidogyne incognita*

(i) Symptoms: The root-knot nematode has been reported to cause damage to coriander. Midha and Trivedi (1991) reported that *M. incognita* was responsible for 52% loss in seed yield of coriander.

(ii) Management

(a) Biological Methods: *P. lilacinus* was effective in increasing the plant growth parameters and in reducing root galling and egg mass production due to *M. incognita* in coriander (Sobita Devi & Pandey, 2001).

20.4. FENUGREEK, *TRIGONELLA FOENUM-GRAECUM*

Fenugreek (*Trigonella foenum-graecum*) seed is a hard lentil seed. Its colour is dark fawn and has astringent aroma. Fenugreek seeds contain 5% bitter fixed oil. The oil has strong celery-like odour. Fenugreek seeds aid in maintaining the blood glucose levels in non-insulin dependent diabetes. The fibre present in the seed may be responsible for this. It is also used with the butter milk in the treatment of dysentery.

20.4.1. Pests

20.4.1.1. Aphid

(i) Damage: Aphid sucks the sap of tender parts of plants and affects the growth, adversely.

(ii) Management

Spraying of 0.03% solution of dimethoate or 0.025% solution of methyl demeton or 0.04% solution of monocrotophos is recommended to control the aphid. If the crop is grown for green (vegetable) purpose then spray malathion.

20.4.2. Diseases

20.4.2.1. Powdery Mildew

(i) Symptoms: This disease appears usually in later stage of crop and becomes serious when pod formation takes place. In this disease, white powdery patches appear on the lower and upper surface of leaves and other parts of plant.

(ii) Management

Crop should be dusted with 300 mesh sulphur dust at 25 kg/ha to control this disease as soon as the symptoms are noticed. Spraying of wettable sulphur or dinocap (karathane or thiovit) can also be used to control the disease at 20–25 g per 10 litre of water at the initial stage of this disease. If needed two more sprays should be given at an interval of 15 days after first spray.

For control of powdery mildew in seed production of fenugreek (Methi), two sprays of propiconazole 0.1% or penconazole 0.05% or carbandazim 0.1% at 15 days interval from the appearance of disease are recommended.

20.4.2.2. Downy Mildew

(i) Symptoms: This disease occurs during February and March. Yellow patches on the upper surface of leaves appear in the infected plants and white cottony mycelium on the lower surface of leaves.

(ii) Management

This disease can be controlled by spraying of 0.2% solution of difolatan or any other copper fungicide.

20.4.2.3. Cercospora *Leaf Spot*

(i) Management

A fenugreek variety '*Rajendra Kanti*' with golden yellow seeds, is resistant to *Cercospora* leaf spot with yield potentials of 1.5–1.8 MT/ha and 40–42% higher yield over local variety. It is widely accepted by the farmers and has a greater coverage in Bihar and other states. It also builds up-soil health because of profuse nodulation.

20.4.2.4. Root Rot, *Rhizoctonia solani*

(i) Management

Soil drenching with 0.1% carbendazim commencing at the time of initiation of the disease and repeated after a month. Soil amendment with antagonistic fungi has minimized the incidence of root rot.

20.4.3. Nematodes

20.4.3.1. Root-knot Nematode, *Meloidogyme incognita*

(i) Management

Seed treatment with *T. viride* (4 g/kg) followed by soil application (5 kg/ha) along with 150 kg/ha of neem cake consistently suppressed root-knot.

20.5. CUMIN, *CUMINUM CYMINUM*

Cumin seed is pungent, sharp and astringent. It contains 2–4% essential oil. The active principle is an aldehyde cumino. It is used as a stimulant and carminative agent. Cumin seeds contain bio-active substances called pthalides which increase the levels of anticancer protective enzymes in the body.

20.5.1. Pests

20.5.1.1. Aphid, *Myzus persicae*

(i) Damage: Aphid is a major pest of cumin crop. It sucks the sap of tender parts and reduces the yield.

(ii) Management

Spraying of 0.03% solution of dimethoate or 0.025% solution of methyl demeton or 0.04% solution of monocrotophos is recommended to control the aphid.

Two sprays with monocrotophos and acephate were effective for the management of aphids in cumin. Honey bees help in pollination of cumin, hence care must be taken in choosing the insecticide that does not affect honey bee activity or at least causes minimum damage. Endosulfan at 0.07% is relatively safe to honey bees and can be sprayed in evening hours when honey bee activity is low.

20.5.1.2. Leaf Eating Caterpillar

(i) Damage: This pest causes damage to the foliage of plants reducing the yield of crop.

(ii) Management

It can be controlled by spraying of 0.02% solution of phosphamidon in the early stage of crop.

20.5.1.3. Mite

(i) Management

Mites can be effectively controlled by spraying the crop with 0.02% ethion. The systemic insecticides normally used against sucking pests might also be used for the control of mites.

20.5.2. Diseases

20.5.2.1. Fusarium Wilt, *Fusarium oxysporum* f. sp. *cumini*

(i) Symptoms: Infected plants show peculiar symptoms of drooping of tips and leaves, epinasty, leading to mortality of the entire plant. Attack of wilt is severe in younger plants. It is both soil-borne and internally seed-borne. Frequent epiphytotics are caused resulting in almost total loss of the crop. A three year cycle of poor crop due to wilt is observed in Rajasthan.

(ii) Epidemiology: Soil temperature of 54.7°F to 57.0°F in the mornings and 61.0 to 66.7°F in the evenings provide congenial conditions for incidence of wilt. Light soils favour growth of the pathogen.

(iii) Management

Crop rotation and use of neem cake are helpful in checking spread of the fungus *vis-à-vis* disease. Seeds collected from disease free plots should only be used for sowing.

Seed treatment with a 1: 1 mixture of carbendazim and captafal at 4 g/kg of seed helps to reduce the incidence of wilt disease.

T. harzianum with neem cake as soil application reduced the incidence of wilt. Seed and soil treatment with *T. harzianum* was also found significantly effective for wilt disease.

20.5.2.2. *Alternaria* Blight, *Alternaria burnsii*

(i) Symptoms: The blight affected plants show very minute brownish necrotic spots, which later turn to blackish. Mostly diseased plants fail to produce seeds. If seeds are produced they remain shriveled, light in weight and dark in colour.

(ii) Management

For the control of this disease, seed treatment and spraying of 0.2% solution of mancozeb 2 times at 15 days interval commencing after flowering is recommended. Add 1 ml soap solution/litre of water for better efficiency of fungicide. Seed treatment with ceresan at 3 g/kg of seed gives control of the disease. The crop should be kept free from weeds. Crops requiring more irrigation and mustard crop should not be grown in vicinity of this crop.

20.5.2.3. Powdery Mildew, *Erysiphe* sp.

(i) Symptoms: Affected plants in early stages show minute whitish spots on leaves, petiole, stem pedicel and seeds. In severe condition, it looks as the plants have been dusted with white powder. At later stages of attack seeds become white and shriveled and light in weight.

(ii) Management

Crop should be dusted with 300 mesh sulphur dust at 25 kg/ha to control this disease as soon as the symptoms are noticed and repeated twice at 15 days interval for the effective management of powdery mildew on cumin.

Spraying of wettable sulphur or dinocap (karathane or thiovit) can also be used at 20–25 g per 10 litres of water at the initial stage of this disease. If needed second spray should be given 15–20 day after first spray.

20.6. CLOVE, *EUGENIA CARYOPHYLLATA*

Clove (*Eugenia caryophyllata*) is the dried flower bud of the clove tree. It contains about 15% essential oil. The chief active principle is eugenol. Cloves are used whole in meats, pickling and fish. It is used ground in cakes, cookies and puddings. Used as flavouring agent in pulav, rice puttu and fruit cake. Cloves contain eugenol which has anti mutagenic effect. It is used for tooth pain.

20.6.1. Pests

20.6.1.1. Stem Borer, *Sahyadrassus malabaricus*

(i) Damage: The stem borer infests the main stem of young trees at the basal region. The larvae of the pest girdle the stem and bores downward into it. The girdled portion and bore hole are covered with a mat-like frass material. The infested trees wilt and succumb to the pest attack subsequently.

(ii) Management

Inspect the base of clove trees regularly for symptoms of pest attack. Spray 0.1% quinalphos around the bore-hole and inject the same into bore-hole after removing the frass. Swab the basal region of the main stem with 0.2% carbaryl and keep the basin free of weeds.

20.6.1.2. Scale Insects

(i) Damage: Many species of scales such as wax scale, *Ceroplastes floridensis*; shield scale, *Pulvinaria psidii*; masked scale, *Mycetaspis personata* and soft scale, *Kilifia accuminata* infest clove seedlings in the nursery and sometimes young plants in the field. The scales are generally seen clustered together on tender stems and lower surface of leaves causing wilting of shoots. The plants present a sickly appearance.

(ii) Management

Spraying 0.05% monocrotophos is effective for the management of scale insects.

20.6.2. Diseases

20.6.2.1. Seedling Wilt, *Cylindrocladium* sp., *Fusarium* sp., *Rhizoctonia* sp.

(i) Symptoms: Leaves of affected seedlings lose their natural luster, droop and ultimately die. Root system and collar region of seedling show varying degree of discolouration and decay.

(ii) Management

Since the infected plants promote further spread of the disease, they are to be removed and the remaining seedlings should be treated with 0.1% carbendazim both as spray and soil drench. Alternatively, foliar spraying with 1% Bordeaux mixture and soil drench with 0.2% copper oxychloride is also effective for the management of the disease.

20.6.2.2. Leaf Rot, *Cylindrocorpon quinqueseptatum*

(i) Symptoms: Leaf rot is noticed in mature trees and seedlings. The infection starts as dark diffuse patches at the leaf tip or margin and later the whole leaf rots, resulting in severe defoliation.

(ii) Management

The foliage of affected trees should be sprayed with 0.1% carbandazim. Prophylactic sprays with 1% Bordeaux mixture also prevent the disease.

20.6.2.3. Leaf Spot and Bud Shedding, *Colletotrichum gloeosporoides*

(i) Symptoms: The disease is characterized by dark brown spots with a yellow halo on leaves. Such spots also appear on the buds resulting in their shedding.

(ii) Management

Prophylactic spraying with 1% Bordeaux mixture or 0.2% copper oxychloride prevents the above diseases.

20.7. FENNEL, *FOENICULUM VULGARE*

The bulb, foliage, and seeds of the fennel (*Foeniculum vulgare*) plant are widely used in many of the culinary traditions of the world. Fennel pollen is the most potent form of fennel, but also the most expensive. Dried fennel seed is an aromatic, anise-flavoured spice, brown or green in colour when fresh, slowly turning a dull grey as the seed ages. For cooking, green seeds are optimal. The leaves are delicately flavoured and similar in shape to those of dill. The bulb is a crisp, hardy root vegetable and may be sautéed, stewed, braised, grilled, or eaten raw.

20.7.1. Pests

20.7.1.1. Yellow Willow Aphid, *Cavariella aegopodii*

(i) Damage: This aphid causes serious problems primarily in the coastal regions. High populations of this aphid can stunt young plants, and can also deposit honeydew, which causes quality problems in fennel.

(ii) Management

For the effective control of aphids on fennel, three sprays with methyl parathion or methyl oxydemeton each at 0.025% or DDVP at 0.05% or endosulfan at 0.07% or phosphamidon at 0.03% at 30 days interval commencing from 30 DAT should be given. Methomyl at

2.25 kg *a.i.*/ha and permethrin at 0.315 kg *a.i.*/ha were also effective in suppressing aphid populations.

20.7.1.2. Beet Army Worm, *Spodoptera exigua*

(i) Damage: This insect causes serious damage to fennel, and is most prevalent during fall production in the desert regions. Reduced stands can result from high insect populations.

(ii) Management

(a) Biological Methods: Parasites, predators, and viral diseases may assist in reducing lepidopterous egg and/or larval populations.

The most common parasites of the beet army worm are wasps (*e.g., Hyposoter exiguae* and *Chelonus insularis*), and the tachinid fly (*Lespesia archippivora*).

Predators also play a role in reducing populations of lepidopterous larvae. Two predators that feed on lepidopterous eggs are minute pirate bugs (*Orius* spp.), and bigeyed bugs (*Geocoris* spp.).

Viral diseases (*e.g.*, nuclear polyhedrosis virus) also play a role in reducing populations of lepidopterous larvae.

(b) Chemical Methods: Foliar spraying with methomyl at 2.25 kg *a.i.*/ha, permethrin at 0.315 kg *a.i.*/ha, spinosad at 0.225 kg *a.i.*/ha, tebufenozide at 0.315 kg *a.i.*/ha or thiodicarb at 1.85 kg *a.i.*/ha gives effective management of the pest.

20.7.1.3. Mite

(i) Damage: Mites sometimes attack fennel. The infestation is more severe on the young inflorescence as well as leaves. The leaves of whole plant become whitish yellow and appear sickly. If the infestation is not checked timely, the yield is reduced and grains become shriveled.

(ii) Management

Mites can be effectively controlled by spraying with 0.02% ethion. The systemic insecticides normally used against the sucking pests can also be used for the control of mites.

20.7.2. Diseases

20.7.2.1. Leaf Blight, *Cercosporidium punctum*

(i) Symptoms: Fennel Leaf blight caused by *Cercosporidium punctum,* is the major disease that affects fennel and can cause serious yield losses if not controlled. The pathogen causes heavy leaf loss and damage to the developing flowers. High humidity during flowering will promote *Cercosporidium* development, such that heavy leaf loss and damage to developing flowers and seed will be sustained.

(ii) Management

This can be managed with early preventative fungicide applications to reduce the level of inoculum. Mancozeb is currently the product used for the control of this disease.

20.7.2.2. Stem Rot, *Sclerotinia minor, S. sclerotiorum*

(i) Symptoms: Stem rot, commonly referred to as "drop," is an extremely important disease of fennel, and is very prevalent in the coastal production areas. The sclerotia of *S. minor* populations may remain viable in the soil for prolonged periods (up to 8 to 10 years). The crop losses from this disease range from 5 to 20%. Individual stalks begin to collapse and wilt. The interior of dead stems frequently reveals small, irregularly-shaped black structures called sclerotia.

(ii) Management

Since the fungus lives in the soil, a two-year rotation with non-susceptible plants, such as corn, to prevent the buildup of pathogenic organisms can reduce the incidence of disease. Foliar application of fosetyl-AL at 6 kg *a.i.*/ha, mancozeb at 4 kg *a.i.*/ha or dicloran at 5 kg *a.i.*/ha gives effective control of the disease.

20.7.2.3. Damping-Off, *Rhizoctonia solani, Pythium* spp.

(i) Symptoms: Seedlings either rot and seeds do not germinate or germinate and collapse. The disease frequently attacks the base of the seedling and leaves a shriveled, dried-up stem at the base of the seedling.

(ii) Management

Control can be achieved by using sterile potting media and fungicide-treated seed. If seedlings become symptomatic, rogue and discard all of the seedlings since asymptomatic seedlings may also be infected.

20.7.2.4. Root and Crown Rot, *Rhizoctonia solani*, *Pythium* spp., *Fusarium* spp.

(i) Symptoms: Symptoms appear as wilting and a slow or rapid collapse of the plant. The roots can appear brown and water-soaked instead of white. A water-soaked lesion can often appear at the base of the stem.

(ii) Management

Control can be achieved by using sterile potting media, clean pots, fungicide-treated seed and using a two-year rotation with non-susceptible plants, such as corn, to prevent the buildup of pathogenic organisms. Affected plants should be rouged out and discarded.

20.7.2.5. Botrytis Blight, *Botrytis cinerea*

(i) Symptoms: Botrytis blight occurs on many different kinds of herbs. The disease appears when plants are grown under conditions of high humidity. The disease appears as tan to brown spots on leaves that can be associated with a grayish mold colonizing the damaged area. Senescing flowers are particularly susceptible and frequently have papery spots on them.

(ii) Management

Botrytis blight can be suppressed by avoiding overhead irrigation and watering in the morning to allow time for leaves and flowers to dry. Proper plant spacing can improve air drainage and rapid leaf drying. Control can also be achieved with the use of fungicide (iprodione) sprays applied as soon as symptoms are visible.

20.7.2.6. Sugary Disease, *Sphacelia* sp.

(i) Symptoms: The sugary disease can be occasionally serious, completely transforming the flowers. The aphids spread the disease from plant to plant.

(ii) Management

Spraying dimecron or dimethoate both at 0.2% can effectively control this disease through aphid control.

20.8. CURRY LEAF, *MURRAYA KOENIGII*

Curry leaf (*Murraya koenigii*) is a fast-growing deciduous shrub or small tree with deep roots and scented leaves. It is cultivated all over south and southeast Asia. The leaves are an important ingredient extensively used in curry in south India and Sri Lanka. They are particularly used in south India cooking to provide a flavouring for curries, vegetable, fish and meat dishes, soups (rasams), pickles, butter milk preparations, chutneys, scrambled eggs and curry powder blends.

Curry leaf is used in South Asian traditional medicine to treat the digestive system, skin conditions and diabetes. Its anti-diabetic properties are supported by scientific research.

20.8.1. Pests

20.8.1.1. Lemon Butterfly, *Papilio demoleus*

(i) Damage: Caterpillars of lemon butterfly feed on leaves (Fig 20.4).

Figure 20.4. Larva and adult of lemon butterfly

(ii) Management

Hand pick and destroy caterpillars.

20.8.1.2. Scale Insect

(i) Damage: Scale insects affect branches of curry leaf plant (Fig 20.5).

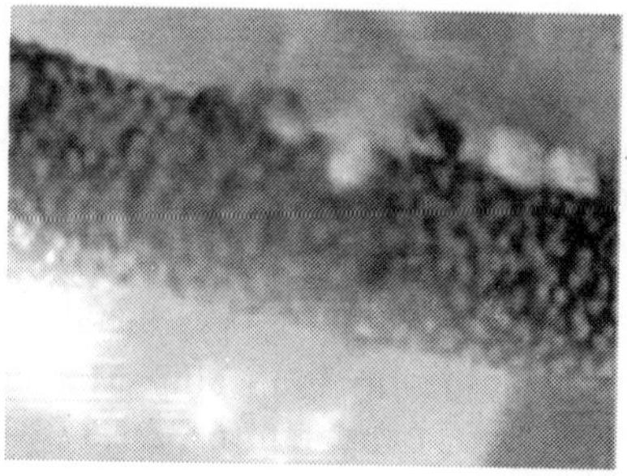

Figure 20.5. Scale infestation on stem of curry leaf plant

(ii) Management

Spraying 4% NSKE gives effective control of the pest.

20.8.2. Diseases

20.8.2.1. Leaf Spot, *Cercospora* sp.

(i) Symptoms: The pathogen causes leaf spots on curry leaves (Fig 20.6) which may drop.

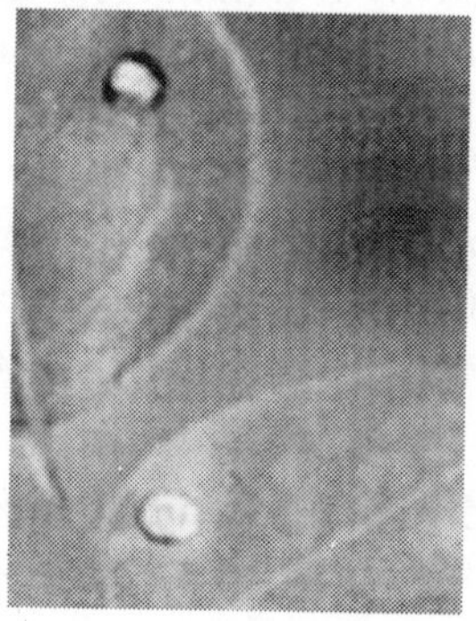

Figure 20.6. *Cercospora* leaf spots on curry leaf

(ii) Management

Spraying copper oxychloride at 0.3% or carbendazim at 0.1% gives effective management of leaf spot disease.

20.9. MUSTARD, *Brassica juncea*

20.9.1. Pests

20.9.1.1. Aphid, *Lipaphis erysimi*

(i) Damage: Aphids are bottlenecks in successful cultivation of mustard crop. They are often seen feeding in large numbers, covering the entire surface of flower buds, shoots, pods, etc. The pest breeds parthenogenetically and the adults give birth to 26–133 nymphs. About 45 generations are completed in a year. Nymphs and adults suck sap from the foliage. The aphid reaches ETL when 30% plants are infested (Bakhetia, 1986) or 50–60 aphids are noticed per 10 cm shoot (Singh *et al.*, 1983).

(ii) Epidemiology: Cloudy and cold weather (20°C or below) is very favourable for the multiplication of the pest. The winged forms are produced in autumn and spring and they spread from field to field and from locality to locality. Mean minimum temperature in the range of 4–13.8°C, mean maximum temperature of 14–30°C and relative humidity of 48–80%, favour quick aphid multiplication (Bakhetia & Sekhon, 1993). At the pod formation stage, rains completely wipe out the aphids, and at bud stage, a small residual population could multiply under favourable conditions (Mahal *et al.*, 1992).

(iii) Management

- Plant the crop early (by 15th–25th October) to reduce the aphid incidence.
- Removal of aphid infested twigs at the initial stage of pest attack at community level to prevent further spread of the pest.
- Spraying of 0.03% dimethoate or 0.025% oxydemeton methyl when pest population crosses ETL.
- Mustard aphid is attacked by a number of natural enemies *viz.*, *Coccinella septempunctata, Chrysoperla carnea,* syrphid fly, etc.

20.9.1.2. Painted Bug, *Bagrada hilaris*

(i) Damage: It is an occasional pest of cruciferous crops appearing at two stages of crop growth, *i.e.* seedling and maturity/harvesting stages. Severe incidence at seedling stage sometimes necessitates the resowing of crops, while at harvesting stage; it causes reduction in oil content and seed weight.

(ii) Epidemiology: Very low and very high temperatures reduce egg viability. It remains active between 21–40°C. Rainfall during February-March proved conducive for its development (Singh *et al.*, 1993).

(iii) Management

- Removal of crop residues of previous crop from the field to check the painted bug incidence.
- Periodical weeding helps in reducing the build up of painted bug population.
- Spraying of 0.05% malathion for the control of painted bug.
- Quick threshing of harvested crop.

20.9.2. Diseases

20.9.2.1. White Rust, *Albugo candida*

(i) Symptoms: Symptoms result from local as well as systemic infection. In case of local infection, symptoms of the disease appear on leaves and are characterized by the appearance of isolated, shiny white, raised pustules or blisters (1–2 mm in diameter), which later coalesce to form patches. The pustules are found scattered on the under surface of the leaves and their corresponding upper surface is yellow. Mature pustules rupture to release a mass of white powder of spores. Affected flowers show malformation, the petals may become green, like sepals, and stamens may be transformed into leaf-like or carpelloid structures. The petals and stamens persist in the flower instead of falling early as in case of normal flower. The ovules and pollen grains are usually atrophied, resulting in complete sterility. The entire plant may remain dwarf with small leaves and the stem often becomes twisted (Saharan *et al.*, 1997).

(ii) Epidemiology: The optimum temperature is about 10°C at which the rate of germination of sporangia and the number of zoospores formed is maximum. Above 25°C, the sporangia do not germinate (Kumar *et al.*, 1995). A temperature of around 12°C, relative humidity of more than 70%, wind velocity from 2.7 to 3.4 km/hr and winter rains are found to be most congenial for epidemic development of the disease under field conditions (Singh *et al.*, 2001).

(iii) Management

- Removal of crop residues of previous crop from the field to check the disease incidence.
- Plant the crop early (by 15th–25th October) to reduce the disease incidence.
- Seed treatment with *Trichoderma viride* at 4 g/kg of seed.
- Removal of heavily disease infected plants from the field.
- Soil application of *T. viride* at 2.5 kg mixed with 62.5 kg FYM/ha.
- Spraying of 0.2% mancozeb whenever required.

Table 20.1. Integrated pest and disease management

Crop stage	Pest/Disease	IPM activities
Pre-sowing		Remove the residues of previous crop from the field to check spread of diseases and to prevent painted bug infestation; grow the variety Varuna (T-59).
Sowing		Plant the crop early (by 15th –25th October) to reduce the incidence of major diseases like white rust and downy mildew and insect pests like aphids; seed treatment with *Trichoderma viride* at 4 g/kg seed for control of white rust.
Seedling & vegetative stage	*Alternaria*, white rust, downy mildew & Powdery mildew diseases	Soil application of *T. viride* at 1 kg mixed with 25 kg FYM/acre; removal of heavily disease infected plants from the field; spray 0.2% mancozeb/0.2% metalaxyl/0.1% karathane whenever required.

Table. 20.1. (Contd.)

Crop stage	Pest/Disease	IPM activities
	Painted bug	Periodical weeding helps in reducing the build up of painted bug population; spray 0.05% malathion for the control of painted bug.
	Mustard sawfly	Collect the larvae during the early morning times & destroy them; apply 0.07% endosulfan or 0.05% quinalphos or 0.05% malathion, if needed.
Flowering & pod stage	Mustard aphid	Remove the aphid infested twigs at the initial stage of pest attack at community level to prevent further spread of the pest; spray 0.03% dimethoate or 0.025% oxydemeton methyl when pest population crosses ETL.
	Stag head infection (white rust + downy mildew)	Spray 0.03% metalaxyl followed by 0.2% mancozeb.
	Leaf miner	Avoid insecticidal spray wherever possible for maintaining natural control.
Maturity	Painted bug	Spray 0.05% malathion, if needed; quick threshing of harvested crop.

21 SOME GENERAL AND USEFUL TIPS IN PLANT PROTECTION

21.1. USE OF POWER SPRAYERS

Power sprayers require less spray fluid/unit area as compared to high volume sprayers like gator, back pack etc. While high volume spray require about 800–1000 litres of spray fluid per hectare, low volume spray with power sprayers require only 200–300 litres. Hence, while using power sprayers, spray solutions should be at least three times more concentrated than high volume sprays and spray solutions also should be reduced three times. For example, if 450 ml of monocrotophos is to be mixed with 300 litres of water for using Gator sprayer, then mix 450 ml of monocrotophos in 100 litres of water for power sprayer.

21.2. USE OF STICKERS/SPREADERS

The addition of stickers/spreaders to the spray solution improves the retention, even spreading of the spray solution over sprayed surface.

Commonly used stickers and spreaders are as follows:

Sandovit	:	0.5 ml/lit
Teepol	:	0.5 ml/lit
Triton	:	0.5 ml/lit
Citowet	:	0.5 ml/lit

21.3. PREPARATION OF 1% BORDEAUX MIXTURE

Dissolve one kg of copper sulphate (98% purity) in 50 litres of water in wooden barrel or plastic drum. Dissolve one kg of fresh hydrated lime in 50 litres of water separately. Both the solutions should be sieved. Pour slowly copper sulphate solution and lime water simultaneously by continuous stirring into tank.

It should be ensured that there is no free copper in the mixture. To check this, a clean iron knife/blade should be held/dipped in the mixture for a few minutes, removed and observed for any reddish brown deposit of copper on it. If the copper deposit is noticed, add a little lime solution to the mixture to neutralize excess copper. The best way is to use litmus paper and check pH. The Bordeaux mixture should be neutral or slightly alkaline, otherwise it will injure the leaves/plants. Bordeaux mixture should always be prepared fresh and should not be stored for more than one day. However, mixture can be stabilized by adding 200 gm of jaggery per 100 litres of mixture.

Preparation of 10% Bordeaux Paste

- Dissolve 1 kg of copper sulphate in 5 litres of water.
- Dissolve 1 kg of quick lime in another 5 litres of water separately.
- Mix the above 2 solutions to get the Bordeaux paste.

Caution while using Bordeaux mixture

- Always spray freshly prepared mixture for effective management. Therefore, prepare only required quantity for the day.
- Use only wooden, earthen or plastic vessels, to avoid the corrosion of the metal vessels by the mixture.
- Do not use the Bordeaux mixture in combination with any other chemical or pesticide.
- To avoid choking of nozzle, it is advisable to strain the Bordeaux mixture through a cloth or a sieve before putting it into spray tank.
- The Bordeaux mixture tends to sediment easily. Therefore, it's stirring while using is desirable.
- In exceptionally hot days, when the plants are showing signs of temporary wilting or when it is raining, the Bordeaux mixture should not be sprayed, particularly on newly emerged tender foliage. Exercise this caution particularly on nursery plants.
- After carrying out spray operations, the appliances should be thoroughly washed with plenty of water to remove any copper deposits.
- The left over Bordeaux mixture should not be dumped in the field as this may prove toxic to the subsequent sowings.

21.4. PREPARATION OF 4% NEEM SEED KERNEL EXTRACT

4 kg kernel of neem seed is required to prepare 100 litres of spray fluid. Grind 4 kg of kernel with 20 litres of water to get a fine paste. For grinding, a mixer or wet grinder may be used. This fine paste is kept for 12 hours and repeatedly filtered, washed and squeezed through a rough cloth and made up to 100 litres, which is used for spraying. About 800–1000 litres of solution is required to spray one hectare of cabbage or other vegetable crops.

21.5. SEED TREATMENT FOR STORAGE

Pests and diseases affect seeds in storage. Following seed treatments may be adopted for protecting seeds during storage (Table 21.1):

Table 21.1. Effect of different pesticides on storage of seeds

Causal organism	Pesticide	Dose
Insects	Carbaryl	4 g/kg seed
Diseases	Subeej/derosal	2 g/kg seed
	Chlorothalonil/captan/iprodione	2 g/kg seed
	Thiride/vitavax	2 g/kg seed

21.6. CONTROL OF RODENTS

There are three main methods to control rodents in the field. Trapping has been employed since the earliest times, and much skill has been developed in its use. Poisoning and environmental alteration have the biggest contribution to make to increasing crop yields.

21.6.1. Control in the Field

21.6.1.1. Poisons

There are three main kinds of poisons for rodent control, multi-dose, single-dose and gassing poisons. Warfarin, coumatetralyl, chlorophacinone, coumachlor, difenacoum and calciferol are multi-dose poisons. All except the last are anticoagulants, which have an effect on the blood system of the animal. These compounds will normally give up to 100% kill. The single dose poison is zinc phosphide. Generally both multi-dose and single-dose poisons have to be mixed with bait and have a lethal effect when eaten, but gassing poisons release poisonous gases which are breathed by the rodents. In most situations where there is a rodent problem in growing crops, the use of bait containing a multi-dose poison is the best method of control.

Detecting Live Burrows (where Rodents are Active): Locate all burrows in the field and close them with mud in the evenings. Next morning, check-up and find out the burrows which have opened. They are live burrows.

Baits: Rats and mice shall usually eat cereals (milled or whole grain) and often the cheapest locally available will be accepted as baits by them. Rodents tend to like sugar and a small amount of sugar is added to the bait to make it more attractive. Sometimes a mixture of cereals will give good bait.

Pre Baiting: To familiarize the rats with the food used for poisoning, pre-baiting is required. For this, prepare plain bait of a mixture of wheat, bajra, sorghum or maize flour and groundnut chops mixed with vegetable oil for 2–3 days. When rats start accepting the bait, they can be poisoned.

Bait Mixing: It is very important that the correct amounts of poison may be mixed throughout the cereal so that it is all poisoned. The concentration of rodenticides has been calculated to produce the maximum kill of rats. The instructions on the label should be followed. Many of the anticoagulants such as warfarin are sold as a master-mix or concentrate which require to be mixed in the proportion of one part to 19 parts of bait. Mixing of zinc phosphide should be done just before use.

Suitable Mixtures: If a cereal flour or meal is chosen as bait, a satisfactory mix can be obtained by mixing the poison and the meal together for about one minute.

Quantities for Warfarin Bait

Fine cereal meal or flour	— 19 parts (1900 g)
Warfarin concentrate (0.5%)	— 1 part (100 g)

Quantities for Zinc Phosphide Bait

Fine cereal meal or flour	— 39 parts (975 g)

Zinc phosphide — 1 part (25 g)

If whole grain bait is chosen, it is necessary to add a 'sticker' to ensure that the poison is evenly distributed. Corn oil or other vegetable oils are the best choice although a mineral oil is also acceptable.

Quantities for Warfarin Bait

Wheat or other grain	— 18 parts (1800 g)
Corn oil	— 1 part (100 ml)
Warfarin concentrate (0.5%)	— 1 part (100 g)

Quantities for Zinc Phosphide Bait

Wheat or other grain	— 36 parts (1800 g)
Corn oil	— 2 parts (50 ml)
Zinc phosphide	— 1 part (25 g)
or	
Zinc phosphide	— 3 parts
Flour	— 80 parts
Groundnut chops	— 16 parts
Edible oil	— Sufficient to make a lump of the bait

Making Wax Blocks: A very useful way of making long lasting baits is by mixing the poison and the bait with melted paraffin wax. The wax sticks the bait and poison together and small blocks can be prepared by the following method:

Fine cereal meal or whole grain	— 10 parts (1000 g)
Sugar	— 1 part (100 g)
Warfarin concentrate (0.5%)	— 1 part (100 g)
Semi-refined paraffin wax (m.p. 60–65°C)	— 8 parts

To prepare the bait blocks, first heat the paraffin wax just enough to melt it. Next, stir the cereal, sugar and poison until all are thoroughly mixed. Then pour the mixture into flat tray and leave it to set. When set hard enough, cut into blocks each weighing about 100 g. The advantage of wax blocks over loose baits lies in their greater resistance to the weather. They remain palatable to rats even when exposed to rain and sun.

Prepare the bait as described above and keep some quantity in each live burrow and close it with mud. Check next day. If burrow remains closed, the rat is dead. If burrow is opened, rat is not dead. Hence, repeat the bating.

Gassing Poisons: Gassing powders act by releasing poisonous gases which are breathed in by the rodents. There are two main ones which have been used against rats; sodium cyanide and aluminium phosphide. They produce poisonous gases such as hydrogen cyanide and aluminium phosphine respectively, when they come in contact with the moisture.

Insert 6 grams (2 tablets) of cellophos (Aluminium phosphide) into each live burrow and close it with mud. Check next day. If burrow remains closed, the rat is dead. If burrow is opened, rat is not dead. Hence, repeat the fumigation.

21.6.1.2. Trapping

Even when most skillfully used, traps will rarely control rats as effectively as poisons. There use is probably confined to larger rodents being caught for food or to the few survivors of a poison treatment. Many different local traps will catch rats and those such as the break back trap which will kill the animal, solve the problem of having to kill rats that are caught in the traps. Any locally available food that the rats eat can be used for bait which should be placed in the centre of the treadle.

21.6.1.3. Weed and Bush Clearance

There is some evidence that weed control both within the crop and around the field edges will limit the numbers of rodents by reducing the sources of food and living spaces. Many species of rodents do not actually live in the crops they attack, but in the undergrowth, banks and grassland surrounding it. Bush clearance at intervals may well be less costly than poisoning. This may be a very useful additional benefit of weed control done primarily for improving crop yields directly.

21.6.2. Control in Storage

Making the store rodent-proof is probably the most satisfactory approach, but it is not always practicable and regular poisoning may be necessary to keep rodents in check. A very important and inexpensive method of minimizing a rodent problem is scrupulous attention to hygiene in and around the store. All potential rodent living areas–piles of unwanted building materials, wooden packing cases, waste paper and sweepings–should be removed completely and their further accumulation prevented. In addition, all undergrowth should be cut down and kept at a height which will enable any new rat holes and runs to be readily detected.

21.6.2.1. Poisons

The use of multi- or single-dose poisons outlined in the field can be used. Aluminium phosphide tablets will kill rodents in stores, as will other fumigating gases.

21.6.2.2. Trapping

Traps may be useful to get rid of the survivors of a poison treatment or to catch a few rodents getting into a store.

21.7. CONTROL OF TERMITES

- Locate termite colony, destroy mound, locate the queen and kill it.
- Most effective method of killing the mound is poisoning the mounds by fumigating with cellophos @ 2 tablets/metre diameter mound and plaster all the openings with wet mud.
- Pour 60 ml Coroban 20 EC/Dursban 20 EC through a hole made in the centre of closed mounds or through open chimney in case of open mounds.

21.8. CONTROL OF ROOT GRUBS

Incorporate in the soil 20 kg phorate10 G or 25 kg carbofuran 3G or 30 kg solvirex 5 G or 25 kg ekalux 5G per hectare or 100 ml dursban/coroban in 10 litres of water (500 litres to cover 1 ha.) after first summer rains (April-May) for maximum kill of root grubs.

21.9. READY RECKONER FOR MAKING PESTICIDE SPRAY SOLUTIONS

Table 21.2. Ready reckoner for making pesticide spray solutions

% *a.i.* of pesticides in commercial products	Quantity of pesticide formulation to be added per litre of water in gram or ml to get the desired strength of solution (%)											
	0.01	**0.02**	**0.03**	**0.04**	**0.05**	**0.06**	**0.07**	**0.08**	**0.09**	**0.1**	**0.2**	**0.3**
2.5	4.00	8.00	12.00	16.00	20.00	24.00	28.00	32.00	36.00	40.00	80.00	120.0
	0.01	0.02	0.03	0.04	0.05	0.06	0.07	0.08	0.09	0.1	0.2	0.3
2.5	4.00	8.00	12.00	16.00	20.00	24.00	28.00	32.00	36.00	40.00	80.00	120.0
5	2.00	4.00	6.00	8.00	10.00	12.00	14.00	16.00	18.00	20.00	40.00	60.0
10	1.00	2.00	3.00	4.00	5.00	6.00	7.00	8.00	9.00	10.00	20.00	30.0
15	0.67	1.30	2.00	2.60	3.30	4.00	4.66	5.20	6.00	6.66	13.30	20.00
20	0.50	1.00	1.50	2.00	2.50	3.00	3.50	4.00	4.50	5.00	10.00	15.00
25	0.40	0.80	1.20	1.60	2.00	2.40	2.80	3.20	3.60	4.00	8.00	12.00
30	0.33	0.66	1.00	1.30	1.65	1.98	2.31	2.64	2.97	3.33	6.60	9.90
35	0.29	0.58	0.87	1.16	1.45	1.74	2.03	2.32	2.61	2.85	5.70	8.55
40	0.25	0.50	0.75	1.00	1.25	1.50	1.75	2.00	2.25	2.50	5.00	7.50
50	0.20	0.40	0.60	0.80	1.oo	1.20	1.40	1.60	1.80	2.00	4.00	6.00
75	0.13	0.27	0.40	0.53	0.66	0.79	0.92	1.07	1.20	1.33	2.66	4.00
80	0.12	0.25	0.37	0.50	0.63	0.75	0.88	1.00	1.13	1.25	2.50	3.75
85	0.12	0.25	0.36	0.48	0.59	0.70	0.82	0.96	1.06	1.18	2.35	3.54
100	0.10	0.20	0.30	0.40	0.50	0.60	0.70	0.80	0.90	1.00	2.00	3.00

Note: The pesticides should be measured in g or ml. For preparation of malathion 0.05% spray solution from 50% malathion, the value at the point of intersection of 10th row and 5th column should be taken *i.e.*, 1.0 ml should be dissolved in 1 litre of water.

Preparation of Spray Solution

The spray solution must be prepared to the exact concentrations as recommended. The quantity of the commercial pesticide formulation required for preparation of the spray solution is calculated by using the following formula:

Quantity of the pesticide formulation

$$= \frac{\text{Quantity of spray solution required} \times \text{Conc. of solution desired}}{\text{\% of a.i. of pesticide in the formulation}}$$

Prepare spray solution first in small quantity and then increase the volume to desired level by adding water. in case of wettable powder, take required quantity of pesticide, add a little quantity of water, mix it thoroughly to prepare the paste and then add remaining quantity of water to this paste with constant stirring.

21.10. COMPATIBILITY CHART

Table 21.3. Compatibility Chart: Precautions to be taken while using insecticides, fungicides and other Agrochemicals

No.	Pesticides/Agrochemicals	1	2	3	4	5	6	7	8	9	10	11	12	13	14	15	16	17	18	19	20	21	22
1	Plant origin insectcides@	a	a	a	a	a	a	a	a	a	a	b	a	a	b	a	a	a	a	b	b	c	a
2	Carbaryl	a	a	a	b	a	c	a	d	a	a	a	d	a	d	a	d	a	a	a	d	a	a
3	DDT Group	a	a	a	a	a	a	a	a	a	a	a	d	a	a	a	b	a	a	a	b	a	a
4	Monocrotophos	a	b	a	a	a	a	a	a	a	a	a	b	a	a	a	a	a	a	a	b	a	a
5	Dimethoate	a	a	a	a	a	a	a	a	a	c	a	c	a	a	a	b	a	a	a	b	a	a
6	Endosulfan	c	a	a	a	a	a	a	d	a	a	a	b	a	c	a	b	a	a	a	c	a	c
7	Fenitrothion	a	a	a	a	a	a	a	a	a	a	a	b	d	a	a	b	a	a	a	b	a	a
8	Kerosene	a	d	a	a	a	d	a	a	a	a	a	b	b	a	b	b	b	b	b	b	a	c
9	Malathion, Parathion, Demeton	a	a	a	a	a	a	a	a	a	a	a	b	d	a	a	a	d	a	a	a	a	a
10	Oxy-demeton methyl	a	a	a	a	c	a	a	a	a	a	a	a	c	a	a	a	c	c	a	a	c	c
11	Phosalane	d	a	a	a	a	a	a	a	a	a	a	a	d	a	a	a	a	a	b	b	b	b
12	Bordeaux mixture	b	b	d	b	b	b	b	b	b	b	d	a	b	d	a	b	b	a	a	a	b	a
13	Captan, Difolatan	a	a	a	a	a	a	d	b	d	a	a	b	a	d	a	b	a	d	a	b	a	a
14	Copper oxychloride	a	b	a	a	a	c	a	a	a	c	a	a	d	d	d	b	b	a	b	a	a	a
15	Dithiocarbamate group#	a	a	a	a	a	a	b	a	a	a	a	a	d	a	b	d	a	a	a	a	a	–
16	Lime sulphur	b	b	b	b	b	b	b	b	b	b	d	b	b	b	b	a	b	a	a	a	a	b
17	Karathane	a	a	a	a	a	a	a	b	a	a	a	b	e	b	a	b	a	b	c	c	b	c
18	Wettable sulphur	a	a	a	a	a	a	a	b	a	a	a	a	d	a	a	a	a	a	a	a	a	a
19	2, 4-D	b	–	a	a	a	c	a	b	a	c	c	a	a	b	a	a	c	a	a	a	c	a
20	Lime	b	b	b	b	b	c	b	b	b	c	a	b	a	a	a	a	a	a	a	a	a	a
21	Streptomycin	c	a	a	a	a	a	a	c	a	c	c	a	a	a	a	a	a	a	c	c	a	c
22	Urea	b	a	a	a	a	c	a	a	a	c	c	a	a	a	b	a	a	a	a	b	c	a
23	Carbendazim	a	a	a	a	–	a	–	–	a	–	–	a	a	–	a	–	–	a	a	–	–	a

@ = Plant origin insecticides: Rotenene, Pyrethrum. # = Dithiocarbamates: Maneb, Zineb, Nabam, Thiram, Ziram, Mancozeb.
a = Can be mixed. b = Not to be mixed. c = Information not available. d = Use immediately.
e = Usually not mixed.

21.11. SOIL SOLARIZATION

Soil solarization is a method of killing soil-borne pathogens (fungi, bacteria, nematodes), soil insects and weed seeds by heat. This technique consists of covering moist soil with a plastic film during periods of intense sunshine and heat, thereby capturing radiant heat energy from the sun, causing physical, chemical and biological changes in the soil. The soil temperature is increased (by 5–15°C) to levels lethal to many soil-borne fungi, bacteria, nematodes and weeds. The short wave incident solar radiation penetrates the polyethylene sheet but the long wave radiation is prevented from soil, thus trapping the heat resulting in thermal inactivation and production of heat shock proteins and irreversible heat injury. Solarization also improves plant nutrition by increasing the availability of N and other essential elements. This practice can be profitably used on a small scale and is especially useful for nursery beds.

Soil solarization for 40–45 days using 400 gauge transparent LDPE was proved ideal pre-sowing treatment for the management of nematodes and other soil-borne diseases in cardamom nurseries. Mean soil temperature during the solarization period increased by 8.7°C in solarized beds and the germination was enhanced by 25.5%, while weed growth was suppressed by 82%. *Pythium vexans* was totally eliminated, while populations of *Rhizoctonia solani, Phyllosticta elettaria* and nematodes were suppressed to varying levels. Solarization was also found to stimulate the growth of cardamom seedlings (Eapen, 1995).

White polyethylene mulching for 15 days on beds prepared for planting betel vine revealed high reduction in nematode populations due to increased soil temperature to 44.1°C (Sivakumar & Marimuthu, 1987).

Soil solarization using thin transparent polyethylene mulches for six weeks caused as much as 72.6 and 88.5% decrease in the population densities of *Tylenchorhynchus vulgaris* and *Hoplolaimus indicus*, respectively in nursery beds. Application of mahua cake before solarization usually caused slightly greater reduction in nematode population (Kamra & Gaur, 1998).

21.12. MASS PRODUCTION OF BIOAGENTS

21.12.1. *Pseudomonas fluorescens*

A loopful of *P. fluorescens* should be transferred to 100 ml of King's B broth in a 250 ml conical flask under aseptic conditions. The flask can be incubated at room temperature for 48 hours. The population of *P. fluorescens* in the broth can be assessed (a minimum of 9×10^9 cfu/ml is necessary).

The pH of the substrate (talc powder) can be adjusted to 7.0 by addition of 150 g calcium carbonate/kg of talc powder and sterilized at 1.1 kg/cm^2 for 30 min for 2 successive days. One kg of the sterilized substrate (talc powder) and 5 g of carboxy methyl cellulose should be placed in a polythene bag or any sterile container under aseptic conditions, 400 ml of *P. fluorescens* suspension to be added and mixed thoroughly. The formulation can be stored in a polythene bags for one month. The product should contain a minimum of 2.5×10^8 cfu/g.

The product should have the following quality parameters:

- Fresh product should contain 2.5×10^8 cfu/g.
- Three months after storage at room temperature, the population should be 8–9 × 10^7 cfu/g.
- Storage period is 3–4 months.
- Minimum population load for use is 1×10^8 cfu/g.
- The product should be packed in white polythene bags.
- Moisture content of the final product should not be more than 20%.
- Population in broth should be $9 \pm 2 \times 10^8$ cfu/ml.

The composition of King's B medium is as follows:

Peptone	—	20.00 g
Dipotassium hydrogen phosphate	—	1.50 g
Magnesium sulphate	—	1.50 g
Glycerol	—	10.00 ml
Agar	—	15.00 g
Distilled water	—	1000 ml
Cycloheximide	—	100 ppm
Ampicillin	—	50 ppm
Chloramphenicol	—	12.50 ppm

(King's B broth is prepared without the agar and antibiotics)

Liquid Formulation Technology to Extend the Shelf Life of *Pseudomonas fluorescens*

Polyvinyl Pyrrolidone (PVP) (2%) was tested as an effective preservation agent in protecting the cells of *P. fluorescens* for more than a year without affecting the natural properties of bacteria. The antagonistic efficacy of preserved cells was also confirmed *in vitro* and *in vivo* (on black pepper, vanilla and ginger).

21.12.2. *Trichoderma viride, T. harzianum*

Among the various fungal biological control agents which are effective against nematode and fungal diseases of horticultural crops, *Trichoderma viride* and *T. harzianum* have a good market. They can be easily mass produced on several organic substrates.

21.12.2.1. Small Scale

The mycelial discs grown in potato dextrose agar medium was transferred to 250 ml conical flasks containing sterilized molasses yeast medium (Yeast extract –5 g, Molasses –30 g, Distilled water –1 litre sterilized at 1.42 kg/cm^2 for 20 min) at 100 ml/flask and incubated for 10 days. The mycelial mat along with the broth was homogenized and mixed with talc powder (500 mesh) at 1:2 ratio. Carboxy methyl cellulose (CMC) was added at 5 g/kg of the product as sticker. The product was shade dried until it attained 12% moisture level. The clumps were broken, homogenized and packed in polythene bags of 100 g capacity. The required population level at the time of packing should be 28×10^7 cfu/g of the product.

The shelf life of the product was 4 months. At the end of expiry, the cfu should not be less than 20×10^6 cfu/g of the product.

21.12.2.2. Large Scale

Add 70 ml of molasses yeast medium in each 250 ml conical flasks and plug with non-absorbent cotton wool. Autoclave at 1.42 kg/cm^2 for 1 hr. Transfer 8 ml mycelial disc of *T. viride/T. harzianum* to each flask and incubate for 10 days. This mother culture is used for inoculating the medium in fermentor.

Add 50 litres of molasses yeast medium in 50 litre fermentor, sterilize at 1.42 kg/cm^2 for 20 min and cool to room temperature. Add 2.5 litres of inoculum (from mother culture) into the sterilized medium and incubate for 10 days. Provide aeration in the fermentor for 4–8 hr/day by agitation. Presence of greenish fungal growth indicates *T. viride/T. harzianum*. This broth culture contains fungal mat, spores and antibiotics. Draw samples 10 days after incubation and assess the spore load in the broth which should be at least 10^8 spores/ml. Mix one litre of broth culture with 2 kg talc powder, add CMC at 5 g/kg of the product, air dry (12% moisture content) and sieve. The population should not be less than 20×10^6 cfu/g of the product. It is packed in opaque polythene bags and stored in a cool dry place before use. The product can be safely stored for 4 months without losing its viability.

21.12.3. Arbuscular Mycorrhizal Fungi (*Glomus mosseae, G. fasciculatum*)

21.12.3.1. In Glasshouse

Cement pots or tanks are filled with sterilized sand: soil (1: 1 by volume) or soilrite: peat (1: 1 by volume) mixture. A standard starter culture of AMF obtained from any authentic source is placed as a band 3 cm below the soil surface. The seeds of Rhoades grass/finger millet/maize are sown on them and covered with soil. Pots are irrigated normally. When the plants are about 10–12 weeks old, the top shoots are cut off and the root system colonized by the fungus along with the soil containing chlamydospores is used as the inoculum.

21.12.3.2. On-Farm Production

A suitable land (even waste land) is selected near the field where the fungus inoculum has to be used. The land is brought to the fine tilth by ploughing, harrowing and cultivator operations. A raised bed (20 cm height) of soil: sand (1: 1 by volume) mixture is prepared on an even land surface. The raised bed is sterilized by soil solarization for 2 to 3 weeks using polythene sheet. The starter culture of AMF is applied at a depth of 3–4 cm from the top by removing the top soil. The seeds of Rhoades grass/ finger millet/maize are sown on them and covered with soil. Irrigation is done normally. When the crop attains a height of 10–15 cm, 12 weeks after sowing, the shoot portion is cut off at ground level. The soil is dug up to a depth of 20 cm along with root system is used as the inoculum.

21.12.3.3. Method of Inoculation

Perennial crops are inoculated in the nursery. AMF inoculum is added to FYM + soil + sand mixture (1: 1: 1 by volume) at the rate of 50 g inoculum/500 g soil mixture in polythene bags. Seeds of crops are planted in each bag. Forty five days after sowing, the root system of seedlings would be colonized by the fungus and ready for transplantation in the field (in pits). Before planting, each pit should be inoculated with 250–500 g of the fungus along with FYM and neem cake. Two months after planting, the roots may be tested for colonization by the fungus. The plants get colonized up to 80–90% within 3–8 months. Inoculation of AMF helps in greater uptake of nutrients like P, Zn and Cu. It also increases stem girth, plant height and hastens flower initiation.

Annual crops can be inoculated with AMF in the nursery beds. The AMF inoculum consisting of chlamydospores, soil and infected root bits is either spread uniformly in the nursery beds or placed in rows over which seeds of are sown. The seedlings get colonized by the fungus in about 4-5 weeks after which they are transplanted in the main field. Inoculated plants show better growth and improved nutrient content.

21.12.4. NPV of *Spodoptera litura*

21.12.4.1. Production and Utilization

- Collect 7–9 days old larvae (4th instar) and starve them for 8 hours.
- Prepare *Sl* NPV suspension (10^8 POBs-Polyhedral Occlusion Bodies/ml) in 250 ml of water with teepol in a bottle.
- Dip clean castor leaves in *Sl* NPV suspension, shade dry for 15–20 min and feed the larvae for 2 days.
- Provide fresh non-infected leaves for the larvae thereafter.
- Alternatively, expose larvae to semi-synthetic diet treated with *Sl* NPV suspension.
- Larvae die after 7 days and their under surface turns pink and later white.
- Collect dead larvae and allow it to putrefy in water for a week.
- Grind dead larvae and filter through muslin cloth.
- Remove the supernatant and suspend the polyhedra in water.
- Centrifuge at 500 rpm for 5–10 min and discard the supernatant retaining the POBs.
- Centrifuge further at 2500 rpm for 5–10 min and suspend the white POBs in water.
- Count using Neubauer haemocytometer and express on larval basis or per unit of larval weight.
- Dry the POBs over calcium chloride or precipitate by acetone.
- Formulate by adding permitted spreaders/wetting agents.

21.12.4.2. Nursery and Field Sprays

- Spray *Sl* NPV at 250 larval equivalent (LE) mixed in 125 litres of water added with crude sugar at 1 ml/l of water and teepol at 1 ml/10 litres of water in the evening hours for one ha of nursery.

- Spray thrice at fortnightly intervals from the day neonate larvae are sighted.
- Continue spraying up to three weeks after planting.
- Spray when eggs start hatching in castor at weekly intervals.
- The dosage varies from 250 LE to 450 LE/ha.
- Larvae feeding on sprayed plants turn pinkish white, move to top of the plant, hang upside down and die (Fig 21.1).

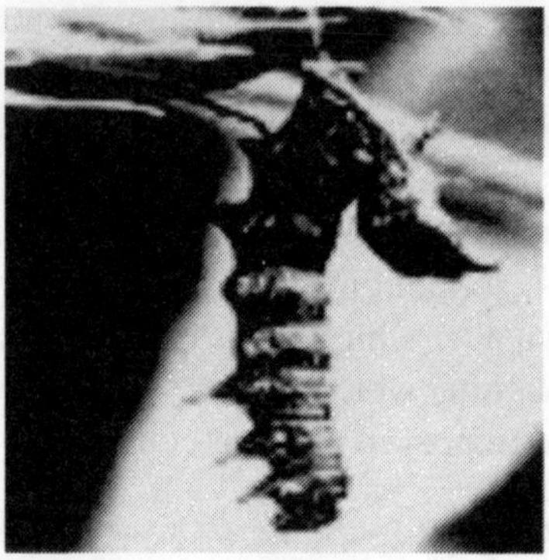

Figure 21.1. Infected larva hanging upside down (Tree top disease)

21.12.4.3. *Spodoptera litura* Culture

- Allow a pair of newly emerged moth in a well ventilated plastic container (20 × 15 cm) lined (inner surface) with paper to obtain eggs.
- Cover the bottom sponge lining with blotting paper.
- Provide 50% honey solution and water in small plastic cups.
- Cut out the egg batches on the paper.
- Dip eggs in 10% formalin for 30 min, wash in running water for 30 min and dry using blotting paper.
- Hatch in sterilized glass vials (20 g capacity).
- Transfer 3 days old larvae in bouquets of sterilized castor leaves for feeding.
- Collect the pupae, sterilize with sodium hypochlorite, wash in water and dry with blotting paper.
- Identify the sex with the help of ovipositor and last abdominal segment and keep over a wet sponge in an adult emergence cage (22 × 15 cm) (Fig 21.2).
- Adults emerge after 10 days.

21.12.5. NPV of *Helicoverpa armigera*

21.12.5.1. Production and Utilization

- Prepare the viral solution (virus at 18×10^6 POBs/ml of distilled water).
- Feed 5–7 days old larvae with virus infected diet.
- Incubate at 26°C for 7 days.

Figure 21.2. *Spodoptera* rearing unit

- Collect the dead infected larvae and macerate in distilled water.
- Standardize the *Ha* NPV with number of POBs/ml in terms of LC_{50}.
- Store POBs in plastic cans/bottles with proper instructions on the containers even up to 6 months without exposure to sun.
- 1 larval equivalent (LE) = 6×10^9 POBs or their equivalent in activity.

21.12.5.2. Field Spray

- *Ha* NPV is sprayed to control *H. armigera* on chickpea, pigeonpea, field bean, and tomato.
- Spray 3–4 times of *Ha* NPV 250 LE/ha with crude sugar at 5 ml/l of water, groundnut oil cake at 1 ml/100 ml of water and chickpea flour at 1 ml/100 ml of water.
- After 3–5 days, dead larvae will be visible hanging upside down.

21.12.5.3. *Helicoverpa armigera* Culture

- Collect the adult insects in light traps or from endemic crop areas.
- Place the adults in oviposition cages (cylindrical iron frame–50 cm height × 30 cm diameter and black or white cloth cover) at 100 pairs/cage.
- Hang 2 cotton swabs dipped in 50% honey and water over a circular plastic mesh, 5 cm from frame base.
- Place the frame on aluminium tray (40 × 40 × 5 cm) with a 3 cm thick sponge at the bottom soaked in water.
- Remove eggs daily from the inner side of the cloth cover.
- Sterilize the eggs with 10% formalin, wash in distilled water for 10 min and dry in laminar air flow chamber.
- Place the eggs in a moist cotton inside a flask and wrap with aluminium foil.
- Larvae emerges after 3–4 days and place them in multi cellular trays.
- Larvae are fed with chickpea based nutritious semi-synthetic diet.
- Harvest pupae after 18–20 days, sterilize, wash and dry.
- Separate the male and female pupae and place in adult emergence cages.
- Adults emerge after 7–8 days.

21.12.6. Trichogrammatid Egg Parasitoids

21.12.6.1. Data Base

- Trichogrammatid egg parasitoids are multiplied on eggs of *Corcyra cephalonica,* a factitious host.
- One cc of *Corcyra* eggs contains 20,000 eggs. This is supplied in a card (Fig 21.3).

Figure 21.3. *Tricho* adult oviposit on *Corcyra* egg, *Tricho* card

- Eight releases of Trichogrammatids and each release with 1,40,000 are recommended to contain first and second generation eggs/ha.
- Releases are spaced at weekly intervals.
- 56 cc of *Corcyra* eggs are to be produced/ha (7 cards/release × 8 releases).
- 56,000 cc of *Corcyra* are required/1000 ha. *i.e.*, 56,000 cards.
- 150 g of sorghum grains are required to produce 1 cc of *Corcyra* eggs.
- 8,400 kg of sorghum grains are required as feed to produce 56,000 cc.
- Approximately 10 t of sorghum grains are required as feed to produce natural enemies to cover 1000 ha.
- The parasitized *Trichogramma* egg card should be handed over to the growers within three days after dispatch.
- It cannot be stored in dealers/retailers shop like antagonistic fungi/ biofertilizers/ pesticides/fertilizers.

21.12.6.2. Production of *Corcera cephalonica*

Ideal Conditions for Rearing

- Temperature: 27 $\pm$ 1°C.
- RH: 60 $\pm$ 5%.
- Light: 200—250 lux and avoid direct sun light.

Rearing

- In each box, put 2.5 kg broken cereal or millets, 10 tablets of dry yeast and 0.5 g streptomycin sulphate.

- Keep 0.5 cc *Corcyra* eggs and secure the lid of the charge box for 40 days.
- On 40th day, moths start to emerge and continue to emerge for 2 months with a peak emergence from 45th to 75 th day.
- Emergence rate is 5–15 moths/box/day in the beginning and it reaches 75–100 moths/box/day during the peak.
- Each box produces 0.5 cc eggs/day during the peak emergence.
- Collect the moths daily using mechanical moth collection devices and transfer them to special oviposition cages.
- Discontinue the use of box and burn the contents properly after 80 days of infestation when less than 10 moths/box emerges.
- Pass the collected eggs through 15, 30 and 40 mesh sieves and run over a slope of paper or by automatic egg cleaner to eliminate the dust particles.
- Treat the eggs with UV rays (30 W UV tube for 45 min at a distance of 60 cm) to prevent eggs from hatching, if they are meant for *Trichogramma* production.

Note: Eggs meant for multiplication of *Corcyra* should not be treated with UV rays.

21.12.6.3. Production and Utilization

- Collect cleaned and UV treated *Corcyra* eggs.
- Glue the eggs to "tricho" cards measuring 15 × 10 cm, which are prepunched into 16 pieces of 4 × 3 cm leaving some space for stapling.
- Expose the eggs to adult female *Trichogramma* in the ratio of 30 : 1 for 24 hr and allow it to parasitize.
- Expose females to the eggs in the ratio of 40 : 1 on cards inside polythene bags, where female oviposits till they die.
- Transfer 3–5 days old parasitized egg cards in safe polythene bags and place dried honey for emerging parasitoids.

21.12.6.4. Field Release

- Cut "tricho cards" into small bits.
- Release cards 8 days after parasitization for emergence of adults.
- Staple the bits to lower side of the leaves at 100 spots/ha in the evening hours at weekly intervals.
- Release is done based on the appearance of eggs of host.

21.12.7. Chrysopids

21.12.7.1. Production and Utilization

Ideal Conditions for Rearing

- Temperature: 27 ± 1°C.
- RH: 70%.
- Light: Normal illumination by a fluorescent tube.
- Room area: 6 × 6 m.

Rearing

- Rear the adults with artificial honey based diet and larvae on *Corcyra* eggs separately.
- Keep 200 adults in the oviposition cage measuring 75 × 30 × 30 cm with a sliding top cover.
- Cover the sides of the cage with smooth nylon lining mesh of '0' size which is unsuitable for egg laying.
- Fit a black cloth on the top cover for laying of eggs.
- Replace the top cover with eggs on alternate days starting from 5th day.
- One batch is kept for 30–35 days and dead adults are removed every alternate day.
- Wait for 24 hr till the eggs harden and gently rub with a soft sponge.
- Expose *Corcyra* eggs to UV rays 30 W tube light for 45 min at 60 cm distance and inactivate them.
- Mix 3 days old 120 chrysopid eggs with 1 cc *Corcyra* eggs in a plastic container (27 × 18 × 6 cm).
- On hatching, *Chrysoperla* larvae feeds on *Corcyra* eggs.
- Remove the larvae on the 3rd day and transfer them into individual 2.5 cm cubical cells of plastic louvers.
- Each louver can hold 192 larvae.
- Sprinkle *Corcyra* eggs in all cells.
- Feed the larvae in 2 doses, first with 1.5 cc and second with 2.5 cc for 100 larvae with a gap of 3–4 days.
- Total quantity of *Corcyra* eggs required for rearing 100 *Chrysopid* larvae is 5 cc.
- Secure each louver on one side by brown paper and after transfer of larvae cover it with acrylic sheet. Brown paper would facilitate clear visibility of *Corcyra* eggs and also formation of cocoon by Chrysopid larvae.
- Collect cocoons after 24 hours of formation by removing paper from one side.
- Place them in adult oviposition cages for emergence (Fig 21.4).
- One set of louver remains in use for 13–15 days, then they are completely cleaned before reuse.

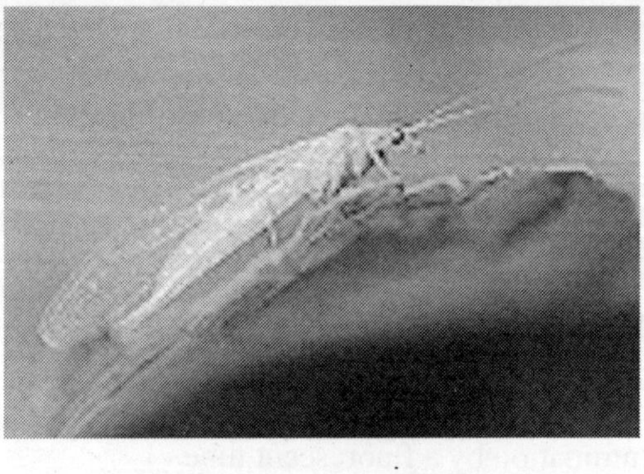

Figure 21.4. Chrysopid adult

21.12.7.2. Field Release

- Mix viable *Chrysoperla* eggs with *Corcyra* eggs. Add saw dust and paper strips. This is done before shipment. Once larvae emerge, they are released in the field.
- Release in the early morning by broadcasting larvae on the plants along with saw dust and paper strips.
- 10,000–20,000 larvae/ha is required in 2 releases which varies with crop, pest predator load.

21.13. ENRICHMENT OF FYM WITH BIOAGENTS (*TRICHODERMA HARZIANUM, PAECILOMYCES LILACINUS*)

Commercial formulation of the bioagent (*T. harzianum/P. lilacinus*) (1 kg/ton) was mixed with moist FYM along with 50 kg of neem/pongamia cake. Mixture was spread 15 cm high heap under shade, covered with polyethylene sheet and incubated for 3 weeks at 25 to 30°C and sprayed with water regularly to maintain moisture.

Colonized FYM can be used at 6 g/kg soil in pot culture experiments (12 g/2 kg soil in 15 cm diameter pots). It can also be used for main field treatment at 5 tonnes/ha before sowing of seeds or transplanting of seedlings. It can be applied at 2 kg/plant of banana, acid lime, papaya, pomegranate, grapes and roses along with 400 g of neem/pongamia cake once in every 6 months after planting for the management of nematodes and other soil-borne pathogens.

T. harzianum colonized compost significantly enhanced germination and growth of both tomato and okra as compared to non-colonized FYM. Population of *T. harzianum* in colonized compost was several folds higher (48.6×10^{12}) than non-colonized FYM (3×10^4).

Colonization of FYM by *T. harzianum* helps to improve the quality of compost. It significantly enhanced total and water soluble humic matter. A number of macro and micro-nutrients like P, K, S, Zn, Cu and Fe (both total and water soluble content) were significantly higher in *T. harzianum* colonized compost. Six times increase in water soluble humic content due to colonization of *T. harzianum* seems quite important as it is reported to enhance the plant growth.

21.14. SEED BIOPRIMING/SOLID MATRIX PRIMING

Seed biopriming (treatment of seeds with a slurry of biocontrol agents mixed with an organic carrier and incubated for 4 days under warm and moist conditions with relative humidity of 80–100% until just prior to radical emergence) has potential advantages over simple coating of seeds as it results in rapid and uniform seedling emergence. *Trichoderma* required less moisture for growth. *Trichoderma* conidia germinate on the seed surface and form a layer around bioprimed seeds. Such seeds tolerate adverse soil conditions better. Biopriming could also reduce the amount of biocontrol agents that is applied to the seed. Population of *T. harzianum* and *T. virens* on the surface of treated seeds increases from 10^4 cfu per seed up to 10^6–10^7 cfu per seed after biopriming.

21.15. FORMULATION OF ALGINATED PELLETS OF BIOCONTROL AGENTS

Alginate is a water soluble polysaccharide gum that has excellent gel-forming properties. Living organisms that are capable of biologically controlling a variety of pathogens can easily be entrapped in a calcium or sodium alginate matrix in aqueous form at room temperature. A pellet formulation of biocontrol agents was prepared by mixing sodium or calcium alginate, clay and streptomycin sulfate. The mixture was homogenized in a blender and then mixed with bioagent propagules. The gel beads, granules or seed coatings that are produced by the process are biodegradable. The encapsulation in Ca or Na alginate maintained the viability of the bioagents for 12 months at 7°C and 5 months at room temperature. Antagonistic fungi (species of *Paecilomyces, Trichoderma, Gliocladium, Talaromyces* and *Pencillium*) and bacteria (species of *Pseudomonas* and *Bacillus*) that can control soil-borne pathogens can be formulated in Ca or Na alginate (Connick, 1988). Alginated pellets can be used for soil application at 400 to 800 pellets/sq.m.

21.16. PRODUCTION OF BIOAGENT COLONIZED SEEDLINGS

The horticultural crops are invariably attacked by soil borne pathogens (nematodes, fungi, bacteria) at the seedling stage itself. Securing healthy seedlings of horticultural crops is essential to ensure freedom from soil-borne pathogens, optimum plant stand, good growth and higher yields. The following methods may be adopted to produce bioagent colonized seedlings of horticultural crops:

21.16.1. Seed Treatment

Seeds have to be treated with talc based formulations of *Pseudomonas fluorescens, Trichoderma harzianum* and *T. viride* at 10 g/kg seed before sowing.

21.16.2. Nursery Bed Treatment

The nursery beds may be treated with *P. fluorescens/T. harzianum/T. viride* at 50 g/m^2 along with neem/pongamia cake at 200 g/m^2 before sowing.

21.16.3. Treatment of Cocoa Peat or any other Substrate

The coco peat or any other substrate used for producing the nursery seedlings in portrays is treated with *P. fluorescens/T. harzianum/ T. viride* at 1 kg/tonne before filling with the substrate.

21.16.4. Treatment of Soil Mixture

The soil mixture used for producing the nursery seedlings in polythene bags in a conventional manner is treated with *P. fluorescens/T. harzianum/ T. viride* at 2 kg/tonne along with neem/ pongamia cake at 50 kg/tonne before sowing.

21.17. TRADITIONAL PEST CONTROL

See Table 21.4.

Table 21.4. Plant products used in traditional pest control

Product	Preparation	Pests controlled
Neem leaf extract	Chop 6 kg fresh neem leaves and make it into a paste; soak the paste in 6 litres of water overnight; on the following morning boil the suspension and filter by pouring over a coarse cloth; add 150 g of detergent soap powder to the filtrate; make up the volume to 60 litres; this is ready to spray.	Most of the leaf feeding caterpillar pests.
Vitex leaf extract (Nochi)	Grind 6 kg fresh leaves into a paste; soak the paste overnight in 3 litres of cow's urine; on the following morning, stir the solution and boil for 3 hours; allow it to cool; pour it over a coarse cloth and squeeze; to the filtrate, add 150 g of detergent soap powder; the filtrate is then made up to 60 litres by adding water; this is ready to spray.	Castor semi looper.
Calotropis leaf extract	Chop1 kg calotropis leaves and make it into a fine paste; mix the paste with 50 litres water and filter the suspension and spray.	Caterpillar pests.
Jatropa leaf extract	Grind 5 kg of leaves and stem of Jatropa into fine paste; soak the paste in 5 litres of cow's urine overnight; on the following morning, add 2 litres of water and boil for 3 hours; the solution is then concentrated to 5 litres and filtered; this is ready to spray.	All pests.
Tobacco leaf extract	Cut 5 kg of tobacco waste into pieces and soak in 5 litres of cow's urine for 1 week; filter the suspension and mix 150 g of detergent soap powder and 250 ml of water; spray at 50 ml/l water.	Aphids, thrips, beetles, lepidopteran caterpillars.
Agave flesh extract	Grind chopped agave leaves and make it to a paste by thoroughly mixing it; soak it in water for 24 hr; mix the suspension and filter it with a coarse cloth; add 10 g of detergent soap powder and 500 ml of cow's urine to every 100 ml of extract and spray.	Leaf folder.
Garlic clove	Chop garlic cloves and make it into a fine paste; soak it in equal volume of kerosene over night; mix and filter it; add 150 g of detergent soap powder.	Brown leaf hopper, green leaf hopper, thrips, diamond back moth.

Table. 21.4. (Contd.)

Product	Preparation	Pests controlled
Mahua oil spray extract	Mix 30 ml of mahua oil and 5 g of detergent soap powder with 1litre of water and spray.	
Nerium seed extract	Seeds of nerium are dried, powdered, sieved and soaked in water over night; filter with coarse cloth and spray the filtrate.	Rhinoceros beetle.
Cow dung extract	Mix 100 g cow dung in 100 ml water; filter it and spray.	
Ipomea leaf	Clip 5 kg of fresh Ipomea leaves and grind to fine paste; mix paste with equal volume of water and boil; add a little of cow's urine to the boiling mixture; when the mixture is ¼ its volume stop boiling and allow it to cool overnight; filter the solution. Repeat the washing with cow's urine to make a volume of 4.5 litres; add 250 g of detergent soap powder in 500 ml water; mix the extract with the soap solution; use 1.25 litres of solution in 25 litres water.	All pests.

22 SOURCES OF CRITICAL INPUTS FOR THE MANAGEMENT OF PESTS AND DISEASES

22.1. INSECTICIDES

Table 22.1. Common and trade names and sources of availability of insecticides

Group	Common Name	Trade Name/s (Source)
Chlorinated Hydrocarbons	Endosulfan	Agrosulfan (16), Dawn (42), Devigor (9), Endocel (14), Endocin (23), Endogaurd (15), Endodhan (13), Endotaf (37), Endonil (11), Endoin (20), Endosul (46), Endosol (28), Hexasulfan (32), Hildan (17), Hysulfan (18), Lusulfan (26), Nagusulphan (29), Parrysulfan (18), Sedostar (41), Speed (31), Thiodan (4), Tagsulfan (50), Thiokill (52), Topsulfan (49).
	Gamma BHC	Lindane.
Organophos-phates	Acephate	Acatin (23), Acehero (39), Acekill (5), Acemil 75 WP (28), Acet (19), Acetop (49), Acifete (26), Aimthene (1), Artin (32), Asataf (37), Dhanraj (11), Hilphate (17), Hythene 75, SP (18), Lancer (52), Laser (20), Orthane (9), Pace (31), Specthane (43), Starthane (41), Tamaron Gold (4), Tarpedo 75 SP (40), Tremor (55), Volace (54).
	Chlorfenvinphos	Chlorfenvinphos.
	Chlorpyriphos	Agro Chlore (16), Blaze (21), Chorobon (52), Chlorgaurd (15), Classic (26), Coroban (7), Corocin (23), Cyphos (19), Dhanwan 20 (11), Doomer (5), Durmet 20 EC (8), Force (31), Hexaban (32), Hilban (17), Hyban (18), Nuklor 20 EC (12), Pyrivol (54), Radar (40), Robust (39), Ruban (9), Scout (1), Specphos20 (42), Starban (41), Strike (55), Sulban (46), Tafaban (37), Tagban 20 EC (50), Tarkash (3).
	Diazinon	Basudin (47), Tik-20 (37).

Table. 22.1. (Contd.)

Group	Common Name	Trade Name/s (Source)
	Dichlorvos (DDVP)	Agro-DDVP (16), Amidos (1), Dicotop (49), Divap 100 (35), Doom (52), Hilfol (17), Luvan (26), Marvex Super (32), Nuvan (47), Specvos (43), Suchlor (45), Vantaf (37), Vapona (3).
	Dimethoate	Hygro (18), Rogar (37), Rogrin (20), Rogar (24), Tara-909.
	Ethion	Acarin (18), Crush (20), Dhanumit (11), Fosmite (35), Indothion (21), Laser (26), MIT-505 (41), Phosmil (21), Tafethion (37), Volthion (54).
	Fenitrothion	Folithion (4), Hexafen (32), Sumithion (37).
	Fenthion	Baytex (4), Lebaycid (4).
	Formathion	Formathion.
	Malathion	Agromala (16), Hilmala (17), Luthion (26), Malamar (32), Malataf (37), Cythion (8), Malatop (49), Mlamal (20), Maltox (1), Specmal (43), Sulmalathion (46).
	Methyl parathion	Devithion (32), Dhanumar (11), Luthion (26), Metacid (4), Milion (28), Prataf (49), Paratox (1), Profit (20), Tagpar (16).
	Monocrotophos	Agro Monark (16), Atom (21), Azodrin (8), Balwan (37), Bilphos (4), Corophos (7), Croton (24), Gaurdian (15), Hilcron (17), Hycrophos (18), Kadett (35), Luphos (26), Macrophos (50), Milphos (20), Moncar (5), Monochem (30), Monocil (9), Monocin (23), Monocron (31), Monodhan (11), Monophos (32), Monosect (2), Monostar (41), Monovol (54), Nuvacron (47), Parryphos (13), Phoskill (52), Specron (42), Sufos (45), Topcil (49).
	Oxy-demeton methyl	Dhanussystox (11), Hexasystox (32), Metasystox (4), Mode (2).
	Phenthoate	Aimsan (1), Amaze (55), Delsan (7), Elsan (52), Phendal (32), Penthomet (11), Tagsan (50).

Table. 22.1. (Contd.)

Group	Common Name	Trade Name/s (Source)
	Phorate	Agro-phorate (16), Dhan 100 (11), Foratox (35), Forcin (23), Granutox (1), Helmet (50), Hilphorate (17), Luphate (26), Milate (28), Specphor (43), Starphor (41), Thimet (8), Umet (52), Volphor (53), Warrant (40).
	Phoslane	Voltas Phosalone (54), Zolone (38).
	Profenophos	Ajanta (33), Banjo (22), Bolero (24), Carina (35), Curacron (47), Proven (27), Prabal (37).
	Quinalphos	Agroquin (16), Basquin (5), Bayruil (4), Chemlux (30), Dhanulux (11), Ekalux (47), Ekatop (49), Flash (21), Hilquin (17), Hyguin (18), Kinalux (52), Knock (42), Krush (55), Miluax (20), Quick (31), Quinal (46), Quinaltaf (37), Quinatox (1), Quingaurd (15), Shakti (13), Smash (40), Starlux (41), Suquin (45), Tagquin (50), Vazara (26), Volquin (54).
	Triazophos	Hostothion (4), Josh (52), Titan (20), Trelka, Truzo (50), Trihero (39).
Carbamates	Carbaryl	Agroryl (16), Dhanuvin (11), Hexavin (32), Kildiryl (25), Killex Cabaryl (34), Parry Win 50 WP (13), Sevidol (38), Sevin 50 WP (38), Sevin D (38), Sevin Flo 42 (38).
	Carbofuran	Carbocil (9), Caardon (52), Diafuran (33), Furadon (37), Furacarb (1), Furatox (1), Fury (31), Hexafuran (32), Sumo (20).
	Carbosulfan	Marshall (37).
	Indoxacarb	Avant (12), Dash (21), Daksha (37), Dhava (11), Fego (52).
	Methomyl	Astra (13), Dunnet (11), Lannata (12).
	Thiodicarb	Larvin (2).
Pyrethroides	Alphacypermethrin	Forward 10 EC (24), Sherpa Alpha 10 EC (4).
	B-Cyfluthrin	Bulldock (4).
	Cypermethrin	Agro-Cyper (16), Angol (18), Ankush (3), Auzar (55), Bilcyp (4), Challenger (50), Cilcord (9), Colt-25 (35), Cybil 25 EC (4), Cymbush (19), Cyper Top (49), Cypercin (23), Cypergaurd (15),

Table. 22.1. (Contd.)

Group	Common Name	Trade Name/s (Source)
		Cypermar (32), Cypermil (28), Cypersul (46), Cyproid (1), Cyrux 25 (52), Hilcyperin (17), Hi-Power (46), Hycyper (18), Indothrin (21), Jawa (12), Laser 10 EC (24), Piothrin (5), Polytrin (47), Ralothrin (37), Ripcord (8), Shakti (26), Sherpa Cyper 25 EC (4), Simper (19), Spec (43), Spec-Cyperin (42), Starcip (41), Superkiller (11), Suraksha (31), Trofy 25 EC (24), Trofy 25 EC (40), Ustad (52), Volcyper (54), White Gold (30).
	Decamethrin	Decakill (18), Decasyn (1), Decis (2), Easytab 25 EC (2), Faulad Rukrin (8).
	Ethofenfrox	Punakaso.
	Fenvalerate	Agrofen (16), Anchor (19), Bhaskarfen (5), Bhasma (55), Bilfen (4), Fen Fen (32), Fenchem (30), Fencron (47), Fenkill (52), Fennock 20 (9), Fenny (31), Fenval (40), Flame (20), Hilfen (17), Lufen (26), Milfen (28), Newfen (15), Starfen (41), Sumitox (1), Trfen (50), Tatafen (37), Topfen (49), Trumpcard (11), Viper (43).
	Lambda-Cyhalothrin	Agent (21), Bravo (20), Cliva (6), Karate (19), Reeva 5 (37).
	Permethrin	Ambush (19), Hilthrin (17), Permasect (32).
	Pyrethrum	
Insect growth regulators	Diflubenzuron	Dimilin 18 EC, 25 WP (15).
	Diafenthuran	Polo (4), Pegasus (47).
	Flufenoxuron	Cascade 10 EC (3).
	Lufenuron	Match 5 EC (47).
	Novaluron	Rimon 10 EC (21), Rim.
	Tiflubenzuron	Nomolt (9).
Combi products	B-Cyfluthrin 12.5 g + Chlorpyriphos 250 g	Bulldock 262.5 EC (4).
	Imidachloprid 50 g + B-Cyfluthrin 50 g	Confider Ultra 100 EC (4).
	Alphamethrin 1% + Chlorpyriphos 16%	Ducord 17 EC (4), Alfatoon, Lethal Plus (20).
	Fenvalerate 3% + Acephate 25%	Koranda 28 EC (29).

Table. 22.1. (Contd.)

Group	Common Name	Trade Name/s (Source)
	Cypermethrin 5% + Ethion 40%	Nagata 45 EC (37), Rimjim (20).
	Cypermethrin 5% + Chlorpyriphos 50%	Noorani 55 EC (37), Nurello D-505, Lrthal Super 50 (20), Synergy 505 (9).
	Cypermethrin 4% + Profenphos 40%	Polytrin 44 EC (36).
	Cypermethrin 3% + Quinalphos 20%	Prachanda 23 EC, Viraat 23 EC (52).
	Deltamethrin 8% + Endosulfan 32%	Sevit 40 EC.
	Deltamethrin 1% + Trizophos 35%	Spark 36 EC (4).
New molecules	Acetamaprid	Lift (21), Mnik (37), Pride (3), Sharp (20).
	Chlorfenapyr	Intrepid (3).
	Emamectin benzoate	Proclaim (47), Prabhav (37).
	Fipronil	Reagent (4).
	Imidachloprid	Admit (24), Atom (21), Confidar (4), Chemida (6), Imida Gold (52), Sabmida (39), Tata Mida (48), Victor.
	Spinosad	Tracer (9), Spintar (4).
	Thiomethoxam	Actra (4), Ananth 25 WP (37), Click (21), Crusier (47), Renova (52), Reno 70 WP.
Acaricides	Abamectin	Vertimec (47).
	Diafenthiuron	Polo, Pegasus.
	Dicofol	Kelthane (9), Colonel-S (21).
	Ethion	Acarin (18), Crush (20), Dhanumit (11), Fosmite (35), Indothion (21), Laser (26), MIT-505 (41), Phosmil (21), Tafethion (37), Volthion (54).
	Fenazaquin	Magister (12).
	Fenpropathrin + Pyriproxifen	Prempt 20 EC.
	Sulphur	Dhanusulf (11), Farmsulf (41), Hexasul (32), Insulf (52), Microsulf (32), Sulcol, Wet-Sulf (14), Sulfin M-20, (16), Sulphosan (1), Sultaf (37), Thiovit (47).
	Propagate	Omite (11).

22.2. FUNGICIDES

Table 22.2. Common and trade names and sources of availability of fungicides

Group	Common Name	Trade Name/s (Source)
Copper fungicides	Copper hydroxide	Kocide 101 (12).
	Copper oxychloride	Blitox (37), Blue Copper (47), Captox (1), Cupramar (32), Cuprastar (41), Devicopper (10), Dhanucop (11), Fytolan (19), Hexamar Copper (32), Hil Copper (17), Blue Vit (25), Trucop (21).
	Cuprous oxide	Cuprous oxide
Di-carboxymides	Iprodione	Rovral (38).
	Vinclozolin	
Heterocyclic nitrogenous fungicides	Captafol	Foltaf.
	Captan	Captra (21), Caftap (37), Deltan (7), Dhanutan (11), Hexacap (32).
	Folpet	
Mercury fungicides	Methoxy ethyl mercuric chloride	Bagalol (52).
	Phenyl mercury acetate	
Sulphur fungicides (Inorganic)	Wettable Sulphur	Dhanusulf (11), Farmsulf (41), Hexasul (32), Insulf (52), Microsulf (32), Sulcol, Wet-Sulf (14), Sulfin M-20 (16), Sulphosan (1), Sultaf (37), Thiovit (47).
Sulphur fungicides (Organic–Carbamates)	Ferbam	Hexferb (32).
	Maneb/Mancozeb	Abie (47), Agromanco (16), Aimcozeb (1), Dhanuka (11), Dithane M-45 (4), Hilthane (17), Indofil (21), Luzen (26), Leo M-45 (20), Zebthane M-45, Flowin-HT, Luzim (26), Saviour (9), Shield (35), Sparsh (55), Spic Mancozeb (44), Uthane M-45 (52), Zeb (31), Zebthane (37).
	Nabam	
	Thiram	Thiride (20), Hexathir (32), Vegfru Thiram (35).
	Vapam	
	Zineb	Discon-Z (1), Devizeb (10), Hexathane (32), Indofil Z-78 (21).

Table. 22.2. (Contd.)

Group	Common Name	Trade Name/s (Source)
	Ziram	Cumin L (47), Hexazir (32), Tagziron (50), Ziride (20), Vegfru Zitox (35), Dhanuka Z-27 (11).
Quinine fungicides	Chloronil	
	Dichlone	
Systemic fungicides Benzimidazoles	Benomyl	Benlate (32), Benofit (13).
	Carbendazim	Agni (21), Agrozim (16), Aimocozin (1), Arrest (24), Bavistin (3), Benfin (21), Bengard (9), Calzin (26), Benzin (5), Carzim (26), Derosal (2), Dhamusten (11), Diafuran (33), Hycarb (18), Nirmool (41), Sahara (20), Stare (32), Volzim (54), Zen (31), Zoom (52).
	Thiabendazole	
Thiophanates	Thiophanate	Thiophanate.
	Thiophanate Methyl	Cercobin, Roko (55), Prism (20).
Oxathin compounds	Carboxin	Carboxin.
	Oxy-Carboxin	Morestein.
Phenylamides	Oxadixyl	
	Metalaxyl	Apron (24), Matco, Himilgard.
Triazoles	Bitertanol (Biloxazol)	Baycor (4).
	Dichlobutrazol	Vigil.
	Difenconazole	Score (47).
	Flusilazol	
	Flutriafol	
	Hexaconazole	Avone (20), Contaf (37), Conquer (52), Danzole(6), Manage (24), Sitara (21).
	Myclobutanil	Systhane (9), Boon (21).
	Penconazole	Topas (47).
	Propiconazole	Radar (37), Tilt (47), Swing, Sona (20), Dhan (21).
	Triadimefon	Bayleton (4).
	Triadimenol	
	Tricyclazole	Beam (21), Force-11 (20).
Pyrimidines	Fenarimol	
Morpholines	Dodemorph	Dodemorph.

Table. 22.2. (Contd.)

Group	Common Name	Trade Name/s (Source)
	Tridemorph	Calixin (3).
Organophosphorus compounds	Fosetyl –AL	Aliette (2).
	Kitazine & Kitazine-P	
	Pyrazophos	Apagon.
Acetamides	Cymoxynil	
Carbamates	Propamocarb	
	Propineb	Antracol (4).
Di-caroxymides	Procymidone	
Imidazoles	Fenapanil	
Piperazine	Triforine	
Phenol derivatives	Chloroneb	
Miscellaneous systemic fungicides	Antibiotics	Aureofungin, Cycloheximide, Griseofulvin, Kasugamycin, Validamycin, Streptocycline.
	Etridiazol	
	Pyroxychlor	
Miscellaneous fungicides	Chinomethionat	
	Clorothalanil	Bravo, Kavach (47).
	Clorothaniz	
	Dinocap	Karathane (4).
	Dodine	Sylil (4), Syllet (28), Noor (21).
	Edifenphos	Hinosan (4).
	Glyodin	
	Morocide	Morocide
	Prochloraz	
	Quintozine	
Combi products	Metalaxyl 8% + Mancozeb 64%	Ridomil MZ (47), Master (48), Himil (20), Matco-8-64 (21).
	Iprodione 25% + Carbendazim 25%	Quintal (4).
	Carbendazim 12% + Mancozeb 63%	Care (20), Companion (21), Top-20 (18), Sixer 75 WP (11), Saaf (52).
	Cymoxanil 8% + Mancozeb 64%	Curzate 72 WP (12).
	Propioconazole + Difenoconazole 30%	Armure 30 EC.

22.3. RODENTICIDES

Table 22.3. Common and trade names and sources of availability of rodenticides

Common name	Trade name/s (Source)
Aluminium phosphide	Synfume, Phosfume, Celphos (14), Quickphos (52), Phostoxin, Delacea (32), Alphos (1), Fumino (52).
Bromadiolone	Lainirat (47), Moosh-Moosh (37).
Coumatetralyl	Racumin (4).
Magnesium phosphide	Magnophos.
Warfarin	Rodafarin, Ratafin, Fumarin, Warfotax (1).
Zinc phosphide	Synphos, Rasol (52), Zincfox (1), Rasox.
D-D 1: 1 mixture (Dichloro propene – Dichloro propane)	Telone, Vidden-D.
EDCT 3: 1 mixture (Ethylene. dichloride carban tetrachloride)	Kiloptera, Fumex, Dichromek.

Source of Pesticides

1. AIMCO
2. Aventis
3. BASF
4. Bayer
5. Bhaskar Agro Chemicals
6. Cheminova
7. Coromandal Indag
8. Cynamid Agro
9. De Nocil
10. Devi Dayal
11. Dhanuka
12. Du Pont
13. EID Parry
14. Excel
15. Gardha Chemicals
16. Gujarat Agro
17. Hindusthan Insecticides
18. Hyderabad Chemicals
19. ICI Zeneca
20. IIL
21. Indofil
22. Insecticides India
23. IOCL
24. IS Agro
25. Kilpest
26. Lupin
27. Meghamani Organics Ltd.
28. Montari
29. Nagarjuna Chemicals
30. New Chemi
31. NFCL
32. Parry Chemicals
33. Parry India Ltd.
34. Paushak
35. Pesticides India
36. Polychem
37. Rallis India
38. Rhone-Poulenc
39. Sabero
40. Searle India
41. Shaw Wallace
42. Southern Pesticides
43. SPEC
44. Spic

45. Sudarshan Chemicals
46. Sulphur Mills
47. Syngenta
48. Tata
49. Thakar Chemicals
50. Tropical Agro
51. T. Stanes
52. United Phosphorus Ltd.
53. Volrho
54. Voltas
55. Wockardt

22.4. NEMATICIDES

Table 22.4. Globally important nematicides and their manufacturers

Active substance	Chemical group	Trade name	Manufacturer
1, 3-Dichloropropene	Halogenated hydrocarbon	Telone	Bayer Crop Science
Dazomet	Methyl isothiocyanate liberator	Basamid	BASF Corporation
Metham sodium	Methyl isothiocyanate liberator	Vapam	Amvac Chemical Corpn.
Ethoprophos	Organophosphorus	Mocap	Bayer Crop Science
Phenamiphos	Organophosphorus	Nemacur	Bayer Crop Science
Fosthiazate	Organophosphorus	Nemathorin	Syngenta
Cadusafos	Organophosphorus	Rugby	FMC Corporation
Aldicarb	Oxime carbamate	Temik	Bayer Crop Science
Oxamyl	Oxime carbamate	Vydate	Du Pont
Carbofuran	Carbamate	Furadan	FMC Corporation

22.5. PREDATORS

Table 22.5. Sources of availability of predators

Predators	State	Sources of availability
Amblyscius sp.	Tamil Nadu	Rajendra Foundn for Agri Res, Chittar
Apertochrysa sp.	Karnataka	Project Directorate of Biol Control, Bangalore
Blastostethus pallescens	Karnataka	Project Directorate of Biol Control, Bangalore
Brumoides suturalis	Karnataka	Project Directorate of Biol Control, Bangalore. Central IPM Centre, Bangalore
	New Delhi	Entomol Divn, Indian Agri Res Inst, New Delhi
Cardiastethus exiguous	Karnataka	Project Directorate of Biol Control, Bangalore
Cheilomenes sexmaculata	Gujarat	Entomol Dept, Anand Agri Univ, Anand
	Karnataka	Project Directorate of Biol Control, Bangalore. Central IPM Centre, Bangalore
	New Delhi	Entomol Divn, Indian Agri Res Inst, New Delhi
Chilocorus nigrita	Himachal Pradesh	Central IPM Centre, Solan. Ent Dept, Y S Paramar Univ of Hort & Forestry, Solan

Table. 22.5. (Contd.)

Predators	State	Sources of availability
	Jammu & Kashmir	Central IPM Centre, Jammu; Central IPM Centre, Srinagar; Entomol Dept, SK Univ Agri Sci & Technol, Srinagar.
	Karnataka	Project Directorate of Biol Control, Bangalore. Bio Control Res Lab, PCI Pvt Ltd., Bangalore
	Tamil Nadu	Rajendra Foundn for Agri Res, Chittar
	West Bengal	Central IPM Centre, Burdwan
Chrysoperla carnea	Andhra Pradesh	Central Tobacco Res Inst, Rajahmundry. Ent Dept, Acharya N G Ranga Agri Univ, Hyderabad. Pest Control (India) Pvt Ltd, Hyderabad. SOM-IPM Systems (I) Ltd, Hyderabad
	Assam	Pest Control (India) Pvt Ltd, Guwahati
	Bihar & Jarkhand	Pest Control (India) Pvt Ltd, Jamshedpur
	Gujarat	Entomol Dept, Anand Agri Univ, Anand. Pest Control (India) Pvt Ltd, Vadodara
	Haryana	Pest Control (India) Pvt Ltd, Chandigarh
	Karnataka	Project Directorate of Biol Control, Bangalore. Central IPM Centre, Bangalore. Univ of Agri Sci, Entomol Dept, Dharwad. Kadur Agro Pvt. Ltd., Bangalore. Pest Control (India) Pvt Ltd, Bangalore
	Kerala	Pest Control (India) Pvt Ltd, Cochin
	Madhya Pradesh	Indore Biotech Inputs & Res (P) Ltd., Indore. Pest Control (India) Pvt Ltd, Bhopal
	Maharashtra	Entomol Dept, Mahatma Phule Agri Univ, Rahuri. Nathkrupa Biocontrol Lab, Nagpur. Pest Control (India) Pvt Ltd Mumbai-400 063. Pest Control (India) Pvt Ltd, Mumbai 400 059. Pest Control (India) Pvt Ltd, Thane-400 601. Pest Control (India) Pvt Ltd, New Mumbai. Pest Control (India) Pvt Ltd, Nashik-422 009. Pest Control (India) Pvt Ltd, Pune-411 033. Pest Control (India) Pvt Ltd, Pune-411 030
	New Delhi	Biotech Intern Ltd., New Delhi. Pest Control (India) Pvt Ltd, New Delhi
	Orissa	Pest Control (India) Pvt Ltd, Bhubaneswar
	Punjab	Pest Control (India) Pvt Ltd, Chandigarh

Table. 22.5. (Contd.)

Predators	State	Sources of availability
	Rajasthan	Pest Control (India) Pvt Ltd, Jaipur
	Tamil Nadu	Entomol Dept, TNAU, Madurai.
		Basarass Biocontrol Res, Vallavar.
		Greentech Agro Prod P Ltd, Coimbatore.
		Pest Control (India) Pvt Ltd, Chennai.
		Rajendra Foundn for Agri Res, Chittar.
		TN Co-op Sugar Fedn Ltd, Chengalpattu
	Uttarakhand	Pest Control (India) Pvt Ltd, Dehradun
	Uttar Pradesh	Pest Control (India) Pvt Ltd, Bareilly.
		Pest Control (India) Pvt Ltd, Lucknow.
		Pest Control (India) Pvt Ltd, Kanpur
	West Bengal	Pest Control (India) Pvt Ltd, Kolkata
Coccinella septumpunctata	Tamil Nadu	Jeypee Bio Techs, Virudhunagar
Cryptolaemus montrouzieri	Andhra Pradesh	Acharya N G Ranga Agri Univ, Hyderabad.
		Pest Control (India) Pvt Ltd, Hyderabad
	Assam	Pest Control (India) Pvt Ltd, Guwahati
	Bihar & Jarkhand	Pest Control (India) Pvt Ltd, Jamshedpur
	Gujarat.	Pest Control (India) Pvt Ltd, Vadodara
	Haryana	Pest Control (India) Pvt Ltd, Chandigarh
	Karnataka	Project Directorate of Biol Control, Bangalore.
		Agri Technol Infmn Centre, IIHR, Bangalore.
		Kadur Agro Pvt. Ltd, Bangalore.
		Pest Control (India) Pvt Ltd, Bangalore
	Kerala	Pest Control (India) Pvt Ltd, Cochin
	Madhya Pradesh	Central IPM Centre, Indore.
		Indore Biotech Inputs & Res (P) Ltd, Indore.
		Pest Control (India) Pvt Ltd, Bhopal
	Maharashtra	Pest Control (India) Pvt Ltd, Mumbai-400 063.
		Pest Control (India) Pvt Ltd, Mumbai-400 005.
		Pest Control (India) Pvt Ltd, Thane.
		Pest Control (India) Pvt Ltd, New Mumbai.
		Pest Control (India) Pvt Ltd, Nashik.
		Pest Control (India) Pvt Ltd, Pune-411 033.
		Pest Control (India) Pvt Ltd, Pune-411 030
	New Delhi	Entomol Divn, Indian Agri Res Inst, New Delhi.
		Pest Control (India) Pvt Ltd, New Delhi
	Orissa	Pest Control (India) Pvt Ltd, Bhubaneswar
	Punjab	Pest Control (India) Pvt Ltd, Chandigarh
	Tamil Nadu	Basarass Biocontrol Res, Vallavar.
		Greentech Agro Prod P Ltd, Coimbatore.

Table. 22.5. (Contd.)

Predators	State	Sources of availability
		Pest Control (India) Pvt Ltd. Rajendra Foundn for Agri Res, Chittar. Shri Durga Agro Services, Coimbatore
	Uttarakhand	Pest Control (India) Pvt Ltd, Dehradun
	Uttar Pradesh	Pest Control (India) Pvt Ltd, Bareilly. Pest Control (India) Pvt Ltd, Lucknow. Pest Control (India) Pvt Ltd, Kanpur
	West Bengal	Pest Control (India) Pvt Ltd, Kolkata
Curinus coeruleus	Karnataka	Project Directorate of Biol Control, Bangalore
Mallada spp.	Gujarat	Entomol Dept, Anand Agri Univ, Anand
	Karnataka	Project Directorate of Biol Control, Bangalore
	New Delhi	Entomol Divn, Indian Agri Res Inst, New Delhi
	Punjab	Entomol Dept, Punjab Agri Univ, Ludhiana
M. astur	Karnataka	Project Directorate of Biol Control, Bangalore
M. boninensis	Karnataka	Project Directorate of Biol Control, Bangalore
Orius tantillus	Karnataka	Project Directorate of Biol Control, Bangalore
Parena nigrolineata	Karnataka	Project Directorate of Biol Control, Bangalore
Pharoscymnus horni	Karnataka	Project Directorate of Biol Control, Bangalore
Scymnus coccivora	Karnataka	Project Directorate of Biol Control, Bangalore

22.6. PARASITOIDS

Table 22.6. Sources of availability of parasitoids

Parasitoids	State	Sources of availability
Apanteles angaleti	Karnataka	Project Directorate of Biol Control, Bangalore
	Maharashtra	Entomol Dept, Agri College, Pune
	New Delhi	Entomol Divn, Indian Agri Res Inst, New Delhi
Botryoideclava bharatiya	Karnataka	Project Directorate of Biol Control, Bangalore
Brachymeria nosatoi	Andhra Pradesh	Ent Dept, Acharya N G Ranga Agri Univ, Hyderabad
	Karnataka	Central IPM Centre, Bangalore
	Kerala	CPCRI Regional Station, Kayangulam
B. nephantidis	Andhra Pradesh	Ent Dept, Acharya N G Ranga Agri Univ, Hyderabad
	Karnataka	Project Directorate of Biol Control, Bangalore Central IPM Centre, Bangalore
	Kerala	CPCRI Regional Station, Kayangulam
Bracon brevicornis	Karnataka	Project Directorate of Biol Control, Bangalore

Table. 22.6. (Contd.)

Parasitoids	State	Sources of availability
	New Delhi	Entomol Divn, Indian Agri Res Inst, New Delhi. Pest Control (India) Pvt Ltd, New Delhi
	Tamil Nadu	Central IPM Centre, Tiruchirapalli. Basarass Biocontrol Res, Vallavar. Rajendra Foundn for Agri Res, Chittar. TN Coconut Parasite Breeding Centre, Coimbatore. TN Co-op Sugar Fedn Ltd, Chengalpattu
B. hebetor	Andhra Pradesh	Ent Dept, Acharya N G Ranga Agri Univ, Hyderabad
	Karnataka	Project Directorate of Biol Control, Bangalore. Central IPM Centre, Bangalore
	Kerala	CPCRI Regional Station, Kayaṅgulam
	Maharashtra	Entomol Dept, Agri College, Pune. Nathkrupa Biocontrol Lab, Nagpur
	New Delhi	Entomol Divn, Indian Agri Res Inst, New Delhi
	Tamil Nadu	Jeypee Bio Techs, Virudhunagar. Rajendra Foundn for Agri Res, Chittar. Sakthi Biocontrol & Biofert Centre, Kallipatti Gobi
B. kirkpatricki	Karnataka	Project Directorate of Biol Control, Bangalore
	Kerala	CPCRI Regional Station, Kayangulam
Campoletis chloridae	Karnataka	Project Directorate of Biol Control, Bangalore
Cephalonomia stephanoderis	Gujarat	Entomol Dept, Anand Agri Univ, Anand
	Karnataka	Coffee Research Sub Station, Chethalli
	Kerala	Entomol Dept, Kerala Agri Univ, Trichur
	Maharashtra	Central Inst of Cotton Res, Nagpur
	Tamil Nadu	Regional Coffee Res Stn, Thandigudi. Centre for Pl Prot Studies, TNAU, Coimbatore
Chelonus blackburni	Karnataka	Project Directorate of Biol Control, Bangalore
Coccidoxenoides peregrinus	Karnataka	Project Directorate of Biol Control, Bangalore
Copidosoma koehleri	Bihar & Jarkhand	Central IPM Centre, Patna
	Karnataka	Project Directorate of Biol Control, Bangalore
	Maharashtra	Entomol Dept, Mahtma Phule Agri Univ, Rahuri
	New Delhi	Entomol Divn, Indian Agri Res Inst, New Delhi
Cotesia flaviceps	Karnataka	Project Directorate of Biol Control, Bangalore
	Orissa	Central IPM Centre, Bhubaneswar

Table. 22.6. (Contd.)

Parasitoids	State	Sources of availability
	Punjab	Entomol Dept, Punjab Agri Univ, Ludhiana
	Tamil Nadu	Divn of Pl Prot, SBI, Coimbatore
	Uttar Pradesh	Entomol Divn, IISR, Lucknow
C. plutellae	Karnataka	Project Directorate of Biol Control, Bangalore
C. africanus	Andhra Pradesh	Central Tobacco Res Inst, Rajahmundry
	Karnataka	Project Directorate of Biol Control, Bangalore
	Maharashtra	Central IPM Centre, Nagpur
Dipha aphidivora	Karnataka	Agri Res Stn, UAS, Bheemarayanagudi
Elasmus nephantidis	Andhra Pradesh	Ent Dept, Acharya N G Ranga Agri Univ, Hyderabad
	Karnataka	Project Directorate of Biol Control, Bangalore Agri Technol Infmn Centre, IIHR, Bangalore
	Kerala	CPCRI Regional Station, Kayangulam
	Tamil Nadu	Basarass Biocontrol Res, Vallavar
Encarsia spp.	Himachal Pradesh	Central IPM Centre, Solan
	Jammu & Kashmir	Entomol Dept, SK Univ Agri & Technol, Srinagar
	Karnataka	Central IPM Centre, Bangalore
Encarsia guadeloupae	Karnataka	Project Directorate of Biol Control, Bangalore. Agri Technol Infmn Centre, IIHR, Bangalore
Epiricania melanoleuca	Maharashtra	Vasanthdada Sugar Institute, Pune
	Tamil Nadu	Central IPM Centre, Tiruchirapalli. Rajendra Foundn for Agri Res, Chittar
	Uttar Pradesh	Entomol Divn, IISR, Lucknow. Central IPM Centre, Gorakhpur
Eriborus argenteopilosus	Karnataka	Project Directorate of Biol Control, Bangalore
Goniozus nephantidis	Andhra Pradesh	Ent Dept, Acharya N G Ranga Agri Univ, Hyderabad
	Karnataka	Project Directorate of Biol Control, Bangalore Pest Control (India) Pvt Ltd, Bangalore
	Kerala	CPCRI Regional Station, Kayangulam
	Tamil Nadu	Central IPM Centre, Tiruchirapalli Greentech Agro Prod P Ltd, Coimbatore Rajendra Foundn for Agri Res, Chittar TN Coconut Parasite Breeding Centre, Coimbatore
Leptomastix dactylopii	Karnataka	Project Directorate of Biol Control, Bangalore Agri Technol Infmn Centre, IIHR, Bangalore Entomol Divn, Central Coffee Res Inst, Chikmagalur

Table. 22.6. (Contd.)

Parasitoids	State	Sources of availability
	Kerala	Entomol Dept, Kerala Agri Univ, Trichur
	Tamil Nadu	Greentech Agro Prod P Ltd, Coimbatore
Micromus igarotus	Karnataka	Regional Res Stn, UAS, Raichur
Oomyzus sokolowskii	Karnataka	Project Directorate of Biol Control, Bangalore
Sturmiopsis inferens	Karnataka	Agri Technol Infmn Centre, IIHR, Bangalore
Telenomus spp.	Karnataka	Project Directorate of Biol Control, Bangalore
T. remus	Andhra Pradesh	Central Tobacco Res Inst, Rajahmundry
	Karnataka	Project Directorate of Biol Control, Bangalore
	Madhya Pradesh	Central IPM Centre, Indore
Tetrastichus howardi	Karnataka	Project Directorate of Biol Control, Bangalore
T. israeli	Gujarat	Entomol Dept, Anand Agri Univ, Anand
	Karnataka	Project Directorate of Biol Control, Bangalore. Central IPM Centre, Bangalore
	Kerala	CPCRI Regional Station, Kayangulam
	Tamil Nadu	Jeypee Bio Techs, Virudhunagar. Rajendra Foundn for Agri Res, Chittar. TN Coconut Parasite Breeding Centre, Coimbatore
Trichogramma achaeae	Karnataka	Project Directorate of Biol Control, Bangalore. Agri Technol Infmn Centre, IIHR, Bangalore. Central IPM Centre, Bangalore
	Punjab	Entomol Dept, Punjab Agri Univ, Ludhiana
	Tamil Nadu	Rajendra Foundn for Agri Res, Chittar. Shri Durga Agro Services, Coimbatore. Sun Agro Ind India, Chennai
T. brasiliensis	Karnataka	Project Directorate of Biol Control, Bangalore. Agri Technol Infmn Centre, IIHR, Bangalore
	Tamil Nadu	Khatau Agro Tech, Mumbai. National Biocontrol Laboratory, Nagpur
T. chilonis, *T. japonicum*	Andhra Pradesh	Central Tobacco Res Inst, Rajahmundry. Ent Dept, Acharya N G Ranga Agri Univ, Hyderabad. Pest Control (India) Pvt Ltd. SOM-IPM Systems (I) Ltd., Hyderabad
	Assam	State Biocontrol Laboratory, Guwahati. Pest Control (India) Pvt Ltd, Guwahati
	Bihar & Jarkhand	Pest Control (India) Pvt Ltd, Jamshedpur
	Goa	Central IPM Centre, Madgaon
	Gujarat	Pest Control (India) Pvt Ltd, Vadodara

Table. 22.6. (Contd.)

Parasitoids	State	Sources of availability
	Haryana	Entomol Dept, CCS Haryana Agri Univ, Hisar Regional Agri Centre, CCSHAU, Karnal State Biocontrol Laboratory, Sirsa Biocontrol Laboratory, Yamuna Nagar
	Karnataka	Project Directorate of Biol Control, Bangalore Agri Technol Infmn Centre, IIHR, Bangalore. Central IPM Centre, Bangalore. Entomol. Dept, Univ of Agri Sci, Dharwad. Inst of Pulses & Oilseeds Res, Gulbarga. Regional Res Stn, UAS, Raichur. AAO, Parasite Laboratory, Bailhongal. AO, Parasite Laboratory, Gangavathi. AAO, Parasite Laboratory, Mandya. Bio Solujan, Bangalore. Hospet Sugar Mills, Hospet. Kadur Agro Pvt. Ltd., Bangalore. Karnataka Agri Res Foundn, Sameerwadi. Pest Control (India) Pvt Ltd, Bangalore
	Kerala	Pest Control (India) Pvt Ltd, Cochin
	Madhya Pradesh	Indore Biotech Inputs & Res (Pvt.) Ltd., Indore. Pest Control (India) Pvt Ltd, Bhopal
	Maharashtra	Central Inst of Cotton Res, Nagpur. Entomol Dept, Mahtma Phule Agri Univ, Rahuri. Entomol Dept, Marathwada Agri Univ, Parbhani. Entomol Dept, Konkan Agri Univ, Dapoli. KNS Biotech, Nanded. Nathkrupa Biocontrol Lab, Nagpur. Pest Control (India) Pvt Ltd, Mumbai-400 063. Pest Control (India) Pvt Ltd, Mumbai-400 059. Pest Control (India) Pvt Ltd, Thane-400 601 Pest Control (India) Pvt Ltd, New Mumbai. Pest Control (India) Pvt Ltd, Nashik-422 009. Pest Control (India) Pvt Ltd, Pune-411 030. Pest Control (India) Pvt Ltd, Pune-411 033. R.S. Kherdekar, Talpathri. Sandoz (I) Ltd., Mumbai
	New Delhi	Entomol Divn, Indian Agri Res Inst, New Delhi. Biotech Intern Ltd., New Delhi. Navdanya (NGO), New Delhi. Pest Control (India) Pvt Ltd, New Delhi.

Table. 22.6. (Contd.)

Parasitoids	State	Sources of availability
	Orissa	Pest Control (India) Pvt Ltd, Bhubaneswar
	Punjab	Pest Control (India) Pvt Ltd, Chandigarh
	Rajasthan	Biocontrol Lab, ARS, RAU, Sri Ganganagar.
		JDA (Pl Path), Biocontrol Lab, Durgapura.
		DDA (Agro), IPM Lab, Ajmeer.
		DDA (Agro), IPM Lab, Malikpur.
		DDA (Agro), IPM Lab, Rampura.
		DDA (Agro), IPM Lab, Cjjatrapura.
		DDA (Agro), IPM Lab, Chittorgarh.
		DDA (Agro), IPM Lab, Hanumangarh.
		DDA (Pl Path), IPM Lab, Banaswara.
		Adarsh Biocontrol Laboratories, Alwar.
		Pest Control (India) Pvt Ltd, Jaipur
	Tamil Nadu	Divn of Pl Prot, SBI, Coimbatore.
		Regional Coffee Res Stn, Thandigudi.
		Bio-control Lab., Papparapatty.
		Bio-control Lab, Coimbatore.
		Bio-control Lab., Salem.
		Bio-control Lab, Villupuram.
		Bio-control Lab., Trichy.
		Bio-control Lab., Panjupettai.
		Bio-control Lab., Kuruppanaichenpalyam.
		Bio-control Lab., Vinayagapuram.
		Bio-control Lab., Tirunelveli.
		Bio-control Lab., Tanjavur.
		Bio-control Lab., Namakkal.
		Agro Res Lab, Coimbatore.
		Basarass Biocontrol Res, Vallavar.
		Biocontrol Res Centre, Kootham Poondi.
		Greentech Agro Prod P Ltd, Coimbatore.
		Jeypee Bio Techs, Virudhunagar.
		Main Biocontrol Res Lab, Chengalpattu.
		Pest Control (India) Pvt Ltd, Chennai.
		Rajendra Foundn for Agri Res, Chittar.
		Sakthi Biocontrol & Biofert Centre, Kallipatti, Gobi.
		Shri Durga Agro Services, Coimbatore.
		Sinagro Industries India, Chennai.
		Sun Agro Ind India, Chennai.
		Vigneswara Biocontrol, Pollachi

Table. 22.6. (Contd.)

Parasitoids	State	Sources of availability
	Uttarakhand	Pest Control (India) Pvt Ltd, Dehradun
	Uttar Pradesh	Entomol Divn, IISR, Lucknow. Biocontrol Lab, SBBP Agri Univ, Meerut. Bioved Res Soc, Allahabad. Crop Health Bioprod Res Centre, Ghaziabad. Pest Control (India) Pvt Ltd, Bareilly. Pest Control (India) Pvt Ltd, Lucknow. Pest Control (India) Pvt Ltd, Kanpur
	West Bengal	Pest Control (India) Pvt Ltd, Kolkata
T. dendrolimi	Karnataka	Project Directorate of Biol Control, Bangalore. Agri Technol Infmn Centre, IIHR, Bangalore
	Tamil Nadu	Rajendra Foundn for Agri Res, Chittar
T. embryophagum	Himachal Pradesh	Central IPM Centre, Solan. Ent Dept, Y S Paramar Univ of Hort & Forestry, Solan
	Jammu & Kashmir	Central IPM Centre, Jammu. Central IPM Centre, Srinagar. Entomol Dept, SK Univ Agri Sci & Technol, Srinagar
	Karnataka	Project Directorate of Biol Control, Bangalore
T. evanescens	Himachal Pradesh	Central IPM Centre, Solan. Y S Paramar Univ of Hort & Forestry, Solan
	Jammu & Kashmir	Central IPM Centre, Jammu. Central IPM Centre, Srinagar. Entomol Dept, SK Univ Agri & Technol, Srinagar
	Karnataka	Project Directorate of Biol Control, Bangalore
T. maidis	Karnataka	Project Directorate of Biol Control, Bangalore
T. pretiosum	Himachal Pradesh	Central IPM Centre, Solan. Ent Dept, Y S Paramar Univ of Hort & Forestry, Solan
	Jammu & Kashmir	Central IPM Centre, Jammu. Central IPM Centre, Srinagar. Entomol Dept, SK Univ Agri Sci & Technol, Srinagar
	Karnataka	Project Directorate of Biol Control, Bangalore
Trichogrammatoidea armigera	Karnataka	Project Directorate of Biol Control, Bangalore
T. bactrae	Karnataka	Project Directorate of Biol Control, Bangalore

Table. 22.6. (Contd.)

Parasitoids	State	Sources of availability
Trichospilus pupivorus	Karnataka	Central IPM Centre, Bangalore
	Kerala	CPCRI Regional Station, Kayangulam
Xanthopimpla punctata	Karnataka	Project Directorate of Biol Control, Bangalore. Central IPM Centre, Bangalore
	Kerala	CPCRI Regional Station, Kayangulam

22.7. PATHOGENS

22.7.1. Arbuscular Mycorrhizal Fungi

Table 22.7. Trade names and source of availability of Arbuscular Mycorrhizal Fungi

Arbuscular Mycorrhizal Fungi	Trade Name	State	Source of Availability
Glomus mosseae	Symbion VAM	Karnataka	Agri Technol Infmn Centre, IIHR, B'lore
		Maharashtra	Bharatiya Agro Industries Foundation, Pune
		Tamil Nadu	T Stanes & Co, Coimbatore
G. fasciculatum	Symbion VAM	Karnataka	Agri Technol Infmn Centre, IIHR, B'lore
		Maharashtra	Bharatiya Agro Industries Foundation, Pune
		Tamil Nadu	T Stanes & Co, Coimbatore
G. aggregatum		Maharashtra	Bharatiya Agro Industries Foundation, Pune
Gigaspora margarita		Maharashtra	Bharatiya Agro Industries Foundation, Pune
Arbuscular mycorrhizal fungi	Multiplex Trishul	Karnataka	Karnataka Agro Chemicals, B'lore

22.7.2. Antagonistic Fungi Against Plant Diseases and Nematodes

Table 22.8. Trade names and sources of availability of antagonistic fungi against Plant Diseases and nematodes

Antagonistic Fungi/ Trade Name	State	Source of Availability
Ampelomyces quisqualis (Bio-Dewcon)	Tamil Nadu	T. Stanes & Co, Coimbatore
Arthrobotrys cladodes var. *macroides*	Karnataka	Project Directorate of Biol Control, Bangalore
Arthrobotrys oligospora	Karnataka	Project Directorate of Biol Control, Bangalore
Aspergillus nizer (Kalisena SD, Kalisena SL, Sanjeevni)	Gujarat	Cadila Pharmaceuticals Ltd., Ahmedabad

Table. 22.8. (Contd.)

Antagonistic Fungi/ Trade Name	State	Source of Availability
	New Delhi	Plant Path Divn, Indian Agri Res Inst, New Delhi
Dactylella brochophaga	Karnataka	Project Directorate of Biol Control, Bangalore
Fusarium oxysporum	Karnataka	Project Directorate of Biol Control, Bangalore
Fusarium sporotrichoides	Karnataka	Project Directorate of Biol Control, Bangalore
Gliocladium catenulatum	Karnataka	Project Directorate of Biol Control, Bangalore
G. deliquescens	Karnataka	Project Directorate of Biol Control, Bangalore
G. roseum	Karnataka	Project Directorate of Biol Control, Bangalore
Phoma glomerata	Karnataka	Project Directorate of Biol Control, Bangalore
Pochonia chlamydosporia (Bionema)	Karnataka	Project Directorate of Biol Control, Bangalore. Agri Technol Infmn Centre, IIHR, Bangalore. Central Sericultural Res &Trng Inst, Mysore
Paecilomyces lilacinus Bio-Nemator, Biocon, Niyantran, Multiplex Sparsha, Abtec Paecilomyces)	Karnataka	Project Directorate of Biol Control, Bangalore. Agri Technol Infmn Centre, IIHR, Bangalore
T. aureoviride	Karnataka	Project Directorate of Biol Control, Bangalore
T. hamatum	Karnataka	Project Directorate of Biol Control, Bangalore
T. harzianum (Binab-T, Ecoderma, F-Stop, Supravit, Tricodex, Trichoderma)	Andhra Pradesh	ADA (BCL), P.M. Palem. ADA (BCL), Kakinada. ADA (BCL), Nidadavole. ADA (BCL), Ibrahimpatnam. ADA (BCL), Ongole. ADA (BCL), Nellore. DDA (FTC), Nandyal. ADA (BCL), Anantapur. DDA (FTC), Rajendranagar. ADA (Oilseeds) (BCL), Mahabubanagar. ADA (BCL), Nalagonda. ADA (BCL), Mulugu Road. ADA (BCL), Karimnagar. ADA (BCL), Adilabad. Pest Control (India) Pvt Ltd, Hyderabad
	Assam	State Biocontrol Laboratory, Guwahati. Pest Control (India) Pvt Ltd, Guwahati
	Bihar & Jarkhand	Pest Control (India) Pvt Ltd, Jamshedpur
	Gujarat	Pest Control (India) Pvt Ltd, Vadodara
	Haryana	State Biocontrol Laboratory, Sirsa Pest Control (India) Pvt Ltd, Chandigarh

Table. 22.8. (Contd.)

Antagonistic Fungi/ Trade Name	State	Source of Availability
	Karnataka	Project Directorate of Biol Control, Bangalore.
		Agri Technol Infmn Centre, IIHR, Bangalore.
		Central IPM Centre, Bangalore.
		Central Sericultural Res & Trng Inst, Mysore.
		Coffee Research Sub Station, Chethalli.
		Central Coffee Res Inst, Entomol Divn, Chikmagalur.
		Univ of Agri Sci, Entomol Dept, Dharwad.
		Inst of Pulses & Oilseeds Res, Gulbarga.
		Regional Res Stn, UAS, Raichur.
		Agri Res Stn, UAS, Bijapur.
		Kadur Agro Pvt. Ltd., Bangalore.
		Pest Control (India) Pvt Ltd, Bangalore.
		P.J.Margo Pvt. Ltd., Bangalore
	Kerala	Travancore Organic Fertilizers, Kottayam.
		VRM Biotech Products, Changanacherry
	Madhya Pradesh	Pest Control (India) Pvt Ltd, Bhopal
	Maharashtra	Bio Era Technologies, Nagpur.
		Hoechst Schering Agr. Envt, Mumbai.
		KNS Biotech, Nanded.
		Nathkrupa Biocontrol Lab, Nagpur.
		Om Agro Organics, Yavatmal.
		Pest Control (India) Pvt Ltd, Mumbai-400 063.
		Pest Control (India) Pvt Ltd, Mumbai-400 059.
		Pest Control (India) Pvt Ltd, Thane.
		Pest Control (India) Pvt Ltd, New Mumbai.
		Pest Control (India) Pvt Ltd, Nashik.
		Pest Control (India) Pvt Ltd, Pune-411 033.
		Pest Control (India) Pvt Ltd, Pune-411 030.
		Sanvardhini Agro Pvt. Ltd., Satara.
		Soman Biofertilizers, Pune.
		Vidyas Biotech Lab, Nagpur.
		West Coast Herbochem P Ltd, Mumbai
	New Delhi	Biotech Intern Ltd., New Delhi.
		Pest Control (India) Pvt Ltd, New Delhi
	Orissa	Pest Control (India) Pvt Ltd, Bhubaneswar
	Punjab	Pest Control (India) Pvt Ltd, Chandigarh
	Rajasthan	Biocontrol Lab, ARS, RAU, Sri Ganganagar.
		JDA (Pl Path), Biocontrol Lab, Durgapura.
		DDA (Agro), IPM Lab, Ajmeer.

Table. 22.8. (Contd.)

Antagonistic Fungi/ Trade Name	State	Source of Availability
		DDA (Agro), IPM Lab, Malikpur.
		DDA (Agro), IPM Lab, Rampura.
		DDA (Agro), IPM Lab, Cjjatrapura.
		DDA (Agro), l IPM Lab, Chittorgarh.
		DDA (Agro), IPM Lab, Hanumangarh.
		DDA (Pl Path), IPM Lab, Banaswara.
		Pest Control (India) Pvt Ltd, Jaipur
	Tamil Nadu	Centre for Pl Prot Studies, TNAU, Coimbatore.
		Bio-control Lab, Papparapatty.
		Bio-control Lab, Coimbatore.
		Bio-control Lab, Salem.
		Bio-control Lab, Villupuram.
		Bio-control Lab, Trichy.
		Bio-control Lab, Panjupettai.
		Bio-control Lab, Kuruppanaichenpalyam.
		Bio-control Lab, Vinayagapuram.
		Bio-control Lab, Tirunelveli.
		Bio-control Lab, Tanjavur.
		Bio-control Lab, Namakkal.
		Basarass Biocontrol Res, Vallavar.
		Greentech Agro Prod P Ltd, Coimbatore.
		Jeypee Bio Techs, Virudhunagar.
		Pest Control (India) Pvt Ltd, Chennai.
		Rajendra Foundn for Agri Res, Chittar.
		Shri Durga Agro Services, Coimbatore.
		Shrishti Bioprod P Ltd, Coimbatore.
		Sun Agro Ind India, Chennai.
		TN Co-op Sugar Fedn Ltd, Chengalpattu.
		TAP Industries, Chennai
	Uttarakhand	Pl Path Dept, GB Pant Uni Agri & Technol, Pantnagar.
		Pest Control (India) Pvt Ltd, Dehradun
	Uttar Pradesh	Crop Health Bioprod Res Centre, Ghaziabad.
		Pest Control (India) Pvt Ltd, Bareilly.
		Pest Control (India) Pvt Ltd, Lucknow.
		Pest Control (India) Pvt Ltd, Kanpur
	West Bengal	Pest Control (India) Pvt Ltd, Kolkata

Table. 22.8. (Contd.)

Antagonistic Fungi/ Trade Name	State	Source of Availability
T. koningii	Karnataka	Project Directorate of Biol Control, Bangalore
T. longibactriatum	Karnataka	Project Directorate of Biol Control, Bangalore
T. polysporum	Karnataka	Project Directorate of Biol Control, Bangalore
T. pseudokoningii (Nursery Guard)	Karnataka	Project Directorate of Biol Control, Bangalore. Central Sericultural Res &Trng Inst, Mysore
T. virens (Gligard, WRC-GV)	Karnataka	Project Directorate of Biol Control, Bangalore
T. viride (Bio-Cure, Bio-Cure F, Agroderma, Antogon TV, Bioderma, Dermapak, Ecofit, Monitor, Multiplex Nisarga, Rakshak, Trichosan, Trichoft, Trichoderm, Trichonik, Tricho-X, Trichodermin)	Andhra Pradesh	ADA (BCL), P.M. Palem. ADA (BCL), Kakinada. ADA (BCL), Nidadavole. ADA (BCL), Ibrahimpatnam. ADA (BCL), Ongole. ADA (BCL), Nellore. DDA (FTC), Nandyal ADA (BCL), Anantapur. DDA (FTC), Rajendranagar. ADA (Oilseeds) (BCL), Mahabubanagar. ADA (BCL), Nalagonda. ADA (BCL), Mulugu Road. ADA (BCL), Karimnagar. ADA (BCL), Adilabad. Pest Control (India) Pvt Ltd, Hyderabad
	Assam	State Biocontrol Laboratory, Guwahati. Pest Control (India) Pvt Ltd, Guwahati
	Bihar & Jarkhand	Pest Control (India) Pvt Ltd, Jamshedpur
	Gujarat	Pest Control (India) Pvt Ltd, Vadodara
	Haryana	State Biocontrol Laboratory, Sirsa. Pest Control (India) Pvt Ltd, Chandigarh
	Karnataka	Project Directorate of Biol Control, Bangalore. Coffee Research Sub Station, Chethalli. Central Coffee Res Inst, Entomol Divn, Chikmagalur. Kadur Agro Pvt. Ltd., Bangalore. Pest Control (India) Pvt Ltd, Bangalore. P.J.Margo Pvt. Ltd., Bangalore. Travancore Org Fert Co P Ltd., Bangalore
	Kerala	Travancore Organic Fertilizers, Kottayam. VRM Biotech Products, Changanacherry

Table. 22.8. (Contd.)

Antagonistic Fungi/ Trade Name	State	Source of Availability
	Madhya Pradesh	Pest Control (India) Pvt Ltd, Bhopal
	Maharashtra	Bio Era Technologies, Nagpur.
		Hoechst Schering Agr. Envt, Mumbai.
		KNS Biotech, Nanded.
		Nathkrupa Biocontrol Lab, Nagpur.
		Om Agro Organics, Yavatmal.
		Pest Control (India) Pvt Ltd, Mumbai-400 063.
		Pest Control (India) Pvt Ltd, Mumbai-400 059.
		Pest Control (India) Pvt Ltd, Thane.
		Pest Control (India) Pvt Ltd, New Mumbai.
		Pest Control (India) Pvt Ltd, Nashik.
		Pest Control (India) Pvt Ltd, Pune-411 033.
		Pest Control (India) Pvt Ltd, Pune-411 030.
		Sanvardhini Agro Pvt. Ltd., Satara.
		Soman Biofertilizers, Pune.
		Vidyas Biotech Lab, Nagpur.
		West Coast Herbochem P Ltd, Mumbai
	New Delhi	Biotech Intern Ltd., New Delhi.
		Pest Control (India) Pvt Ltd, New Delhi
	Orissa	Pest Control (India) Pvt Ltd, Bhubaneswar
	Punjab	Pest Control (India) Pvt Ltd, Chandigarh
	Rajasthan	Biocontrol Lab, ARS, RAU, Sri Ganganagar.
		JDA (Pl Path), Biocontrol Lab, Durgapura.
		DDA (Agro), IPM Lab, Ajmeer.
		DDA (Agro), IPM Lab, Malikpur.
		DDA (Agro), IPM Lab, Rampura.
		DDA (Agro), IPM Lab, Cjjatrapura.
		DDA (Agro), I IPM Lab, Chittorgarh.
		DDA (Agro), IPM Lab, Hanumangarh.
		DDA (Pl Path), IPM Lab, Banaswara.
		Pest Control (India) Pvt Ltd, Jaipur
	Tamil Nadu	Centre for Pl Prot Studies, TNAU, Coimbatore.
		Bio-control Lab, Papparapatty.
		Bio-control Lab, Coimbatore.
		Bio-control Lab, Salem.
		Bio-control Lab, Villupuram.
		Bio-control Lab, Trichy.
		Bio-control Lab, Panjupettai.

Table. 22.8. (Contd.)

Antagonistic Fungi/ Trade Name	State	Source of Availability
		Bio-control Lab, Kuruppanaichenpalyam.
		Bio-control Lab, Vinayagapuram.
		Bio-control Lab, Tirunelveli.
		Bio-control Lab, Tanjavur.
		Bio-control Lab, Namakkal.
		Basarass Biocontrol Res, Vallavar.
		Greentech Agro Prod P Ltd, Coimbatore.
		Jeypee Bio Techs, Virudhunagar.
		Pest Control (India) Pvt Ltd, Chennai.
		Rajendra Foundn for Agri Res, Chittar.
		Shri Durga Agro Services, Coimbatore.
		Shrishti Bioprod P Ltd, Coimbatore.
		Sun Agro Ind India, Chennai.
		TN Co-op Sugar Fedn Ltd, Chengalpattu.
		TAP Industries, Chennai.
		T. Stanes & Co, Coimbatore
	Uttarakhand	Pl Path Dept, GB Pant Univ Agri & Technol, Pantnagar.
		Pest Control (India) Pvt Ltd, Dehradun
	Uttar Pradesh	Crop Health Bioprod Res Centre, Ghaziabad.
		Pest Control (India) Pvt Ltd, Bareilly.
		Pest Control (India) Pvt Ltd, Lucknow.
		Pest Control (India) Pvt Ltd, Kanpur
	West Bengal	Pest Control (India) Pvt Ltd, Kolkata
Verticillium lecanii	Karnataka	Project Directorate of Biol Control, Bangalore.
		Pest Control (India) Pvt Ltd, Bangalore
	Maharashtra	Sio Agro Research Lab, Mumbai
	Tamil Nadu	Sakthi Biocontrol & Biofert Centre, Kallipatti Gobi
V. suchlosporium	Karnataka	Project Directorate of Biol Control, Bangalore
Arthrobotrys sp. + *Plilacinus* + *Verticillium* sp. (Nemastin)	Maharashtra	Kumar Krishimitra Bioproducts, Pune
T. harzianum + *T. viride* (Ecoderma, NIPROT)	Karnataka	Bio Control Res Lab, PCI Pvt. Ltd., Bangalore.
		P.J. Margo Pvt. Ltd., Bangalore
T. harzianum + *T. viride* + *T. virens* (Combat)	Maharashtra	Kumar Krishimitra Bioproducts, Pune

22.7.3. Entomopathogenic Fungi

Table 22.9. Trade names and sources of availability of entomopathogenic fungi against insect pests

Entomopathogenic Fungi	State	Source of Availability
Beauveria bassiana (Bio-Power, Tracobea, Dispel, Larvocel, Don Muscardin, BABA, Abtech Beauveria)	Karnataka	Project Directorate of Biol Control, Bangalore. Entomol Dept, Univ of Agri Sci, Dharwad. Travancore Org Fert Co P Ltd., Bangalore
	Maharashtra	Vidyas Biotech Lab, Nagpur
	New Delhi	Defence Inst of Physiol, New Delhi. Navdanya (NGO), New Delhi
	Rajasthan	Pesticides India Ltd., Udaipur.
	Tamil Nadu	Divn of Pl Prot, SBI, Coimbatore. Basarass Biocontrol Res, Vallavar. Greentech Agro Prod P Ltd, Coimbatore. Jeypee Bio Techs, Virudhunagar. Rajshree Sugars & Chem Ltd, Varadaraj Nagar. Shri Durga Agro Services, Coimbatore. T. Stanes & Co, Coimbatore
Botryodiplodia theobromae	Karnataka	Project Directorate of Biol Control, Bangalore
Fusarium cocophilum	Karnataka	Project Directorate of Biol Control, Bangalore
F. moniliformae	Karnataka	Project Directorate of Biol Control, Bangalore
F. roseum	Karnataka	Project Directorate of Biol Control, Bangalore
Metarhizium anisopliae (Bio-Magic)	Gujarat	Entomol Dept, Anand Agri Univ, Anand
	Karnataka	Project Directorate of Biol Control, Bangalore. Entomol Dept, Univ of Agri Sci, Dharwad
	Kerala	CPCRI Regional Station, Kayangulam
	Maharashtra	Bio Era Technologies, Nagpur. National Biocontrol Laboratory, Nagpur
	New Delhi	Biotech Intern Ltd., New Delhi. Navdanya (NGO), New Delhi
	Tamil Nadu	Divn of Pl Prot, SBI, Coimbatore. Basarass Biocontrol Res, Vallavar. Biocontrol Res Centre, Kootham Poondi. Rajendra Foundn for Agri Res, Chittar. Shri Durga Agro Services, Coimbatore.

Table. 22.9. (Contd.)

Entomopathogenic Fungi	State	Source of Availability
Nomuraea rileyi	Andhra Pradesh	Directorate of Oil Seeds Res, Hyderabad
	Karnataka	Project Directorate of Biol Control, Bangalore. Agri Technol Infmn Centre, IIHR, Bangalore. Entomol Dept, Univ of Agri Sci, Dharwad
Paecilomyces fumosoroseus (Priority)	Tamil Nadu	Jeypee Bio Techs, Virudhunagar. T. Stanes & Co, Coimbatore
Scopulariopsis brevicaulis	Karnataka	Project Directorate of Biol Control, Bangalore
Verticillium lecanii (Bio-Catch, Tracovert, Verticel, Varsha, Abtech Verticillium)	Karnataka	Project Directorate of Biol Control, Bangalore. Entomol Dept, Univ of Agri Sci, Dharwad. Agri Res Stn, UAS, Bijapur. Travancore Org Fert Co P Ltd., Bangalore
	Maharashtra	Wockhardt Ltd., Mumbai
	Tamil Nadu	T. Stanes & Co, Coimbatore

22.7.4. Bacterial Biopesticides

Table 22.10. Trade names and sources of availability of bacterial biopesticides

Bacterial Biopesticides/ Trade Name/s	State	Source of Availability
Botryodiplodia theobromae	Karnataka	Project Directorate of Biol Control, Bangalore
B. sphaericus	New Delhi	Biotech Intern Ltd, New Delhi
	Tamil Nadu	Sinagro Industries India, Chennai
B. subtilis (Kill Dew DP Epic, Kodiac)	Maharashtra	Kumar Krishimitra Bioproducts, Pune
B. thuringiensis var. *kurstaki* (Biolep, Biosap, Delfin, Dipel, Bibot, Spicturin, Halt)	Andhra Pradesh	Indian Inst of Chem Technol, Hyderabad
	Madhya Pradesh	Indore Biotech Inputs & Res (Pvt.) Ltd., Indore
	Maharashtra	Lupin Laboratories (I) P Ltd., Mumbai. Sandoz (I) Ltd., Mumbai. Wockhardt Ltd., Mumbai
	New Delhi	Biotech Intern Ltd., New Delhi. Hindustan Insecticides, New Delhi
	Tamil Nadu	Tuticorn Alkali Chem & Fert Ltd, Chennai
Pseudomonas fluorescens (Bio-Cure B, Bioshield, Tracomonas, Sudocel, Dagger-G)	Assam	State Biocontrol Laboratory, Guwahati

Table. 22.10. (Contd.)

Bacterial Biopesticides/ Trade Name/s	State	Source of Availability
	Haryana	State Biocontrol Laboratory, Sirsa
	Karnataka	Agri Technol Infmn Centre, IIHR, Bangalore.
		Travancore Org Fert Co P Ltd., Bangalore
	Kerala	Plantrich Chem & Fert Ltd, Kottayam
	New Delhi	Biotech Intern Ltd., New Delhi
	Tamil Nadu	Bio-control Lab, Papparapatty.
		Bio-control Lab, Coimbatore.
		Bio-control Lab, Salem.
		Bio-control Lab, Villupuram.
		Bio-control Lab, Trichy.
		Bio-control Lab, Panjupettai.
		Bio-control Lab, Kuruppanaichenpalyam.
		Bio-control Lab, Vinayagapuram.
		Bio-control Lab, Tirunelveli.
		Bio-control Lab, Tanjavur.
		Bio-control Lab, Namakkal.
		Basarass Biocontrol Res, Vallavar.
		Greentech Agro Prod P Ltd, Coimbatore.
		Jeypee Bio Techs, Virudhunagar.
		Rajendra Foundn for Agri Res, Chittar.
		Sun Agro Ind India, Chennai.
		T. Stanes & Co, Coimbatore

22.7.5. Combination Products of Biocontrol Agents

Table 22.11. Source of availability of combination products of biocontrol agents

Combination Products of Biocontrol Agents/ Trade Name/s	State	Source of Availability
B. pumilis + *P. chlamydosporia* (Bacillus Plus-P)	Karnataka	Agri Technol Infmn Centre, IIHR, Bangalore
B. subtilis + *P. chlamydosporia* (Bacillus Plus-S)	Karnataka	Agri Technol Infmn Centre, IIHR, Bangalore
P. fluorescens + *P. chlamydosporia* (Pseudomonas Plus)	Karnataka	Agri Technol Infmn Centre, IIHR, Bangalore
Arthrobotrys sp. + *P.lilacinus* + *Verticillium* sp. (Nemastin)	Maharashtra	Kumar Krishimitra Bioproducts, Pune

Table. 22.11. (Contd.)

Combination Products of Biocontrol Agents/ Trade Name/s	State	Source of Availability
Aspergillus niger + *T. viride* + *P. lilacinus* (Multiplex Shock)	Karnataka	Karnataka Agro Chemicals, B'lore
P. chlamydosporia + *B. pumilis* (Bacillus Plus-P)	Karnataka	Agri Technol Infmn Centre, IIHR, Bangalore
P. chlamydosporia + *B. subtilis* (Bacillus Plus-S)	Karnataka	Agri Technol Infmn Centre, IIHR, Bangalore
P. chlamydosporia + *P.lilacinus* (Biovert Plus)	Karnataka	Agri Technol Infmn Centre, IIHR, Bangalore
P. chlamydosporia + *P.fluorescens* (Pseudomonas Plus)	Karnataka	Agri Technol Infmn Centre, IIHR, Bangalore
P. chlamydosporia + *T. harzianum* (Trichorich)	Karnataka	Agri Technol Infmn Centre, IIHR, Bangalore
P. chlamydosporia + *T. viride* (Trichovert)	Karnataka	Agri Technol Infmn Centre, IIHR, Bangalore
T. harzianum + *T. viride* (Ecoderma, NIPROT)	Karnataka	Bio Control Res Lab, PCI Pvt Ltd, Bangalore. Margo Pvt Ltd, Bangalore
T. harzianum + *T. viride* + *T. virens* (Combat)	Maharashtra	Kumar Krishimitra Bioproducts, Pune

22.7.6. Entomopathogenic Viruses

Table 22.12. Trade names and sources of availability of entomopathogenic viruses

Entomopathogenic Viruses/Trade Name/s	State	Source of Availability
Ha NPV of *Helicoverpa armigera* (Helivax, Heliokill, Helicide, Helinash, Heliocel)	Andhra Pradesh	ADA (BCL), P.M. Palem. ADA (BCL), Kakinada. ADA (BCL), Nidadavole. ADA (BCL), Ibrahimpatnam. ADA (BCL), Ongole. ADA (BCL), Nellore. DDA (FTC), Nandyal. ADA (BCL), Anantapur. DDA (FTC), Rajendranagar.

Table. 22.12. (Contd.)

Entomopathogenic Viruses/Trade Name/s	State	Source of Availability
		ADA (Oilseeds) (BCL), Mahabubanagar. ADA (BCL), Nalagonda. ADA (BCL), Mulugu Road. ADA (BCL), Karimnagar. ADA (BCL), Adilabad. Kamal Herbal & Pheromone Exp (P) Ltd., Hyderabad. Pest Control (India) Pvt Ltd, Hyderabad. Sneha Biotech, Vijayawada
	Assam	State Biocontrol Laboratory, Guwahati. Pest Control (India) Pvt Ltd, Guwahati
	Bihar & Jarkhand	Pest Control (India) Pvt Ltd, Jamshedpur
	Gujarat	Pest Control (India) Pvt Ltd, Vadodara
	Karnataka	Project Directorate of Biol Control, Bangalore. Agri Technol Infmn Centre, IIHR, Bangalore. Entomol Dept, Univ of Agri Sci, Dharwad. Inst of Pulses & Oilseeds Res, Gulbarga. Regional Res Stn, UAS, Raichur. Agri Res Stn, UAS, Bheemarayanagudi. Bio-Pest Management P Ltd, Bangalore. Cee Kay Associates, Bangalore. Innovative Pest Control Lab, Bangalore. Kadur Agro Pvt. Ltd., Bangalore. Pest Control (India) Pvt Ltd, Bangalore. P.J.Margo Pvt. Ltd., Bangalore
	Kerala	Pest Control (India) Pvt Ltd, Cochin
	Madhya Pradesh	Indore Biotech Inputs & Res (Pvt.) Ltd., Indore. Pest Control (India) Pvt Ltd, Bhopal. Samriddi Bioculture Pvt. Ltd., Indore
	Maharashtra	Central Inst of Cotton Res, Nagpur. Entomol Dept, Mahtma Phule Agri Univ, Rahuri. Entomol Dept, Marathwada Agri Univ, Parbhani. Entomol Dept, Punjabrao Desmukh Agri Univ, Akola. Entomol Dept, Konkan Agri Univ, Dapoli. Ajay Biotech Lab (P) Ltd, Pune. Anand Biocontrol Lab, Jalgaon. Ankur Seeds Pvt. Ltd., Ratnagiri. Arag Biotech, Nagpur. Bio Era Technologies, Nagpur.

Table. 22.12. (Contd.)

Entomopathogenic Viruses/Trade Name/s	State	Source of Availability
		Chintawar Biocontrol Lab, Yeotmal.
		Harit Biocontrol Laboratories, Yeotmal.
		JR Biocontrol Lab, Yeotmal.
		KNS Biotech, Nanded.
		Nathkrupa Biocontrol Lab, Nagpur.
		National Biocontrol Laboratory, Nagpur.
		Pest Control (India) Pvt Ltd, Mumbai-400 063.
		Pest Control (India) Pvt Ltd, Mumbai-400 059.
		Pest Control (India) Pvt Ltd, Thane.
		Pest Control (India) Pvt Ltd, New Mumbai.
		Pest Control (India) Pvt Ltd, Nashik.
		Pest Control (India) Pvt Ltd, Pune-411 033.
		Pest Control (India) Pvt Ltd, Pune-411 030.
		Sanvardhini Agro Pvt. Ltd., Satara.
		Tirupati Agro Inputs, Latur.
		Universal Biotech, Nagpur
	New Delhi	Biotech Intern Ltd., New Delhi.
		JK Traders, Delhi.
		Pest Control (India) Pvt Ltd, New Delhi
	Orissa	Pest Control (India) Pvt Ltd, Chandigarh
	Punjab	Pest Control (India) Pvt Ltd, Chandigarh
	Rajasthan	Biocontrol Lab, ARS, RAU, Sri Ganganagar.
		Adarsh Biocontrol Laboratories, Alwar.
		Pest Control (India) Pvt Ltd, Jaipur
	Tamil Nadu	Centre for Pl Prot Studies, TNAU, Coimbatore.
		Entomol Dept, TNAU, Madurai.
		Bio-control Lab, Papparapatty.
		Bio-control Lab, Coimbatore.
		Bio-control Lab, Salem.
		Bio-control Lab, Villupuram.
		Bio-control Lab, Trichy.
		Bio-control Lab, Panjupettai.
		Bio-control Lab, Kuruppanaichenpalyam.
		Bio-control Lab, Vinayagapuram.
		Bio-control Lab, Tirunelveli.
		Bio-control Lab, Tanjavur.
		Bio-control Lab, Namakkal.

Table. 22.12. (Contd.)

Entomopathogenic Viruses/Trade Name/s	State	Source of Availability
		Basarass Biocontrol Res, Vallavar.
		Greentech Agro Prod P Ltd, Coimbatore.
		Jeypee Bio Techs, Virudhunagar.
		Pest Control (India) Pvt Ltd, Chennai.
		Rajendra Foundn for Agri Res, Chittar.
		Shri Durga Agro Services, Coimbatore.
		Spinco Biotech Pvt. Ltd, Chennai.
		Sun Agro Ind India, Chennai.
		T. Stanes & Co, Coimbatore
	Uttarakhand	Pest Control (India) Pvt Ltd, Dehradun
	Uttar Pradesh	Biocontrol Lab, SBBP Agri Univ, Meerut.
		Crop Health Bioprod Res Centre, Ghaziabad.
		Pest Control (India) Pvt Ltd, Bareilly.
		Pest Control (India) Pvt Ltd, Lucknow.
		Pest Control (India) Pvt Ltd, Kanpur
	West Bengal	Pest Control (India) Pvt Ltd, Kolkata
Sl NPV of *Spodoptera litura* (Spodovax, Spodokill, Spodonash)	Andhra Pradesh	Central Tobacco Res Inst, Rajahmundry
		ADA (BCL), P.M. Palem.
		ADA (BCL), Kakinada.
		ADA (BCL), Nidadavole.
		ADA (BCL), Ibrahimpatnam.
		ADA (BCL), Ongole.
		ADA (BCL), Nellore.
		DDA (FTC), Nandyal.
		ADA (BCL), Anantapur.
		DDA (FTC), Rajendranagar.
		ADA (Oilseeds) (BCL), Mahabubanagar.
		ADA (BCL), Nalagonda.
		ADA (BCL), Mulugu Road.
		ADA (BCL), Karimnagar.
		ADA (BCL), Adilabad.
		Kamal Herbal & Pheromone Exp (P) Ltd., Hyderabad.
		Pest Control (India) Pvt Ltd, Hyderabad
	Assam	Pest Control (India) Pvt Ltd, Guwahati
	Bihar & Jarkhand	Pest Control (India) Pvt Ltd, Jamshedpur
	Gujarat	Pest Control (India) Pvt Ltd, Vadodara

Table. 22.12. (Contd.)

Entomopathogenic Viruses/Trade Name/s	State	Source of Availability
	Karnataka	Project Directorate of Biol Control, Bangalore.
		Agri Technol Infmn Centre, IIHR, Bangalore.
		Entomol Dept, Univ of Agri Sci, Dharwad.
		Regional Res Stn, UAS, Raichur.
		Bio-Pest Management Pvt. Ltd., Bangalore.
		Pest Control (India) Pvt Ltd, Bangalore.
	Kerala	Pest Control (India) Pvt Ltd, Cochin
	Madhya Pradesh	Pest Control (India) Pvt Ltd, Bhopal
	Maharashtra	National Biocontrol Laboratory, Nagpur.
		Pest Control (India) Pvt Ltd, Mumbai-400 063.
		Pest Control (India) Pvt Ltd, Mumbai-400 059.
		Pest Control (India) Pvt Ltd, Thane-400 601.
		Pest Control (India) Pvt Ltd, New Mumbai.
		Pest Control (India) Pvt Ltd, Nashik-422 009.
		Pest Control (India) Pvt Ltd, Pune-411 030.
		Pest Control (India) Pvt Ltd, Pune-411 033
	New Delhi	Biotech Intern Ltd., New Delhi.
		Pest Control (India) Pvt Ltd, New Delhi
	Orissa	Pest Control (India) Pvt Ltd, Chandigarh
	Punjab	Pest Control (India) Pvt Ltd, Chandigarh
	Rajasthan	Biocontrol Lab, ARS, RAU, Sri Ganganagar.
		Adarsh Biocontrol Laboratories, Alwar.
		Pest Control (India) Pvt Ltd, Jaipur
	Tamil Nadu	Centre for Pl Prot Studies, TNAU, Coimbatore.
		Entomol Dept, TNAU, Madurai.
		Basarass Biocontrol Res, Vallavar.
		Jeypee Bio Techs, Virudhunagar.
		Pest Control (India) Pvt Ltd, Chennai.
		Shri Durga Agro Services, Coimbatore.
		Sun Agro Ind India, Chennai
	Uttarakhand	Pest Control (India) Pvt Ltd, Dehradun
	Uttar Pradesh	Pest Control (India) Pvt Ltd, Bareilly.
		Pest Control (India) Pvt Ltd, Lucknow.
		Pest Control (India) Pvt Ltd, Kanpur
	West Bengal	Pest Control (India) Pvt Ltd, Kolkata

22.7.7. Entomopathogenic Nematodes

Table 22.13. Trade names and sources of availability of entomopathogenic nematodes

Entomopathogenic Nematodes/Trade Name/s	State	Source of Availability
Heterorhabditis spp. (Green Cammandos)	Karnataka	Project Directorate of Biol Control, Bangalore
	Maharashtra	Bio Sense Crop Protection, Mumbai
Steinernema spp. (Soil Commandos)	Karnataka	Project Directorate of Biol Control, Bangalore
	Maharashtra	Bio Sense Crop Protection, Mumbai

22.8. PHEROMONES

Table 22.14. Source of availability of pheromones of fruit crop pests

Pheromone	State	Source of Availability
Oriental fruit moth	Maharashtra	Agri Sense BCS, Mumbai. Bio Sense Crop Protection, Mumbai
Gypsy moth	Maharashtra	Bio Sense Crop Protection, Mumbai
Codling moth	Maharashtra	Agri Sense BCS, Mumbai. Bio Sense Crop Protection, Mumbai
San Jose scale	Maharashtra	Agri Sense BCS, Mumbai
Apple leaf roller	Maharashtra	Agri Sense BCS, Mumbai
Apple fruit miner	Maharashtra	Agri Sense BCS, Mumbai
Red banded leaf roller	Maharashtra	Agri Sense BCS, Mumbai
Pear leaf roller	Maharashtra	Agri Sense BCS, Mumbai
Peach tree borer	Maharashtra	Agri Sense BCS, Mumbai
Peach twig borer	Maharashtra	Agri Sense BCS, Mumbai
Grape torteix	Maharashtra	Agri Sense BCS, Mumbai
Mediterranean fruit fly	Maharashtra	Agri Sense BCS, Mumbai
California red scale	Maharashtra	Agri Sense BCS, Mumbai

22.9. NEEM-BASED AND OTHER BOTANICALS

Table 22.15. Source of availability of neem–based and other botanicals

Neem-Based & other Botanicals	State	Source of Availability
Achook	Maharashtra	Bahar Agrochemicals, Ratnagiri Mahagrapes, Pune
Aphidin	Maharashtra	Eco-Max Agrosystems Ltd, Mumbai
Avana	Tamil Nadu	EID Parry (I) Ltd., Bioprod Divn, Chennai

Table. 22.15. (Contd.)

Neem-Based & other Botanicals	State	Source of Availability
Aza	Andhra Pradesh	SOM-IPM Systems (I) Ltd., Hyderabad
Azadit	Rajasthan	Pesticides India Ltd., Udaipur
Biogold	Tamil Nadu	SPIC, Bio-Tech Divn, Chennai
Bioneem	Maharashtra	Ajay Biotech Lab (P) Ltd, Pune
Bionema	Uttar Pradesh	Bioved Res Soc, Allahabad
Biopest	Uttar Pradesh	Bioved Res Soc, Allahabad
Biosol	Tamil Nadu	A.V. Thomas Pvt. Ltd., Chennai
Eco-Neem	Karnataka	Murkumbi Bio Agro P Ltd., Belgaum. Pest Control (India) Pvt Ltd, Bangalore
Eco-Neem Plus	Karnataka	Murkumbi Bio Agro P Ltd., Belgaum. Pest Control (India) Pvt Ltd, Bangalore
Ecotin	Karnataka	Monarch Biofert & Res Centre, Bangalore
Field Marshal	Gujarat	Ganesh Biocontrol System, Gondal
Fortuna Aza	Andhra Pradesh	Fortune Bio-tech Limited, Secunderabad
Jai Neem	Haryana	Jai Chemicals, Faridabad
Jawan	Maharashtra	Mc DA Agro Pvt. Ltd., Mumbai
Jeevan Crop Protector	Maharashtra	Mc DA Agro Pvt. Ltd., Mumbai
Juerken	Tamil Nadu	Madurai Chem & Agro Ind Ltd., Madurai
Limonool	Karnataka	Bio-Multi Tech Pvt. Ltd., Bangalore
Margocide-CK	Karnataka	Monofix Agro Products Ltd., Hubli
Margosan-O	Karnataka	Monofix Agro Products Ltd., Hubli
Neemactin	Maharashtra	Wockhardt Ltd., Mumbai
Neemak	Karnataka	West Coast Herbochem P Ltd, Bangalore
	Maharashtra	West Coast Herbochem P Ltd, Mumbai
Neemarin	New Delhi	Biotech Intern Ltd., New Delhi
Neemark	Karnataka	West Coast Herbochem P Ltd, Bangalore
Neemasol	Tamil Nadu	EID Parry (I) Ltd., Pesticides Divn, Chennai
Neemax	Maharashtra	Eco-Max Agrosystems Ltd, Mumbai
Neemazal-F	Tamil Nadu	EID Parry (I) Ltd., Bioprod Divn, Chennai
Neemazal-T/S	Tamil Nadu	EID Parry (I) Ltd., Bioprod Divn, Chennai
Neemocide	West Bengal	Vinayak Fats & Proteins P Ltd, Kolkata
Neemol	Andhra Pradesh	Ramson Laboratories, Vijayawada
	Tamil Nadu	Agro Links, Trichy

Table. 22.15. (Contd.)

Neem-Based & other Botanicals	State	Source of Availability
Neemolin	Maharashtra	Khatau Agro Tech, Mumbai
Neemosan	Tamil Nadu	Agronule Industries, Trichy
Neemox	Maharashtra	Prakash Farm Chemicals, Mumbai
Neempourn	Tamil Nadu	Prabhakar Oil Mills, Trichy
Neemta	Gujarat	A.J. Chemicals, Ahmedabad
Neethrin	Uttar Pradesh	Amitul Agro Chem Pvt. Ltd., Gorakhpur
Neem Base	New Delhi	Agrochem Divn, Indian Agri Res Inst, New Delhi
Neem Based pesticides	Maharashtra	Bahar Agrochemicals, Ratnagiri. KNS Biotech, Nanded. Om Agro Organics, Yavatmal
	Tamil Nadu	Ecobiocides & Botanicals Pvt. Ltd., Theni. Jeypee Bio Techs, Virudhunagar
Neem Cake	Uttarakhand	Sri Ram Solvent Extraction Plant, Jaspur
Neem Gold	Tamil Nadu	EID Parry (I) Ltd., Bioprod Divn, Chennai. SPIC, Bio-Tech Divn, Chennai
Neem Guard	Maharashtra	Gharda Chemicals Pvt. Ltd., Mumbai
Neem Hit	Maharashtra	Skylark Agro Chemicals, Thane
Neem Oil	Karnataka	Agro Extracts Ltd., Bangalore
	Maharashtra	Mc DA Agro Pvt. Ltd., Mumbai. National Organic Chem Ind Ltd., Mumbai. Sio Agro Research Lab, Mumbai
	Uttarakhand	Sri Ram Solvent Extraction Plant, Jaspur
Neem Plus	Maharashtra	BD Keathen & Co., Mumbai
Neem Rich	Maharashtra	National Chem Lab, Pune
Neem seed	Maharashtra	Cheminova Agro-Tech, Nandubar
Neem Soap	Karnataka	Agri Technol Infmn Centre, IIHR, Bangalore
Neem Top	Tamil Nadu	Sri Krishna Co, Coimbatore
Nim-76	New Delhi	Defence Inst of Physiol, New Delhi
Nimba	New Delhi	Entomol Divn, Indian Agri Res Inst, New Delhi
Nimbasal	New Delhi	Nimba Foods & Agrochem, New Delhi. Pesto-Chem India Ltd, New Delhi
Nimbicidine	Tamil Nadu	T. Stanes & Co, Coimbatore
Nimbitor	Maharashtra	Zandu Pharmaceutical Works, Mumbai
Nimbosol	Tamil Nadu	Victoria Laboratories, Salem

Table. 22.15. (Contd.)

Neem-Based & other Botanicals	State	Source of Availability
Nimitox	Karnataka	Rallis India Ltd., Bangalore
Nimlin	Maharashtra	Sunline Agro Chem P Ltd, Dhulia
Nimbo Bas	Tamil Nadu	Basarass Biocontrol Res, Vallavar
Ozoneem	Haryana	Ozone Biotech, Faridabad
Phytowin	Tamil Nadu	Phyto Products Ltd., Thiruthuripoondi
Pongamia oil-based biopesticide	Tamil Nadu	Jeypee Bio Techs, Virudhunagar
Pongamia Soap	Karnataka	Agri Technol Infmn Centre, IIHR, Bangalore
Rakshak	Karnataka	Murkumbi Bio Agro P Ltd., Belgaum. P.J. Margo Pvt. Ltd., Bangalore
RD-9 Repellin	Andhra Pradesh	ITC Ltd., Hyderabad
Replin 555	Tamil Nadu	Micro Chemicals Ltd., Madurai
Shaktiman	Gujarat	Khetiwadi Corner, Vadodara. Krishna Biotech Pvt. Ltd., Veravel. Oceon Agro India Ltd., Baroda
Sukrina	Tamil Nadu	Canster Chemicals Ltd., Chennai
Suneem	Maharashtra	Sunida Exports Ltd., Mumbai
Swaticure	Maharashtra	Swathi Industries Ltd., Mumbai
Tric	Uttar Pradesh	Amitul Agro Chem Pvt. Ltd., Gorakhpur
Tricure	Karnataka	Murkumbi Bio Agro P Ltd., Belgaum. P.J. Margo Pvt. Ltd., Bangalore
Vapacide	Andhra Pradesh	Indian Inst of Chem Technol, Hyderabad

REFERENCES

Ahmed, M. 1990. Deviation in feeding behaviour of *Achaea janata* Linn. (Lepidoptera: Noctuidae). *Ann. Ent.* **8** (2): 15–17.

Alam, M.M., Khan, A.M. and Saxena, S.K. 1973. Some new host records of root-knot nematode, *Meloidogyne incognita* (Kofoid & White, 1919) Chitwood, 1949. *Indian J. Nematol.* **3**: 159–160.

Ali, S.S. 1982. Occurrence of root-knot nematodes in cardamom plantations of Tamil Nadu. *Proc. of Fifth Symp. on Plantation Crops*, Kasaragod, Kerala, India, pp. 615–20.

Ali, S.S. 1984. Preliminary observations on the effect of some systemic nematicides and neem oil cakes in a cardamom field infected with root-knot nematodes (Abstr.). *Proceedings of the Sixth Symposium on Plantation Crops*, Kottayam, Kerala, India, pp. 215–23.

Ali, S.S. 1986. Root-knot nematode problem in cardamom and its management. *Proceedings of the Second Group Discussion on the Nematological Problems of Plantation Crops,* Central Coffee Res. Inst., Balehonnur, pp. 10–12.

Ali, S.S. and Koshy, P.K. 1982. Occurrence and distribution of root-knot nematodes in cardamom plantations of Kerala. *Nematol. Medit.* **10**: 107–10.

Ali, S.S. and Venugopal, M.N. 1992. *Nematol. Medit.* **20**: 65–66.

Ali, S.S. and Venugopal, M.N. 1993. *Curr. Nematol.* **4**: 19–24.

Anandaraj M., Devasahayam, S., Krishnamoorthy, B., Mathew, P.A. and Rema, J. 2001. *Cinnamon.* Extn. Pamphlet, Indian Inst. of Spices Res., Calicut, Kerala, India.

Anandaraj, M. and Devasahayam, S. 2004. Pests and diseases of cinnamon and cassia, *In. Cinnamon and Cassia* (P.N. Ravindran, K.N. Babu and M. Shylajah, *eds.*), pp.139–258. CRC Press, New York.

Anon. 1979. *Tropical Legumes: Resources for the Future.* National Acad. Sci., Washington DC, pp. 117–21.

Anon, 1987. Improvement of betel vine cultivation in Mahoba. *Progress Report 1986–87.* National Botanical Res. Inst., Lucknow, India.

Anon. 1996. *Cinnamon; Cultivation and Processing.* Tech. Bull. 5, Dept. of Export Agriculture, Ministry of Agriculture, Lands and Forestry, Sri Lanka, pp. 7–8.

Awasthi, M.D. 2001. Pesticide residues in fruits and vegetables. *In: Integrated Pest Management in Horticultural Ecosystems* (P. Parvatha Reddy, A. Verghese and N.K. Krishna Kumar, *eds.*), pp. 263–78. Capital Publishing Co., New Delhi.

Bakhetia, D.R.C. 1986. Pest management in rapeseed and mustard. *Pestcides* **20**: 32–39.

Bakhetia, D.R.C. and Sekhon, B.S. 1993. Ecological factors influencing the incidence of the aphid, *Lipaphis erysimi* (Kalt.) on *Brassica* spp. in Punjab. *In: National Seminar on Oilseeds Research and Development in India: Status and Strategies,* Hyderabad, pp. 183–84.

Basu, S.D. and Gope, B. 1985. Control of root-knot nematode *Meloidogyne incognita* (Kofoid & White) Chitwood in tea seedlings with some systemic nematicides. *Two Leaves and a Bud* **32**: 31–35.

Bhat, M.N., Hegde, R.K., Hiremath, P.C. and Naidu, K.S. 1988, A note on the occurrence of a die-back on cinnamon in Karnataka, *Curr. Res.*, Univ. Agric. Sci., Bangalore **17** (11): 13.

Blair, G.P. 1966. The use of immature nuts of *Cocos nucifera* for studies on *Rhadinaphelenchus cocophilus*. *Nematologica* **11**: 590–92.

Brasier, C.M., and Griffin, M.J. 1979. Taxonomy of *Phytophthora palmivora* of cocoa. *Trans. Br. Mycol. Soc.* **72**: 111–43.

Butani, D.K. 1979. *Insects of Fruits*. Periodical Export Agency, Vivek Vihar, Delhi, 415 pp.

Butani D.K. 1983. Spices and pest problems 2. Cinnamon. *Pesticides* **17** (9): 32–33.

Cardet, C., Kanji, T., Delobel, A. and Dantu, P. 1998. Efficacy of neem and groundnut oil in protecting leguminous seeds against seed beetles in the Sahel. *Agroforestry Systems* **40**: 39–40.

Chakravarthy, A.K. 1993. *Vertebrate Pest Management*. Final Report of ICAR ad-hoc Project, Univ. of Agri. Sci., Bangalore, 98 pp.

Chakravarthy, A.K., Srihari, K. and Chandra Prakash, J. 1998. Management of bird pests in oil palm plantations in Karnataka. *In: Advances in IPM for Horticultural Crops* (P. Parvatha Reddy, N.K. Krishna Kumar and A. Verghese, *eds.*), pp. 183–87. Assoc. for Advancement of Pest Mangmt. in Hort. Ecosystems, Bangalore.

Chandra Mouli, B. and Parthiban, M. 1991. *UPASI Tea Sci. Dept. Bull.* **44**: 14–16.

Chang, T.T. 1992. Decline associated with brown *Phellinus noxius*. *Plant* **2**: 90–95.

Charles, J.S.K. and Kurian, K.J. 1979. Studies on nematode incidence in ginger. *Proceedings of the Second Symposium on Plantation Crops*, pp. 50–57.

Chaturvedi, A.N. 1985. *Firewood Farming on the Degraded Lands of the Gangetic Plain*. U.P. Forest Bull. No. 50, Lucknow, Govt. of India Press (I), 286 pp.

Coronel, R.E. 1991. *Tamarindus indica* L. *In: Plant Resources of South East Asia, Wageningen; Pudoc. No. 2. Edible Fruits and Nuts* (E.W.M. Verheij and R.E. Coronel, *eds.*), pp. 298–301. PROSEA Foundation, Bogor, Indonesia.

Da Graca, J.V., Vuuren, S.P. and Van. 1980. Transmission of avocado sunblotch disease to cinnamon. *Plant Dis.* **64**: 475.

D'Souza, G.I., Kumar, A.C., Viswanathan, P.R.K. and Shamanna, H.V. 1970. Relative distribution and prevalence of plant parasitic nematodes in the coffee tracts of South-Western India. *Indian Coffee* **34**: 329–30 & 342.

Eapen, S.J. 1987. Plant parasitic nematode problem in small cardamom. *Proceedings of the Third Group Discussion on the Nematological Problems of Plantation Crops*, Sugarcane Breeding Institute, Coimbatore, pp. 43–44.

Eapen, S.J. 1995. *Investigations on Plant Parasitic Nematodes Associated with Cardamom*. Final Report of the Project Nema I (813), Indian Inst. of Spices Research, Calicut, 39 pp.

Eapen, S.J., Geetha, S.M. and Leemoll, M. 1987. A note on the association of *Radopholus similis* with the yellow disease of betel vine. *Indian J. Nemat.* **17**: 137–38.

Eapen, S.J., Ramana, K.V. and Sarma, Y.R. 1997. Evaluation of *Pseudomonas fluorescens* isolates for control of *Meloidogyne incognita* in black pepper (*Piper nigrum* L.). *In: Biotechnology of Spices, Medicinal & Aromatic Plants* (S. Edison, K.V. Ramana, B. Sasikumar, K.N. Babu and S.J. Eapen, *eds.*), pp. 129–33. Indian Inst. of Spices Res., Calicut.

Eapen, S.J. and Venugopal, M.N. 1995. Field evaluation of *Paecilomyces lilacinus* and *Trichoderma* spp. in cardamom nurseries for the control of root-knot nematodes and rhizome rot disease. *National Symp. on Nematode Problems of India–An Appraisal of the Nematode Mangmt. with Eco-friendly Approaches and Biocomponents,* Indian Agri. Res. Inst., New Delhi.

Erwin, D.C., Bartnicki-Garcia, S., and Tsao, P.H. (*eds.* 1983. Phytophthora: *Its Biology, Taxonomy, Ecology, and Pathology*. American Phytopathological Society, St. Paul, MN.

Erwin, D.C., and Ribeiro, O.K. 1996. Phytophthora *Diseases Worldwide*. American Phytopathological Society, St. Paul, MN.

Esser, R.P. 1969. *Rhadinaphelenchus cocophilus* a potential threat to Florida palms. *Nematology Circular No. 9,* Division of Plant Industry, Florida Department of Agriculture, Gainesville, Florida, USA.

Evans, H.C. 1981. *Pod Rot of Cacao Caused by* Moniliophthora (Monilia) roreri. Phytopathol. Pap. 24. Commonw. Mycol. Inst., Kew, Surrey, England.

Evans, H.C. 2001. Risk assessment of fungal biocontrol agents of weeds and plant diseases. *BMS International Symposium on Bioactive Fungal Metabolites*, University of Swansea.

Evans, H.C., and Prior, C. 1987. Cocoa pod diseases: causal agents and control. *Outlook on Agriculture* **16**:35–41.

Evans, H.C., Stalpers, J.A., Sampson, R.A., and Benny, G.L. 1978. On the taxonomy of *Monilia roreri*, an important pathogen of *Theobroma cacao* in South America. *Can. J. Bot.* **56**: 2528–2532.

Freitas, O.M.B.L. De and Moura, R.M. 1986. Response of cultivars of cassava (*Manihot esculenta* Crantz.) to *Meloidogyne incognita* and *M. javanica* (Nematoda: Heteroderidae) and comparison with hydrogen cyanide content. *Nematol. Medit.* **10**: 109–31.

Fu and Chang, T.T. 1999. Anthracnose of cinnamon, *Taiwan J. Sci.* **14**: 513–15.

Fulton, R.H. 1989. The cacao disease trilogy: black pod, monilia pod rot, and witches'-broom. *Plant disease* **73**: 601–603.

Gaspin, R.M. and Valdez, R.B. 1979. Pathogenicity of *Meloidogyne* spp. and *Rotylenchulus reniformis* on sweet potato. *Ann. Trop. Res.* **1**: 20–25.

Gnanapragasam, N.C. 1982. Effect of potassium fertilization and of soil temperature on the incidence and pathogenicity of the root-lesion nematode, *Pratylenchus loosi* Loof on tea. *Tea Quarterly* **51**: 169–74.

Gnanapragasam, N.C. 1987. Integrated strategies to manage nematodes in tea. *Proceedings of the International Conference on Tea Quality and Human Health*, Hangzhou, China, pp. 117–83.

Gnanapragasam, N.C. 1991. Influence of soil amendments in reducing pathogenicity to tea by the root lesion nematode, *Pratylenchus loosi*. *Proceedings of the International Symposium on Tea Science,* Shizuoka, Japan, pp. 684–88.

Gnanapragasam, N.C., Dharmasena, W.A.M. and Liyanage, D.D. 1989. Development of age resistance in young tea to different species of *Meloidogyne*. *Int Nematol. Network Newsl.* **6**: 4.

Gnanapragasam, N.C. and Herath, U.B. 1990. Pathogenicity of the burrowing nematode, *Radopholus similis* to young tea (susceptible clone, TRI 2025) at different density. *Sri Lanka Journal* **58**: 83–86.

Gnanapragasam, N.C. and Manuelpillai, M.E.K. 1990. Influence of inoculum levels and temperature on the population build-up and pathogenicity of the root-lesion nematode, *Pratylenchus loosi* Loof on tea. *Tea Quarterly* **53**: 19–22.

Gowda, P.V. and Nambiar, K.K.N. 1988. *J. Plantn. Crops* **18** (Suppl.): 233–38.

Gregory, P.H. (*ed.*). 1974. Phytophthora *Diseases of Cocoa*. Longman, London.

Gregory, P.H., and Maddison, A.C., *eds.* 1981. *Epidemiology of* Phytophthora *on Cocoa in Nigeria*. Commonw. Mycol. Inst., Kew., Phytopathol. Pap. No. 25.

Griffith, R. 1977. A species of bacterium pathogenic to the palm weevil, *Rhynchophorus palmarum*. 1976 Ann. Rept., Red Ring Res. Divn., Minst. of Agri. Land and Fisheries, Trinidad and Tobago, pp. 31–35.

Griffith, R. 1987. Red ring disease of coconut palm. *Plant Dis.***71**: 193–96.

Griffith, R. and Koshy, P.K. 1989. Nematode parasites of coconut and other palms. *In: Plant Parasitic Nematodes in Subtropical and Tropical Agriculture* (M. Luc, R.A. Sikora & J. Bridge, *eds.*), pp. 363–86. CAB International, Wallingford, Oxon, U.K.

Gunasekaran, C.R., Vadivelu, S. and Jayaraj, S. 1987. Experiments on nematodes of turmeric–A review. *Proceedings of the Third Group Discussions on the Nematological Problems of Plantation Crops*, Sugarcane Breeding Institute, Coimbatore, pp. 45–46.

Gunasena, H.P.M. and Hughes, A. 2000. *Tamarind*. International Centre for Underutilized Crops, Southampton, UK, 171 pp.

Hagley, E.A.C. 1962. The palm weevil *Rhynchophorus palmarum* L. a probable vector of red ring disease of coconuts. *Nature* **193**: 499.

Handa, D.K. and Mathur, B.N. 1981. Control of root-knot of chillies with nematicides. *Second Nematology Symposium*, Tamil Nadu Agri. Univ., Coimbatore, p. 48.

Hazarika, K., Dutta, P.K. and Saikia, L. 1998. Field efficacy of *Paecilomyces lilacinus* against root-knot nematode in betel vine. *J. Phytol. Res.* **11**: 171–73.

Hocking, D. 1993. *Trees for Drylands*, Tamarindus indica *L. Family Leguminosae: Caesalpinodeae*. Oxford & IBH Publishing Co., New Delhi, pp. 305–07.

Hoyle, J.C. 1968. A simple method for poisoning coconut trees. *J. Agri. Sci. Trinidad* **68**: 459–65.

Huang, C.S. 1966. Host-parasite relationships of the root-knot nematode in edible ginger. *Phytopathology* **56**: 755–59.

Ichniohe, M. 1980. Studies on the root-knot nematode of black pepper plantation in Amazon. *Annual Report of the Society of Plant Protection, North Japan* **31**: 1–8.

Jagdale, G.B., Pawar, A.B. and Darekar, K.S. 1985a. Pathogenicity of *Meloidogyne incognita* on betel vine (*Piper betle* L.). *Indian J. Nematol.* **15**: 244.

Jagdale, G.B., Pawar, A.B. and Darekar, K.S. 1985b. Effect of organic amendments on root-knot nematodes infecting betel vine. *Int. Nematol. Network Newsl.* **2** (3): 7–10.

Jahagirdar, S. and Siddaramaiah, A.L. 2000. Estimation of avoidable loss due to foot rot of black pepper. *Karnataka J. Agril. Sci.* **13**: 1019–1021.

Jonathan, E.I., Arulmozhiyan, R., Muthusamy, S. and Manuel, W.W. 2000. Field application of *Paecilomyces lilacinus* for the control of *Meloidogyne incognita* on betel vine, *Piper betle*. *Nematol. Medit.* **28**: 131–33.

Jonathan, E.I., Cannayane, I., Bommiraju, P., Amutha, G. and Samiyappan, R. 2004. Biological control of root-knot nematode, *Meloidogyne incognita* and *Phytophthora* wilt complex in betel vine with plant growth promoting rhizobacteria. *National Symposium on Paradigms in Nematological Research for Biodynamic Farming*, Univ. of Agri. Sci., Bangalore, p. 73–74.

Jonathan, E.I., Nagalakshmi, S. and Padmanabhan, D. 1990. Estimation of yield losses in betel vine due to *Meloidogyne incognita*. *Int. Nematol. Network Newsl.* **7** (4): 26.

Jonathan, E.I., Padmanabhan, D. and Ayyamperumal, A. 1995. Biological control of root-knot nematode on betel vine, *Piper betel* by *Paecilomyces lilacinus. Nematol. Medit.* **23**: 191–93.

Joseph, K.V. and Oommen, P. 1960. Notes on some insect pests infesting dry tamarind fruits in Kerala State. *Indian J. Ent.* **22**: 172–80.

Kairu, G.M. 1991. *The Coffee Research Foundation Ann. Rept.* 1991–92.

Kalita, S. and Bora, B.C. 2006. Effect of nematicides and organic amendments in the management of *Meloidogyne incognita* in nursery tea. *Indian J. Nematol.* **36**: 148–49.

Kandasamy, K. and Sivakumar, C.V. 1981. Effect of *Meloidogyne arenaria* on the germination of chilli (*Capsicum annum* Linn.) seeds. *Second Nematol. Symp.*, Tamil Nadu Agri. Univ., Coimbatore, p. 16.

Karunakaran, P. and Nair, M.C. 1980. Leaf spot and die back disease of *Cinnamomum zeylanicum* caused by *Colletotrichum gloeosporioides, Plant Dis.* **64**: 220–21.

Karunakaran, P., Nair, M.C. and Das, L. 1993. Grey blight disease of cinnamon (*Cinnamomum verum* Bercht. & Presl.) leaves. *J. Spices Aromatic Plants* **2**: 66–67.

Kaur, D.J. 1987. *Studies on Nematodes Associated with Ginger* (Zingiber officinale *Rosc.*) *in Himachal Pradesh*. Ph. D. thesis, Himachal Pradesh Univ., Shimla, India, 198 pp.

Kellam, M.K., and Zentmyer, G.A. 1986. Morphological, physiological, ecological, and pathological comparisons of *Phytophthora* species isolated from *Theobroma cacao. Phytopathology* **76**: 159–64.

Kerr, A. and Vythilingam, M.K. 1966. Replanting eelworm infested areas. *Tea Quarterly* **37**: 67–72.

Khan, A.M., Saxena, S.K., Siddiqi, Z.A. and Upadhyay, R.S. 1975. Control of nematodes by crop rotation. *Indian J. Nematol.* **5**: 214–21.

Khan, M.R. and Ghadipur, M.H. 2004. Management of root-knot disease of some solanaceous vegetables by soil application of fly ash. *National Symposium on Paradigms in Nematological Research for Biodynamic Farming,* Univ. of Agri. Sci., Bangalore, p. 92.

Khan, S.N., Misra, B.M. and Rehill, B.S. 1988. Two new species of *Cercospora*. *Indian J. Forestry* **11**: 188–90.

Kore, S.S. and Dhaliwal, A.L. 1981. Epidemiology of leaf spot disease of chilli caused by *Xanthomonas campestris pv. vesicatoria. Indian Phytopath.* **34**: 113.

Koshy, P.K. 1986. The burrowing nematode, *Radopholus similis* (Cobb, 1893) Thorne, 1949. *In: Plant Parasitic Nematodes of India-Problems and Progress.* (G. Swarup and D.R. Dasgupta, *eds.*), pp. 223–248. Indian Agri. Res. Inst., New Delhi, India.

Koshy, P.K. 2002. Nematode pests of spices and condiments and their control. *In: Nematode Control in Crops* (M.M. Alam & N. Sharma, *eds.*), pp. 187–200. International Book Distributing Company, Lucknow, India.

Koshy, P.K. and Geetha, S.M. 1992. Nematode pests of palms and cocoa. *In: Nematode Pests of Crops* (D.S. Bhatti and R.K. Walia, *eds.*), pp. 214–227. CBS Publishers & Distributors, Delhi.

Koshy, P.K. and Jasy, T. 1990. Effect of the burrowing nematode, *Radopholus similis* on sweet potato (*Ipomea batatas* L.) (Abstr.). *National Symposium on Recent Advances in the Production and Utilization of Tropical Tuber Crops,* ISRC, Central Tuber Crops Res. Inst., Thiruvanathapuram, India, p. 47.

Koshy, P.K. and Jasy, T. 1991. Host preference of the burrowing nematode, *Radopholus similis* populations from India. *Indian J. Nematol.* **21**: 39–51.

Koshy, P.K., Nair, R.C.P., Sosamma, V.K. and Sundararaju, P. 1976. On the incidence of root-knot nematode in cardamom nurseries. *Indian J. Nematol.* **6**: 174–75.

Koshy, P.K. and Sosamma, V.K. 1975. Host-range of the burrowing nematode *Radopholus similis* (Cobb, 1893) Thorne, 1949. *Indian J. Nematol.* **5**: 255–57.

Koshy, P.K. and Sosamma, V.K. 1977. Identification of the physiological races of *Radopholus similis* populations infesting coconut, areca nut and banana in Kerala, South India. *Indian Phytopath.* **30**: 415–16.

Koshy, P.K. and Sosamma, V.K. 1978. Studies on the population fluctuations of *Radopholus similis* in coconut and areca nut roots. *Indian Phytopath.* **31**: 180–85.

Koshy, P.K. and Sosamma, V.K. 1979. Control of the burrowing nematode, *Radopholus similis* (Cobb, 1893) Thorne, 1949 on coconut seedlings with DBCP. *Indian J. Nematol.* **9**: 32–33.

Koshy, P.K. and Sosamma, V.K. 1987. Pathogenicity of *Radopholus similis* on coconut (*Cocos nucifera* L.) seedlings under greenhouse and field conditions. *Indian J. Nematol.* **17**: 108–18.

Koshy, P.K., Sosamma, V.K. and Samuel, R. 1999. Interactive effect of vesicular arbascular mycorrhizae (VAM) with *Radopholus similis* on coconut seedlings. *In: Nematology–Challenges and Opportunities in 21st Century, (*Usha K. Mehta, *ed.*), pp. 106–10. Sugarcane Breeding Institute, Coimbatore.

Koshy, P.K., Sosamma, V.K. and Sundararaju, P. 1977. Screening of plants used as pepper standards against root-knot nematode. *Indian Phytopath.* **30**: 128–29.

Koshy, P.K., Sosamma, V.K. and Sundararaju, P. 1979. Reactions of thirty one areca germplasm collections to the burrowing nematode, *Radopholus* similis. *Plant Dis. Reptr.* **63**: 433–35.

Koshy, P.K., Sundararaju, P. and Sosamma, V.K. 1978. Occurrence and distribution of *Radopholus similis* (Cobb, 1893) Thorne, 1949 in South India. *Indian J. Nematol.* **8**: 49–58.

Krishnamoorthy, P.N. and Krishna Kumar, N.K. 2002. Advances in the use of botanicals in the IPM of major vegetable crops. *Proc. Intn. Conf. Vegetables,* Bangalore, pp.

Krauss, U., and Hebbar, P. 1999. *Research Methodology in Biocontrol of Plant Diseases. Workshop Manual.* CATIE, Turrialba, Costa Rica, 28 June-4 July, 1999.

Krauss, U., and Soberanis, W. 2000. Biological control of frosty pod (*Moniliophthora roreri*) and other pod pathogens in Peru. Paper presented at *13th Int. Cocoa Research Conf.*, Kota Kinabalu, Sabah, Malaysia, 9–14 October, 2000.

Krauss, U., Soberanis, W., and Matthews, P. 1999. The use of antagonist mixtures in biocontrol. *In: Research Methodology in Biocontrol of Plant Diseases with Special Reference to Fungal Diseases of Cocoa* (U. Krauss and K. P. Hebbar, *eds.*), pp.112–22. CATIE, Costa Rica.

Kumar, A.C. 1982. Evaluation of Nemacur 5G against coffee root-lesion nematode, *Pratylenchus coffeae*. *J. Coffee Res.* **12**: 1–7.

Kumar, A.C. 1984. The symptoms and diagnosis of the disorder 'spreading decline' (Cannoncadoo dieback) with a note on spread and control of the causal agent. *J. Coffee Res.* **14**: 156–59.

Kumar, A.C. 1985. Investigations on the three species of *Hemicriconemoides* (Nematoda) associated with 'Crinkle leaf' disorder in coffee. *J. Coffee Res.* **15**: 90–98.

Kumar, A.C. 1988. Nematode problem of coffee and its management. *Indian Coffee* **52**: 12–19.

Kumar, A.C., Viswanathan, P.R.K. and D'Souza, G.I. 1971. A study of plant parasitic nematodes of certain commercial crops in coffee tracts of South India. *Indian Coffee* **36**: 1–3.

Kumara, K.L.W. 1999a, Study of leaf galls, leaf spots and scab-like condition of cinnamon *(Cinnamomum zeylanicum* Blume) in Matara District, *Proc.* SLAAS, Colombo, Sri Lanka, 65 p.

Kumara, K.L.W. 1999b. Effect of some agronomic practices on the incidence of leaf spot disease *(Colletotrichum gloeosporioides)* in cinnamon *(Cinnamomum zeylanicum* Blume) in southern region of Sri Lanka. SLAAS-University of Ruhuna Seminar on "*Pest Control in the Next Millennium*", SLAAS, Colombo.

Lawrence, J.S., Luz, E.D.M.N., and Resnik, F.C.Z. de. 1982. The relative virulence of *Phytophthora palmivora* and *P. capsici* on cacao in Bahia, Brazil. *Revista Theobroma* **12**: 189–98.

Lazarovits, G., Conn, K., and Tenuta, M. 2000. Control of *Verticillium dahliae* with soil amendments: Efficacy and mode of action. *In: Advances in Verticillium Research and Disease Management* (E.C. Tjamos, R.C. Rowe, J.B. Heale, and D.R. Fravel, *eds.*), PP. 274–91. American Phytopathological Society, St. Paul, Minn.

Luz, E.D.M.N., and Campêlo, A.M.F.L. 1984. Equilíbrio populacional das espécies de *Phytophthora* responsáveis pela podridao-parda na Bahia, Brasil. Pages 279–86 in: *Proc. 9th Int. Cocoa Res. Conf.*, Lome, Togo, 12–18 Feb., 1984.

Mahal, M.S., Singh, R., Bakhetia, D.R.C. and Singh, B. 1992. Strategies for the supervised control of mustard aphid, Lipaphis erysimi (Kalt.) on raya, Brassica juncea Coss. *In: Proceedings of National Symposium on Recent Advances in Integrated Pest Management,* Punjab Agri. Univ., Ludhiana, pp. 13–14.

Malaguti, G. 1953. Pudricon del cogello de la palmera de accite Africana (*Elaeis guineensis* Jacq.). *Agr. Trop. Venezuela* **3**: 13–21.

Mammen, K.V. 1974. On a wilt disease of betel vine in Kerala caused by root-knot nematode. *Agric. Res. J. Kerala* **12**: 76.

Mani, M. and Krishnamoorthy, A. 1994. Selectivity of pesticides to the parasitoid *Aphelinus* sp. of the green peach aphid, *Myzus persicae* infesting sweet pepper. *J. Aphidol.* **8**: 79–87.

Mani, M.S. 1973. *Plant Galls of India.* Macmillan Co. India Ltd, New Delhi, India, 354 pp.

Medhane, N.S., Jagdale, G.B., Pawar, A.B. and Darekar, K.S. 1985. Effect of *Tagetes erecta* on root-knot nematodes infecting betel vine. *Int. Nematol. Network Newsl.* **2** (3): 11–12.

Menon, K.K. 1949. The survey of pollu and root diseases of pepper. *Indian J. Agric. Sci.* **19**: 89–136.

Menon, K.S. and Nair, T.V.R. 1978. *Proc. PLACROSYM* **2**: 416–24.

Menon, M.R., Sajoo, B.V., Ramakrishnan, C.K. and Remadevi, L. 1972. *Curr. Sci.* **41**: 231.

Midha, R.L. and Trivedi, P.C. 1991. Estimation of losses caused by *Meloidogyne incognita* on coriander, cumin and fennel. *Curr. Nematol.* **2**: 159–62.

Midha, R.L., Trivedi, P.C. and Yadav, B.D. 2001. *Paecilomyces* spp. as biocontrol agents against *Meloidogyne incognita* on coriander. *Natl. Cong. on Centenary of Nematol. in India–Appraisal & Future Plans*, Indian Agri. Res. Inst., New Delhi, pp.146–47.

Mital, V.P. and Khanna, H.C.L. 1978. Efficacy of some fumigants against adults of tamarind bruchid, *Caryedon serratus* F. *Bull. Grain Technol.* **16**: 147–48.

Mohandas, C. and Ramana, K.V. 1987. Slow wilt disease of black pepper and its control. *Indian Cocoa, Areca nut and Spices J.* **11**: 10–11.

Mohandas, C. and Ramana, K.V. 1987. *Third Group Discussion on the Nematological Problems of Plantation Crops*, Sugarcane Breeding Institute, Coimbatore, pp. 43–44.

Mohandas, C. and Ramana, K.V. 1991. Pathogenicity of *Meloidogyne incognita* and *Radopholus similis* on black pepper (*Piper nigrum* L.). *J. Plantation Crops* **19**: 41–43.

Mohanty, K.C., Sahoo, N.K. and Ray, S. 1972. Occurrence of *Radopholus similis* (Cobb, 1893) Thorne, 1949 on banana and pepper in wide areas of Orissa, India. *Afro-Asian Nematol. Network Newsl.* **1** (1): 25–26.

Mollison, J.W. 1900. *Cultivation of the Betel Palm, Cardamom and Pepper in the Karnataka District of Bombay Presidency with Notes on the Manures used in the Spice Gardens of the Tract*. Bull. No. 20, Dept. Land Rec. and Agri., Bombay, 24 pp.

Muraleedharan, N. 1991. Non-arthropod pests–Nematodes and rodents. *In: Pest Management in Tea* (Director, *ed.*), pp. 77–81. United Planter's Association of Southern India, Tea Res. Inst.

Muralidharan, A., Nair, M.G. and Jayasankar, N.P. 1986. *Indian Coco. J.* **17**: 3–6.

Naidu, R. andVenugopaal, M.N. 1982. *Proc. PLACROSYM* **5**: 563–571. Indian Soc. for Plantn. Crops, CPCRI, Kasaragod, Kerala, India.

Nakat, R.V., Acharya, A., Jonathan, E.I., Hazarika, K., Jha, S., Singh, U.S., Wasnikar, A.R., Subba Rao, D.V. and Sitaramaiah, K. 1995. Evaluation of ***Paecilomyces lilacinus*** for the control of root-knot nematodes *Meloidogyne incognita* on betel vine. *National Symp. on Nematode Problems of India–An Appraisal of the Nematode Mangmt. with Eco-friendly Approaches and Biocomponents*, Indian Agri. Res. Inst., New Delhi, p. 44.

Nambiar, K.K.N. and Rethinam, P. 1986. *Thanjavur Wilt/Ganoderma Disease of Coconut*. Pamphlet No. 30, Central Plantation Crops Res. Inst., Kasaragod.

Nambiar, K.K.N. and Rethinam, P. 1987. *Proc. Int. Symp. on Ganoderma Wilt Disease of Palms and other Perennial Crops,* Thanjavur, India, p. 19.

Nambiar, K.K.N. and Sarma, Y.R. 1976. *J. Plantn. Crops* **4**: 21–22.

Okioga, D.M. 1978. *Kenya Coffee* **43**: 221–30.

Orchard, J., Collin, H.A., Hardwick, K., and Isaac, S. 1994. Changes in morphology and measurement of cytokinin levels during the development of witches' brooms on cocoa. *Plant Pathology* **43**: 65–72.

Pandey, R.C. and Trivedi, P.C. 1992. Biological control of ***Meloidogyne incognita*** by ***Paecilomyces*** *lilacinus* in *Capsicum annum. Indian Phytopath.* **45**: 134–35.

Parvatha Reddy, P. and Nagesh, M. 2002. A*vermectins–Isolation, Fermentation, Preliminary Characterization and Screening for Nematicidal Activity.* Tech. Bull.17, Indian Inst. of Hort. Res., Bangalore, 28 pp.

Patel, B.A., Patel, R.G., Patel, H.R., Vyas, R.V. and Patel, B.N. 2001. Management of root-knot nematodes in turmeric. *Natl. Cong. on Centenary of Nematol. in India–Appraisal & Future Plans*, Indian Agri. Res. Inst., New Delhi, p. 165.

Patel, D.J., Makadia, B.M. and Shah, H.M. 1982. Occurrence of root-knot on turmeric (*Curcuma longa* L.) and its chemical control. *Indian J. Nematol.* **12**: 168–71.

Patel, J.R., Patel, R.C. and Amin, P.R. 1983. Effect of systemic insecticides on thrips and mites and its relation to leaf curl disease of chilli. *Gujarat Agri. Univ. Res. J.* **8**: 129–33.

Patel, N.C., Patel, J.J., Jayani, D.B. and Patel, J.R. 1997. Evaluation of insecticide schedules for controlling chilli thrip, *Scirtothrips dorsalis* Hood infesting chilli, *Capsicum annuum* L. *Indian J. Agric. Sci.*

Patel, P.N. *et al.* (*eds.*). 1972. *Plant Bacteriology Vol. I. Bacterial Diseases of Plants in India*. Indian Council of Agri. Res., New Delhi, 261 pp.

Patnaik, P.R., Mohanty, K.C., Mahapatra, S.N. and Sahoo, S. 2004. Evaluation of chilli varieties against root-knot nematode, *Meloidogyne incognita. National Symposium on Paradigms in Nematological Research for Biodynamic Farming,* Univ. of Agri. Sci., Bangalore, p. 100.

Pereira, J.L. 1985. Chemical control of Phytophthora pod rot of cacao in Brazil. *Cocoa Growers Bull.* **36**: 23–38.

Pereira, J.L. 1996. *Management of Witches' Broom disease of Cocoa: A Contemporary Retrospective*. Cocoa Producers' Alliance, 8/10 Broad Street, P.O. Box 1718, Lagos, Nigeria.

Perera, H.A.S., Sritharan, R., and Perera, K.P. 1985. Some studies of cinnamon galls in Sri Lanka. *Sri Lankan J Agric. Sci.* **22**: 23–27.

Perveen, S., Ehteshamul-Haque, S. and Ghaffar, A. 1998. Efficacy of *Pseudomonas aeruginosa* and *Paecilomyces lilacinus* in the control of root rot-root knot disease complex on some vegetables. *Nematol. Medit.* **26**: 209–12.

Piper, S., Martínez, A., Hidalgo, E., and Krauss, U. 2000. Effect of formulation on population dynamics of mycoparasites on the surface of cocoa pods. Paper presented at the *INCOPED 3rd Int. Seminar on Cocoa Pest and Diseases*, Kota Kinabalu, Sabah, Malaysia, 16–17 October, 2000.

Poornima, K. and Vadivelu, S. 1998. Pathogenicity of *Meloidogyne incognita* to turmeric (*Curcuma longa* L.). *In: Nematology-Challenges and Opportunities in 21st Century* (Usha K. Mehta, *ed.*), pp. 29–33. Sugarcane Breeding Institute, Coimbatore.

Prakasam, V. 1991. Red leaf spot of cinnamon in Lower Pulney hills of Tamil Nadu. *Indian Cocoa, Arecanut and Spices J.* **14**: 123 p.

Purdy, L.H., and Schmidt, R.A. 1996. Status of cacao witches' broom: biology, epidemiology, and management. *Annu. Rev. Phytopathol.* **34**: 573–94.

Radha, K., Rethinam, P. and Menon, K.P.V. 1962. *Indian Cocon. J.* **16**: 3–11.

Rajagopal, V., Patil, K.D. and Amma, B.S.K. 1987. *Indian Cocon. J.* **18**: 3–6.

Rajapakse, R.H.S. and Ratnasekera, D. 1997. Studies on the distribution and control of leaf galls in cinnamon caused by *Trioza cinnamoni* Boselli. in Sri Lanka. *Int. J. Trop. Agric.* **15**: 53–56.

Rajapakse, R.H.S. and Kulasekera, V. L. 1982, Some observations on insect pests of cinnamon in Sri Lanka. *Entomon* **7**: 221–23.

Raju, K.S. and Rao, G.S. 1984. Studies to fix up optimum spray intervals against bacterial leaf spot and fruit rot disease of chilli. *South Indian Hort.* **32**: 319–20.

Ramachandran, N., Sarma, Y.R., Anandraj, M. and Jose, A. 1988. Effect of climatic factors on *Phytophthora* leaf infection in black pepper mixed cropping system. *J. Plantn. Crops* **16**: 140–48.

Ramana, K.V. 1986. Slow wilt disease of black pepper and the role of plant parasitic nematodes in its etiology (Abstr.). *Proceeding of the Second Group Discussion on Nematological Problems of Plantation Crops*, Central Coffee Research Station, Balehonnur, pp. 17–21.

Ramana, K.V., Eapen, S.J. and Sarma, Y.R. 1998. Effect of *Meloidogyne incognita, Pythium aphanidermatum* and *Ralstonia solanacearum* alone and in combinations in ginger. *In: Nematology-Challenges and Opportunities in 21st Century* (Usha K. Mehta, *ed.*), pp. 87 93. Sugarcane Breeding Institute, Coimbatore.

Ramana, K.V. and Mohandas, C. 1987. Plant parasitic nematodes associated with black pepper (*Piper nigrum* L.) in Kerala. *Indian J. Nematol* **17**: 62–66.

Ramana, K.V., Mohandas, C. and Balakrishnan, R. 1987. Role of plant parasitic nematodes in the slow wilt disease complex of black pepper (*Piper nigrum* L.) in Kerala. *Indian J. Nematol* **17**: 225–30.

Rands, R.D. 1922. Stripe canker of cinnamon caused by *Phytophthora cinnamomi* n. sp. *Meded. Inst. Voor Plantenzieketen* **54**: 53

Rethinam, P. 1984. *Indian Cocon. J.* **15** (2): 3–11.

Rethinam, P., Rajamannar, M. and Narasimhachari, C.L. 1989. *Indian Cocon. J.* **20**: 1–4.

Ristaino, J.B., and Johnston, S.A. 1999. Ecologically based approaches to management of *Phytophthora* blight on bell pepper. *Plant Dis.* **83**:1080–1089.

Rudgard, S.A., and Butler, D.R. 1987. Witches' broom disease in Rondonia, Brazil: pod infection in relation to pod susceptibility, wetness, inoculum, and phytosanitation. *Plant Pathology* **36**: 515–22.

Saharan, G.S., Verma, P.R. and Nashaat, N.I. 1997. *Monograph on Downy Mildew of Crucifers.* Agri. And Agri. Food Canada, Saskatoon Res. Centre, Tech. Bull., 150 pp.

Sahasranaman, K.N., Pillai, N.G., Jayasankar, N.P., Potty, V.P., Varkey, T., Amma, P.G.K. and Radha, K. 1983. *In: Coconut Research and Development* (N.M. Nayar, *ed.*), pp. 160–65). Wiley Eastern, New Delhi, 518 pp.

Saikia, B. 1992. *Pathogenicity, Crop Loss Assessment and Biological Control of* Meloidogyne incognita *(Kofoid and White, 1919) Chitwood, 1949 on Betel Vine (*Piper betle *L.)*. M. Sc. (Agri.) thesis, Assam Agri. Univ., Jorhat.

Samuels, G.J., Pardo-Schultheiss, R., Hebbar, K.P., Lumsden, R.D., Bastos, C.N., Costa, J.C., and Bezerra, J.L. 2000. *Trichoderma stromaticum*, sp. nov., a parasite of the cacao witches broom pathogen. *Mycol. Res.* **104**: 760–64.

Sarma, Y.R., Nambiar, K.R.N. and Nair, C.P.R. 1974. Brown rot of turmeric. *J. Plantation Crops* **2**: 33–34.

Saxena, S.K., Alam, M.M. and Khan, A.M. 1974. Chemotherapeutic control of nematodes with Vydate oxamyl, VC-13 and Dazomet on chilli plants. *Indian J. Nematol.* **4**: 235–38.

Schieber, E. and Sosa, O.N. 1960. Nematodes on coffee in Guatemala. *Plant Dis. Reptr.* **44**: 722–23.

Seenivasan, N., Parameswaran, S., Sridar, P., Gopalakrishnan, C. and Gnanamurthy, P. 2001. Application of bioagents and neem cake as soil application for the management of root-knot nematode in turmeric. *Natl. Cong. on Centenary of Nematol. in India–Appraisal & Future Plans*, Indian Agri. Res. Inst., New Delhi, pp. 164.

Shah, J.J. and Raju, E.C. 1977. Histopathology of ginger (*Zingiber officinale*) infected by soil nematode *Meloidogyne* spp. *Phyton* (Austria) **16**: 79–84.

Shekhawat, P.S. and Chakravarti, B.P. 1976. Factors affecting development of bacterial leaf spot of chillies caused by *Xanthomonas vesicatoria. Indian Phytopath.* **29**: 392–97.

Shekhawat, P.S. and Chakravarti, B.P. 1977. Assessment of loss, symptomatology and occurrence of bacterial leaf spot of chillies caused by *Xanthomonas vesicatoria* in Rajasthan. *Indian J. Mycol. Pl. Path.***7**: 1–14.

Shekhawat, P.S. and Chakravarti, B.P. 1978. Survival of *Xanthomonas vesicatoria* the incitant of chilli leaf spot and a medium for its isolation. *Indian J. Pl. Path.***14**: 11–19.

Siddaramaiah, A.L. and Kulkarni, S. 1982. Control of powdery mildew disease of tamarind seedlings. *Indian Forester* **108**: 361–64.

Singh, H., Singh, B., Gupta, S.K. and Singh, K. 1993. Response of mustard aphid to different cultivars of *Brassica* under protected and unprotected conditions. *J. Insect Sci.* **6**: 117–18.

Singh, S., Kumar Surender, Singh Amerika. 2001. *Integrated Pest Management for Rapeseed and Mustard.* Tech. Bull. (in Hindi), NCIPM, New Delhi, 35 pp.

Singh, V., Dubey, O.P., Nair, C.P.R. and Pillai, G.B. 1978. Biology and bionomics of insect pests of cinnamon. *J. Plan. Crops* **6**: 24–27.

Sivakumar, M. and Marimuthu, T. 1986. Efficacy of different organic amendments against phytonematodes associated with betel vine. *Indian J. Nematol.* **16**: 278.

Sivakumar, M. and Marimuthu, T. 1987. Preliminary studies on the effect of soil solarization on phytonematodes of betel vine. *Indian J. Nematol.* **17**: 58–59.

Sivakumar, M., Natarajan, S. and Balasubramaniam, M. 1987. Nematode pests of betel vine (*Piper betle* L.) and their management. *Proceedings of the Third Group Discussions on the Nematological Problems of Plantation Crops*, Sugarcane Breeding Institute, Coimbatore, pp. 34–35.

Sivapalan, P. 1972. Nematode pests of tea. *In: Economic Nematology* (J.M. Webster, *ed.*), pp. 285–310. Academic Press, New York and London.

Sivaprasad, P. and Sheela, M.S. 1999. Distribution of phytonematodes and VAM fungi in the rhizosphere of pepper and its impact on nematode management. *In: Nematology – Challenges and Opportunities in 21st Century* (Usha K. Mehta, *ed.*), pp. 102–105. Sugarcane Breeding Institute, Coimbatore.

Soberanis, W., Ríos, R., Arévalo, E., Zúniga, L., Cabezas, O., and Krauss, U. 1999. Increased frequency of phytosanitary pod removal in cacao (*Theobroma cacao*) increases yield economically in eastern Peru. *Crop Protection* **18**: 677–85.

Sosamma, V.K. 1984. *Studies on the Burrowing Nematode of Coconut.* Ph. D. thesis, Kerala University, Trivandrum, Kerala, India, 166 pp.

Sosamma, V.K., Geetha, V.K. and Koshy, P.K. 1994. Effect of the fungus *Paecilomyces lilacinus* on the burrowing nematode, *Radopholus similis* infesting betel vine. *Indian J. Nematol.* **24**: 50–54.

Sosamma, V.K. and Koshy, P.K. 1978. A note on the association of *Cylindrocarpon* spp. with *Radopholus similis* in coconut. *Indian Phytopath.* **31**: 381–82.

Sosamma, V.K. and Koshy, P.K. 1995. Effect of *Pasteuria penetrans* and *Paecilomyces lilacinus* on population build up of root-knot nematode, *Meloidogyne incognita* on black pepper. *National Symposium on Nematode Problems of India – An Appraisal of the Nematode Mangmt. with Eco-friendly Approaches and Biocomponents,* Indian Agri. Res. Inst., New Delhi, p. 47.

Sosamma, V.K., Koshy, P.K. and Bhaskar Rao, E.V.V. 1980. Susceptibility of coconut cultivars and hybrids to *Radopholus similis* in the field. *Indian J. Nematol.* **10**: 250–52.

Sosamma, V.K., Koshy, P.K. and Samuel, R. 1998. Effect of vesicular arbuscular mycorrhizae, *Glomus mosseae* on nematodes infecting banana under coconut based high density multispecies cropping system. *In: Nematology – Challenges and Opportunities in 21st Century* (Usha K. Mehta, *ed.*), pp. 94–97. Sugarcane Breeding Institute, Coimbatore.

Sosamma, V.K., Koshy, P.K. and Sundararaju, P. 1980a. *Nematodes Associated with Cocoa–A review*. Tech. Doc. No. 123, Plant Prot. Committee for the South East Asia and Pacific Region, 14 pp.

Sosamma, V.K., Koshy, P.K. and Sundararaju, P. 1980b. Plant parasitic nematodes associated with cocoa. *Cocoa Growers Bull.* **29**: 27–30.

Sosamma, V.K., Sundararaju, P. and Koshy, P.K. 1979. Effect of *Radopholus similis* on turmeric. *Indian J. Nematol.* **9**. 27–31.

Srihari, K. and Chakravarthy, A.K. 1998. Ecological approaches for the management of vertebrate pests in Western Ghats region of Karnataka. *Indin J. Ecol.* **25**. 55–61.

Srinivasan, C.S. and D'souza, G.I. 1965. A note on the occurrence of Brazilian pyroid coffee nematode, *Meloidogyne exigua* Goeldi in South India. *Indian Coffee* **29** (6): 13.

Sudha, S. 1998. Screening of coconut cultivars against the burrowing nematode, *Radopholus similis*. *In: Advances in IPM of Horticultural Crops* (P. Parvatha Reddy, N.K. Krishna Kumar and A. Verghese, *eds.*), pp. 332–33. Association for Advancement of Pest Management in Horticultural Ecosystems, Bangalore.

Sudha, S. and Sundararaju, P. 1998. Effect of neem oil cake and nematicide for the control of burrowing nematode, *Radopholus similis* in the areca nut based cropping system. *In: Nematology-Challenges & Opportunities in 21st Century* (Usha K. Mehta, *ed.*), pp. 251–57. Sugarcane Breeding Inst., Coimbatore.

Sukumaran, S. and Sundararaju, P. 1986. Pathogenicity of *Meloidogyne incognita* on ginger (*Zingiber officinale* Rosc.). *Indian J. Nematol.* **16**: 258.

Sukumaran, S. and Sundararaju, P. 1986. Effect of root-knot nematode, *Meloidogyne incognita* on the growth of turmeric (Abstr.). *Seventh Symposium of Plantation Crops*, Coonoor, Nilgiris, Tamil Nadu, India, p. 33.

Sundararaju, P. and Koshy, P.K. 1982. Response of Areca collections against *Radopholus similis. Indian J. Nematol.* **12**: 404–05.

Sundararaju, P. and Koshy, P.K. 1986. Control of *Radopholus similis* on areca nut seedlings with aldicarb, aldicarb sulfone, carbofuran and fensulfothion. *Indian J. Nematol.* **16**: 4–7.

Sundararaju, P. and Koshy, P.K. 1988a. Screening of areca nut cultivars and hybrids for resistance to *Radopholus similis. Indian J. Nematol.* **18**: 376–77.

Sundararaju, P. and Koshy, P.K. 1988b. Histopathology of areca nut roots infected with *Radopholus similis. Indian J. Nematol.* **18**: 378–80.

Sundararaju, P., Koshy, P.K. and Sosamma, V.K. 1979. Plant parasitic nematodes associated with spices. *J. Plantation Crops* **7**: 15–26.

Sundararaju, P., Sosamma, V.K. and Koshy, P.K. 1979. Pathogenicity of *Radopholus similis* on ginger. *Indian J. Nematol.* **9**: 91–94.

Suseela Bhat, R., Joseph Thomas and Naidu, R. 1992a. *Proc. PLACROSYM* **10**. Indian Soc. for Plantn. Crops, CPCRI, Kasaragod, Kerala, India.

Suseela Bhat, R., Joseph Thomas and Naidu, R. 1992b. *J. Plantn. Crops* **20** (Suppl.): 90–91.

Tarjan, A.C., Jimenez, M.G. and Soria, J. 1973. Increasing yields of cocoa by application of nematocides. *Turialba* **23**: 138–42.

Thyagaraj, N.E., Chakravarthy, A.K. and Srihari, K. 1996. Vertebrate pests of cocoa and their management practices in hill region of Karnataka. *IPM and Sustainable Agriculture* **6**: 214–18.

Timyan, J. 1996. *BWA Yo: Important Trees in Haiti*. Southeast Consortium for International development, Washington DC, USA.

Trivedi, P.C. and Tiagi, A. 1984. Effect of some plants on *Meloidogyne incognita* population in chilli soil. *Indian J. Bot. Sci.* **63**: 416–19.

Van Hoof, H.A. and Seinhorst, J.W. 1962. *Rhadinaphelenchus cocophilus* associated with little leaf of coconut and oil palm. *Tijdschr. Plziekt.* **68**: 251–56.

Varma, P.M. 1962. *Arecanut J.* **10**: 7–16.

Venkata Ram, C.S. 1963. *The Root-knot Eelworm of Mature Tea* Meloidogyne brevicauda, *its Importance in South India and an Evaluation of Certain Chemical Treatments for its Control.* Tea Scientific Report for 1962–63, United Planters Association of South India, Appendix, pp. 31–36.

Venkata Ram, C.S. 1975. *Planters' Chron.* **70**: 217–18.

Venkitesan, T.S. 1976. *Studies on the Burrowing Nematode*, Radopholus similis *(Cobb, 1893) Thorne, 1949 on Pepper (Piper nigrum L.) and its Role in Slow Wilt Disease.* Ph. D. thesis, Univ. of Agri. Sci., Bangalore.

Venkitesan, T.S. and Jacob, A. 1985. Integrated control of root-knot nematode, infecting pepper vines in Kerala (Abstr.). *Indian J. Nematol.* **15**: 261–62.

Venkitesan, T.S. and Setty, K.G.H. 1977. Pathogenicity of *Radopholus similis* to black pepper (*Piper nigrum*). *Indian J. Nematol.* **7**: 17–26.

Venugopal, M.N. and Naidu, R. 1984. *Proc. PLACROSYM* **6**: 115–119. Indian Soc. for Plantn. Crops, CPCRI, Kasaragod, Kerala, India.

Victoria, K.J., Sanchez, P.A. and Barriga, O.R. 1970. Eradication de palms de cocotero afectadas por el anillo rojo (*Rhadinaphelenchus cocophilus* (Cobb, 1919) Goodey, 1960, Nematoda: Aphelenchoididae) mediante la utilization de sustancias quimicas. *Revista Inst. Colomb. Agropec.* **5**: 185–97.

Vijayan, K.M. and Natarajan, S. 1972. *Coconut Bull.* **2** (12): 2–4.

Viswanathan, P.R.K., Kumar, A.C. and D'Souza, G.I. 1974. Nematode association of cardamom in South India. *Cardamom News* **6**: 5 & 19.

Weiss, E.A. 2002. *Essential Oil Crops.* CAB international, 191 pp.

Wheeler, B.E.J., and Suárez, C. 1993. The pathosystem. *In: Disease Management in Cocoa. Comparative epidemiology of witches' broom* (S.A. Rudgard, A.C. Maddison, and T. Andebrhan, *eds.*), pp. 9–19. Chapman & Hall.

SUBJECT INDEX

H

I

K

L

M